Woodhead Publishing Series in Food Science, Technology and Nutrition: Number 280

A Complete Course in Canning and Related Processes

Fourteenth Edition

Volume 1: Fundamental Information on Canning

Revised by

Susan Featherstone

ELSEVIER

AMSTERDAM • BOSTON • CAMBRIDGE • HEIDELBERG
LONDON • NEW YORK • OXFORD • PARIS • SAN DIEGO
SAN FRANCISCO • SINGAPORE • SYDNEY • TOKYO
Woodhead Publishing is an imprint of Elsevier

WOODHEAD
PUBLISHING

Woodhead Publishing is an imprint of Elsevier
80 High Street, Sawston, Cambridge, CB22 3HJ, UK
225 Wyman Street, Waltham, MA 02451, USA
Langford Lane, Kidlington, OX5 1GB, UK

Thirteenth Edition © 1996 published by CTI Publications, Inc.
Fourteenth Edition © 2015 published by Elsevier Ltd.

British Library Cataloguing in Publication Data
A catalogue record for this book is available from the British Library

Library of Congress Control Number: 2014955925

ISBN 978-0-85709-677-7 (print)
ISBN 978-0-85709-685-2 (online)
ISBN 978-1-78242-500-7 (Three-volume set – print)
ISBN 978-1-78242-501-4 (Three-volume set – online)

For information on all Woodhead Publishing publications
visit our website at http://store.elsevier.com

Typeset by TNQ Books and Journals
www.tnq.co.in

{
This book belongs to

REASEHEATH
COLLEGE

(01270) 613226
library@reaseheath.ac.uk

Date due back:

27-2-17.

WITHDRAWN
For Sale

Thank you for using your
Learning Resources
Centre
}

A Complete Course in Canning and Related Processes

Related titles

Packaging Technology
(ISBN 978-1-84569-665-8)

Trends in Packaging of Food, Beverages and Other Fast-Moving Consumer Goods (FMCG)
(ISBN 978-0-85709-503-9)

Advances in Food and Beverage Labelling: Information and Regulations
(ISBN 978-1-78242-085-9)

Contents

Part Three   Canning operations   185

Woodhead Publishing Series in Food Science, Technology and Nutrition

Preface

This complete course in canning is presented in three parts: Volume 1, Volume 2, and Volume 3. Together they cover all aspects of the production of canned and heat-treated, shelf-stable foods.

This 14th Edition of these books builds on the solid foundations laid down by the canners and authors who contributed to the original articles that were published in the early 1900s in the magazine, *The Trade*, later *The Canning Trade*, published in Baltimore in the United States of America. *A Complete Course in Canning and Related Processes* has been published in 13 editions and over the years there have been many contributions by canners and food regulators. More latterly they have been edited by Dr Anthony Lopez (1958–1987) and Dr Douglas L. Downing (1996). The U.S. Food and Drug Administration and National Food Processors Association, now known as Grocery Manufacturers' Association, deserve special recognition for the excellent information they make available to food processors and that has been used in many of the editions. For this edition, information from *Codex Alimentarius* and Food and Agricultural Organisation of the United Nations was also referenced. In Volume II the chapters on Thermal Processing and Energy Saving were updated and written by Dr Gary Tucker, Campden BRI.

I would like to thank my employers, Nampak Research and Development, for my fascinating job that has enabled me to gain experience and expertise in the field of food processing and packaging and for allowing me to edit this series of books. I would also like to thank my family, especially my children, Nicholas and Katherine, for their patience and understanding during the preparation of the manuscripts.

Susan Featherstone
Cape Town, South Africa
June 2014

Introduction

1 Why this series of books?

The name of this series of books, *A Complete Course in Canning and Related Processes*, indicates that it is intended as a source of information on canned foods. The reader will find here factual and reliable data on all the important facets of canned foods, such as product formulations, manufacturing procedures, food laws, sanitation, sterilisation, spoilage, containers, food plant characteristics, warehousing, and others.

This 14th edition of these books builds on the excellent foundations laid by the authors who contributed to the original articles that were published in the early 1900s in the magazine *The Trade*, published in Baltimore in the United States of America (it was later called *The Canning Trade* and is now called *Food Production Management*). In 1906 the articles were published in a book entitled *A Complete Course in Canning: Being a Thorough Exposition of the Best Practical Methods of Hermetically Sealing Canned Foods, and Preserving Fruits and Vegetables: Originally Republished from the Serial Articles Appearing in 'The Canning Trade,' the Canned Food Authority*. This book was regularly updated with contributions from various authors, and between 1969 and 1987 the 9th through the 12th editions were edited by Dr Anthony Lopez. Originally there was only one book, but to live up to the name of "complete" in 1975 the 10th edition was expanded into two volumes, and in 1987 the 12th edition was further expanded into three volumes. In 1996 the 13th edition was edited by Dr Douglas L. Downing.

When this work first appeared as a serial article in the pages of *The Trade*, the claim "complete" seemed boastful, if not questionable. At the beginning of the twentieth century, the canning industry was, if not in its infancy, at best in its kindergarten age and the amount of production was a fraction of what it is today. Factory equipment and layout were crude and just commencing to develop; quality and grades of products were as varied and as numerous as the producers, because food laws were then nonexistent. Science, as applied to canning and food preservation, was just looming on the horizon. There were no set, definite formulae, except those that experience had taught through dint of heavy cost and that were accordingly carefully nursed and protected by their possessors, the "expert processors." These "expert processors" lorded over the work and the men who employed them and refused to divulge their "secrets." Losses from spoilage, as well as from poor quality, were accepted as normal. At the turn of the century, the industry had little or no scientific knowledge or assistance to depend upon.

The fact that there were no definite formulae obtainable, in printed form or otherwise, at that time, brought the canners of that day, and the new men wishing to enter the industry in particular, to *The Canning Trade*, as the sole source of canning information, asking for direction on the canning of the particular product in which they were interested. Since his earliest association with the industry, in the founding of *The Trade*, later called *The Canning Trade*, in 1878, its first editor, Edward S. Judge, began the accumulation of information on processing and handling, keeping these findings in a big black book – a sort of treasure chest. From this source of information, type-written recipes/formulae were given free to inquirers. The demand was so heavy that it forced consideration of publication of the information in the weekly issues of the industry's journal, *The Canning Trade*. To get more recipes, a competition was set up that offered several thousand dollars in prizes for the best, or most complete, formu-lae for the canning, preserving, or pickling of all the various products, the stipulation being that all offerings became the publisher's property, whether or not they won prizes. Responses were prompt and plentiful, coming from all manner of "processors," expert chefs, cooks, etc., including the most famous and most experienced. The awards were paid, and then began the compilation of the work. *A Complete Course in Canning* was, accordingly, the expression of the best experience existent, its formulae as dependable as possible.

As in previous revisions of this book, this 14th edition has been brought up to date. The aim and desire of these revisions has been to help producers advance the safety and success of food production of this kind, to warn against the dangers and the pitfalls, to keep producers upon safe ground, and to make products safe for public consumption. *A Complete Course in Canning*, as the textbook of this industry, used as it is throughout the entire world, affords the opportunity to put information into the hands of the individuals who need it and can make the best use of it. It is intended to be a compendium of the industry's researches and studies. This 14th edition has more detail on food microbiology and a new chapter on microbiological and nonmicro-biological spoilage. To keep abreast of a very important subject for today's canners, there is a new chapter on reducing energy consumption. The series of books has been converted to primarily metric and centigrade and the focus is more on general canning legislation and guidelines and is less specifically aimed at the American canner but does use the excellent base of the U.S. FDA and USDA legislation and guidelines, which have been used as building blocks for much of the text.

Mechanical equipment and construction of the factory itself have so advanced and improved that no canner should fail to check carefully with the latest and best sources of information. To that end, they should consult the builders of factory buildings, mak-ers of canning and preserving machinery, manufacturers of metal and glass containers, commercial horticulturalists, specialists in food labels, etc. Every such firm willingly furnishes detailed information, without obligation, and their recommendations can be relied upon. It is impossible to lay down a uniform factory plan, mechanical equip-ment, or label design. Every individual or firm wants to carry out personal ideas or desires, and it is well that this is so; but we urge all to call in these experts and have confidence in them, as a surety that the best possible job, under the circumstances, in quality, safety, and cost of production, is being done.

Despite the fact that the greatest care has been exercised in the preparation of the recipes, formulae, times, and directions given in this book, they should, nevertheless, be taken largely as suggestive only, as a reliable working basis, to be altered or changed to fit particular conditions. The formulae given herein are practical and ready to use; they have all been tried and proven, but a change in temperature, altitude, or raw material quality or composition; difference in soil or fertiliser used; a wet or dry season; and a hundred and one other causes may necessitate a change in the process. As was said in the first edition of this book, and repeated here: "there is one reservation that goes with this: 'Considerable Common Sense Must Be Added To All Recipes and Formulae.' As a result, "No Large Pack of a New Product Should Ever Be Produced Until a Trial Batch Has Been Made." To quote the last paragraph from the Introduction in the first edition of this work: "If used judiciously, in this manner, these formulae will be found satisfactory, differing possibly with different processors, as is natural, but worthy of the high approval set upon them when they were first published." Keeping careful check upon raw materials received, and on factory operations as they progress, may save heavy losses from spoilage or a lowering of quality. It is too late to check after the product is in the warehouse. Similarly, all the guidance and information given should be considered and then adapted accordingly to the specific factory and processing conditions and product type.

2   A brief history of canning technology

It is more than two centuries (Figure 1) since Nicholas Appert published his book in which he documented his methods for producing heat-preserved foods in hermetically sealed containers. His invention has been incredibly successful and has contributed in a significant way to the improvement of nutrition and health of consumers all over the world. Today approximately 50 billion[1] (Featherstone, 2012) cans of food are manufactured and consumed every year, globally.

3   Nicholas Appert discovers and documents a safe way of heat-preserving food

The sixteenth and seventeenth centuries were a time of great scientific advancement in the areas of chemistry, mathematics, and physics. This was known as the Scientific Revolution, which laid the foundation for the Age of Enlightenment in the eighteenth century, a period in which science became popular with the ordinary person, and an increasingly literate population was hungry for knowledge, for information, and to learn. Evening science lectures, with demonstrations, were very popular as a form of entertainment for the working class. In addition, the Industrial Revolution was taking place in Europe in the eighteenth and nineteenth centuries. Great strides were made in the areas of textiles, steam generation, and metallurgy. Steam generation was much more efficient; fossil fuels were used for the first time, instead of wood-based fuel, resulting in a much more efficient source of energy. Advances in mining techniques and metal-working, especially iron founding, resulted in many new uses for metals such as iron, copper, and tin.

In France, the French Revolution (1789–1799) took place, largely as a result of growing dissatisfaction owing to a shortage of food and increasing malnutrition. The Napoleonic Wars (1803–1815) further influenced the economy and developments in Europe.

---

[1] In British English, a billion used to be equivalent to a million million (i.e. 1,000,000,000,000), while in American English it has always equated to a thousand million (i.e. 1,000,000,000). British English has now adopted the American figure, though, so that a billion equals a thousand million in both varieties of English.

Figure 1   In July 2010, Monaco issued a commemorative stamp to celebrate the bicentenary of Appert's invention.

It was in this time that Nicolas Appert was working as a confectioner. He was born on 17 November 1749 at Charlon-sur-Marne. His family was not wealthy, and young Nicolas acquired an education largely through personal effort. He was trained as a chef and worked and experimented with various kinds of food (fermenting, distilling, and preserving) for his own interest all of his life.

The French army was busy with many military campaigns, and a major concern to the French government was that they were losing more troops to diseases, such as scurvy and malnutrition, than to battle casualties. In 1795, they offered a prize of 12,000 francs (a lot of money at that time) to anyone who could find a way to safely preserve food. This offer caught the attention of Nicolas Appert. He had noticed that the sugar syrups that he used for his confectionary kept almost indefinitely when heated and stored in stoppered glass bottles. He began experimenting with preserving other types of food, also by heating them and storing them in stoppered jars. He worked scientifically and had a keen attention to detail. He started with using champagne bottles but soon had them modified with a wider mouth so that he could fill them more easily.

By 1804 he felt confident enough to test some product on the French navy. The test was a huge success. In 1806 more tests were conducted. His invention was assessed, and early in 1810 he was informed that he could claim the award of 12,000 francs but had to publish the exact details of his discovery, which he did.

Appert is known as the father of canning. Heat sterilisation is also known as appertisation. Although his first products were in glass bottles, the term canning is often used interchangeably with thermal or heat processing of foods. He had to deliver 200 copies of his processing methods (printed at his own expense) to the French government before they would give him his award. The book was very detailed and described the canning process much as it is still today. Appert described the process as follows:

- Enclose the foods to be preserved in bottles.
- Cork the bottles carefully.
- Boil the bottles in water for various lengths of time (depending on the food).
- Remove the bottles and cool them.

In 1811 he had a second edition printed in French, as well as English and Swedish, and in 1823 a third, and in 1831 a fourth. His book was also translated into many languages. Although Appert's methods worked, no-one at the time understood why. Appert thought it was due to the heating and exclusion of air. At that time it was widely believed that air itself was the cause of spoilage (Goldblith, 1971).

4 The Appert food preservation method explained

Louis Pasteur (1822–1895) was a French chemist and microbiologist. In 1862 he proved, by demonstration, that fermentation is caused by the growth of micro-organisms and not by spontaneous generation or exposure to air. In his experiment he heated broth in long swan-necked jars to sterilise it. The jars either had filters on them or had very long necks that allowed only air and not dust and other particles through. Nothing grew in the broth unless the flasks were broken open. He therefore correctly concluded that the living organisms that grew came from outside, as spores on dust, rather than being spontaneously generated within the broth or in the air. He showed that the growth of micro-organisms was responsible for spoiling products such as beer, wine, and milk. He invented a process in which milk was heated to kill most bacteria and molds already present. This process was soon afterward known as pasteurisation.

A little known fact about Pasteur is that he, because of his status as a respected scientist, served on the Board of Public Hygiene and Sanitation in France. One of his responsibilities was food laws. As a result of his discoveries regarding the cause of food spoilage, various methods of preserving foods were researched by many others. Some of these involved adding chemicals to the food. In 1870, Pasteur advocated that the public had "the right to know" what was being put into the food and stated that all additives must be declared on the label.

5 The development of food microbiology

Food microbiology was in its infancy in the nineteenth century. Many people contributed to developing it into the science it is today. A few of these scientists who made significant contributions to the understanding of the science of canning are mentioned below.

All canners had losses due to sporadic swelling and spoilage, but the actual causes of it were unknown, and the canners did not know what they should do to overcome this problem. In 1895 Dr Harry L. Russell published a paper describing swelling spoilage in canned peas. He did experiments in which he processed peas at higher temperatures and longer times and showed that the percentage of spoilage was significantly reduced.

In 1895 William Lyman Underwood, a canner and the grandson of one of the first American canners, wanted to understand canned food spoilage, so he went to the Massachusetts Institute of Technology for help. There he met Samuel Prescott and together they did an enormous amount of work that had a great effect on the development of canning. They published detailed scientific papers describing:

- that bacteria were the cause of spoilage in canned foods;
- that some kinds of food needed to be heated above the boiling point to ensure sterility;
- the use of maximum registering thermometers and the importance of heat penetration;
- recommendations for incubation testing for canned foods; and
- the importance of postprocessing cooling in canned foods.

In 1913, the National Canners Association in the United States was formed, with Dr W.D. Bigelow as its head. Under his guidance they undertook significant research and published many bulletins for the canning industry (Goldblith, 1972).

At the National Canner's Association a lot of good work was not only done but, more importantly, published and shared, the first being "Some Safety Measures in Canning Factories" by Dr A.W. Bitting (1937). Some examples of other important work that was done include:

- 1917: Bigelow used thermocouples to measure the continuous heating of cans of baked beans.
- 1920: Bigelow and J.R. Esty showed that spores die off more quickly at higher temperatures.
- 1921: Bigelow showed that death time curves are logarithmic.
- 1921: Bigelow and P.H. Cathcart described the effect of acidity in lowering the thermal processing requirements.
- 1922: Esty and K.F. Meyer demonstrated the maximum resistance of *Clostridium botulinum* spores using moist heat, laying the foundation for the 12D concept.
- 1923–1927: C. Olin Ball together with Bigelow introduced the direct calculation of processing parameters. In 1927 Ball published the concept of a sterilisation value.
- 1948: C.R. Stumbo introduced the concept of integrating sterilisation over the entire can contents. Using Esty and Meyer's data he calculated a Z value of 10 and an Fo of 2.78.
- 1965: Stumbo published his classic text book entitled *Thermobacteriology in Food Processing* (Stumbo, 1965).

There are many excellent scientists who have contributed invaluable insights into the field of thermal processing, but most of the foundation work was done by those mentioned above.

The early bacteriological studies on spore death kinetics were done by different researchers at various temperatures. This work was done between 1921 and 1948. As could be expected, the lower the temperature the slower the rate of kill. Stumbo took this information and calculated a Z value of 18 °F (10 °C) and an *F* value of 2.78 min on a thermal death time curve that passed though 250 °F (121.1 °C) at

2.78 min (Tucker, 1991). This temperature was appropriate for practical cooking times as well as being safely achievable in the processing vessels of the day. This was the basis of the Fo3 at 250 °F concept.

The general method for calculating scheduled process times was originally described by Bigelow et al. in 1920, but contributions by Ball in 1928 and O.T. Schulz and F.C.W. Olson in 1940 resulted in a much improved general method. M. Patashnik published his improvements, which are the most widely used today, in 1953 (Patashnik, 1953).

The invention of a standardised reliable thermocouple for measurement of real-time heat penetration temperatures was an important contribution made by O.F. Ecklund (Ecklund, 1949).

J.R. Manson, A.A. Teixeira, and K. Purohit were three of Stumbo's graduate students who also contributed significantly to the field of thermal processing. They were engineers and the first to apply engineering mathematics to simulate the coupling of heat transfer with thermal inactivation kinetics in thermal processing of canned foods. Teixeira used this approach to find optimum retort time–temperature combinations that would maximise quality retention while delivering specified target lethality. Manson carried Teixeira's work further by improving the mathematical model to simulate convection as well as conduction heat transfer. Working together, Teixeira and Manson demonstrated how such models could be employed in real-time online control of batch retorts by automatically extending process time to precisely compensate for unexpected process deviations (Teixeira, Dixon, Zahradnik, & Zinsmeister, 1969).

The research mentioned above took place largely in the United States; in the United Kingdom T.G. Gillespy and his team did valuable work on processing times and temperatures of a large range of foods at Campden Experimental Factory (now Campden BRI) (Tucker, 2008). In France, H. Cheftel at Carnaud Research did extensive work on canned foods and published *Principles and methods for establishing thermal processes for canned foods* in French. In South Africa, G.G. Knock did much to increase the understanding, and hence reduce the incidence, of thermophilic "flat sour" spoilage in canned peas (Knock, 1954). These researchers and others from all over the world collaborated to improve the understanding of the mechanisms of canned food spoilage and the requirements necessary to ensure that safe canned food is produced. Improvements in product safety, quality, and energy usage are still continuously being made. G.S. Tucker at Campden BRI further developed numerical techniques for thermal process calculations and a computer program to recalculate process deviations in real time.

6 Packaging for heat-preserved foods

Nicolas Appert's first products were packed in glass. Soon after his discovery was published, Peter Durand, a British merchant, patented the idea of preserving food using tin cans. The patent (No. 3372) was granted on August 25, 1810 by King George III of England. After receiving the patent, Durand did not make any canned food himself, but in 1812 sold his patent to two other Englishmen, Bryan Donkin and John Hall, for £1000. Donkin was involved with tinning of iron from 1808 and was keen to expand it to the food industry. Donkin and Hall set up a commercial canning factory and by 1813 were producing their first canned goods for the British army. In 1818, Durand introduced tin cans in the United States by repatenting his British patent in the United States of America.

The first "canisters" were made from iron that was dipped into molten tin to stop it from corroding. The ends were soldered closed with lead. The metal was thick and the cans were heavy and strong. The cans often weighed more than the food that was in them. They had to be cut open with a hammer and chisel.

Appert also started using cans. He made them himself in his cannery. They had the capacity of between 4 and 45 pounds and could be reused. He also added handles to some of them so that they could be used as cooking pots once opened. Although he preferred round cans, he made oval and rectangular ones, too, at the request of his customers.

Improvements to the can came when steel was invented, allowing a much thinner metal that had the same strength to be used. In 1888 the double seam was invented by Max Ams. This paved the way for automated can lines to be made. Whereas before about six cans per hour were made, the first automated can lines could make about 60 cans per hour. Can-making lines today can run as fast as 1500 cans per minute.

Tin is an expensive metal. In the 1930s hot dipping of tinplate was replaced by electroplating, by which much less tin can be used to perform the same job. Improvements in steel-making technology have resulted in even lighter weight cans. Single-reduced tinplate of 0.19–0.21 mm thickness and double-reduced tinplate as thin as 0.10–0.15 mm are now used to make cans all over the world.

Although improvements in can design can compensate to some extent for the loss of strength due to the thinner metal, many of the down-gauging improvements are possible only because of improvements in can handling.

7 Convenience – the can opener is invented

Only when thinner steel cans came into use could the can opener be invented. Before then, canned food used to come with the written instructions: "Cut round the top near the outer edge with a chisel and hammer." The first can openers were a primitive claw-shaped or "lever-type" design. Robert Yates patented the first can opener in Great Britain in 1855 and Ezra Warner patented another in 1858 in the United States – his looked like a bent bayonet. Its large curved blade was driven into a can's rim, then forcibly worked around its edge. This first type of can opener never left the grocery store as it was deemed to be too dangerous for ordinary people to use; the store clerk had to open each can before it was taken away.

A can opener with a cutting wheel that rolls around the rim was invented by William Lyman of the United States in 1870 but was difficult to operate. A break-through came in 1925 when a second, serrated wheel was added to hold the cutting wheel on the rim of the can. The basic principle of this opener is the same as is used in modern can openers. The first electric can opener was introduced in December 1931.

The easy-opening end is the ultimate in convenience as far as can openers go; it was first patented by Ermal Fraze of Ohio, in 1966.

8   Other forms of packing for "canned foods"

Other forms of packaging for "canned foods" have also become popular and practical: glass, various plastics, and composites. Glass is not new, being the packaging that Appert used to develop his method, but back then it was difficult to seal adequately and cans soon became the packaging of choice. Developments in the closures for glass, starting with the Mason jar in 1858, resulted in glass becoming a popular alternative to cans. Improvements in glass-making technology have resulted in bottles that do not break easily under the high heat and pressure conditions experienced during thermal processing. Improvements in lids, caps, and closures that form hermetic seals, are easy to open and reclose, and have tamper-evident features such as buttons that "pop" on first opening have helped to make glass a viable alternative.

Developments in other packaging types are more recent and restricted to the past 40 years or so. Rigid plastics are useful for ready meals, as they are not breakable and can be heated in the microwave oven. Pouches are flexible, and during processing the flat dimensions result in faster heat penetration and therefore shorter cooking times and better quality product. Pouches and some rigid containers come with their own challenges for thermal processing as their irregular shapes make it more difficult to measure the cold spot during processing. Tetra Recart's "square" format saves space when packing and stacking. All of these options have their advantages and disadvantages. None of the plastics and laminates are as good a gas barrier as glass and metal. Light can also cause deterioration in some products.

The ideal packaging choice depends on the requirements of product type, processing conditions, required shelf life, and target market.

9   Developments in cannery equipment

Seaming: The equipment used in canneries also had to develop significantly. Originally cans were manually soldered closed, and good artisans could do about six an hour. Once the double seam was invented, closing of cans became much faster. Today there are seamers that can close over 1000 cans per minute.

Processing: Processing (heating) of the sealed cans started off as being an all day event. Appert processed most of his products in boiling water. He did experiment with pressure processing, but at that time "digesters" were quite dangerous, and it was not the norm. Around 1863 processors used "chemical baths," in which high concentrations of calcium chloride enabled "water" to boil at up to 121 °C. This allowed for significantly shorter cooking times. By 1870 basic retorts were being used to temperatures up to 121 °C, but they were still quite dangerous and hand operated.

In 1950 the first reel-and-spiral continuous cooker was introduced and was some-thing very innovative. It was the first retort that did not process in batches. It was called the Anderson–Barngrover cooker and was taken over by FMC, which is cur-rently John Bean Technologies.

Around the same time aseptic filling was pioneered. In 1961 flame sterilisation (using direct gas flame heating) was introduced, as was the continuous hydrostatic pressure steriliser. Today, with computers being able to control equipment we have many more options and more precise control. There are combinations of steam and air, raining water, and rotation. All of these developments have as their objective to improve temperature distribution and product heating. Modern retorts can process at temperatures up to 145 °C, yielding faster throughput.

10 Canned foods – a healthy option

One of the negative perceptions about canned food is that it is not as nutritious as other types of food. The original canned foods were made for military rations. At that time anything was better than nothing or rotten food, and it revolutionised the military. The troops were much better nourished than ever before. But the food actually was overprocessed and as a result the nutritional value was slightly compromised. As the understanding of canning grew, the quality of the products became better.

Today canned food is minimally, but adequately, processed, from good quality raw ingredients under strict hygiene conditions. Many studies have been done that show that the nutritional value of canned food is as good as that of its fresh or frozen coun-terparts. Thermally processed foods provide excellent nutrition over extended periods. Most crops, meat, and fish are seasonal. Thermal processing allows seasonal products to be available all year round. There are many studies that have been done on the nutritional content of canned food. This is just one of the very positive statements that have come out of the findings.

Canned food, contrary to popular belief, can form part of a healthy balanced diet. It is often assumed that canned foods are a poor source of vitamins and min-erals. Canned foods in many cases provide amounts of vitamins and minerals that are similar to fresh equivalents and are often a good source of protein and fiber, too.

11 The future of thermally processed foods

Many of the current developments in thermally processed foods are driven by cost saving and an attempt to reduce the carbon footprint. These can be done by reducing the weight of the packaging, optimising the processing, and improving the heat trans-fer into the product either by changing the packing format or shape or by using more efficient forms of heating. To achieve these goals, new or improved packaging must be used (for example, thinner, stronger metals, plastics, or laminates with better barrier properties and strength). More efficient retorts and heat transfer mechanisms are continually being tested. Many canners are also starting to realise that they often overprocess their products for added safety. Understanding the product and proper control of the processing, with properly tested processes, are enough. Lack of knowledge and control often waste time, money, and quality.

We can expect more niche products as developments in food technology allow for new ingredients and improvements in processing give us better quality. We will definitely see innovations in packaging variants. Interestingly, canned food has proven to sell particularly well in times of recession because of the tendency of financially stressed people to stick to what they know and trust and to eat at home.

12 Are canned foods sustainable?

Sustainability is defined as meeting the needs of present generations without jeopardising the needs of future generations. Preservation reduces waste – this and the need for constant availability of food are what drove the invention of canning. For the canned food processor waste can be minimised as products can be tailored to the type and size of the raw ingredients. For the consumer the waste from shelf-stable, canned foods is very low compared with fresh and chilled products.

Thermally processed foods provide excellent nutrition over extended periods, all year round, anywhere in the world. Glass, tinplate, aluminum, and many plastics used in thermally processed products can be recycled. Compared with other types of food processing thermal processing produces low greenhouse gas emissions. Thermally processed foods are sustainable.

References

Bitting, A. W. (1937). *Appertizing or the art of canning; its history and development.* San Francisco, CA: The Trade Pressroom.

Ecklund, O. F. (1949). Apparatus for the measurement of the rate of heat penetration in canned foods. *Food Technology, 3*(7), 231–233.

Featherstone, S. (2012). A review of development in and challenges of thermal processing over the past 200 years—a tribute to Nicolas Appert. *Food Research International, 47,* 56–160.

Goldblith, S. A. (December 1971). A condensed history of the science and technology of thermal processing – part 1. *Food Technology, 25,* 1256–1262.

Goldblith, S. A. (January 1972). A condensed history of the science and technology of thermal processing – part 2. *Food Technology, 26,* 64–69.

Knock, G. G. (1954). Technique for the approximate quantitative prediction of flat-souring in canned peas. *Journal of Food Agriculture, 5,* 113–119.

Patashnik, P. (January 1953). A simplified procedure for thermal process evaluation. *Food Technology, 7*(1), 1–6.

Stumbo, C. R. (1965). *Thermobacteriology in food processing*. New York and London: Academic Press.

Teixeira, A. A., Dixon, J. R., Zahradnik, J. W., & Zinsmeister, G. E. (1969). Computer determination of spore survival distribution in thermally-processed conduction heated foods. *Food Technology, 23*, 352–354.

Tucker, G. S. (2008). *History of the minimum botulinum cook for low-acid canned foods*. Confidential R&D Report No 260, Campden BRI.

Tucker, G. S. (January 1991). Development and use of numerical techniques for improved thermal process calculations and control. *Food Control, 2*(1), 15–19.

Part One

Business planning and regulations for canned foods

Creating a business plan

1

1.1 Introduction

Starting a cannery requires a substantial investment. Depending on the type of products you wish to can, you will need to get permission/licenses from the relevant food authority and will have to be inspected for hygiene compliance by the local health department. The cannery will need to comply with all of the health, fire, and safety codes in your area and may require special licensing, such as food handlers' licenses for you and your employees. Contact your local health department to determine all of the licensing and inspection requirements for your location. Liability insurance can protect your financial position should there be a case of a contamination or a food poisoning incident, provided you can prove that all of the required measures were in place to prove due diligence.

The key to the success of any business is the comprehensive development of a written document called a business plan. A business plan describes the business, its product, market, people, operational characteristics, and financing needs. The process of putting a business plan together forces one to take an objective, critical, unemotional look at the entire business proposal. A well-prepared business plan serves several purposes:

1. For a new business, it helps the owner determine the feasibility and desirability of pursuing the steps necessary to start a business.
2. For a company seeking financing, it is an important sales document for raising capital from outside investors.
3. For a new or existing company, a business plan forms the basis of a more detailed operational plan and thus becomes an important management tool for monitoring the growth of the company and charting future directions.
4. For all companies, it is an operational tool which, when properly used, will help manage a business and work toward its success.

It has been said that many businesses fail without a business plan. No one plan will cover all situations, so it should be tailored to the specific circumstances of the proposed business, should emphasise the strengths of the venture, and address the potential problems and challenges to be faced.

1.2 Proposed outline for a business plan

The overall plan will consist of several components that will probably include the following: business organisation, a financial plan, a marketing plan, a management plan, human resource management, supply chain management, and operations management, among others. It can be helpful to view the business plan as a collection of subplans: one for each of the main business disciplines that have been identified.

A Complete Course in Canning and Related Processes. http://dx.doi.org/10.1016/B978-0-85709-677-7.00001-3

1.2.1   Description of the business organisation

1.2.1.1   Business contact details

Business Name
Street Address
Mailing Address
Telephone/Fax number(s)
Website/email addresses
Owner(s) Name(s)
Business Form (proprietorship, partnership, corporation)
   [If incorporated, state the jurisdictional government body such as state or country.]

Include copies of key subsidiary documents in an appendix. Partnerships require written agreements of terms and conditions to avoid later conflicts and to establish legal entities and equities. Corporations require charters, articles of incorporation, and by-laws.

1.2.1.2   Business purpose and function

In this section, write an accurate concise description of the business. Describe the business in narrative form.

1. What is the principal activity? Will this be a manufacturing or a service business?
2. How will it be started? Will it be a new start-up, the expansion of an existing business, or the purchase of a going business, and what is the actual or projected start-up date?
3. Why will it succeed? Promote your idea here.
4. What is unique about the business? What is its market 'niche', and how and why will this business be successful?

What is your experience in this business? Include a current resume of your career in an appendix and reference it here. If you lack specific experience, explain how you plan to gain it, such as training, apprenticeship, or working with partners who have experience.

1.2.2   Management plan

A management plan describes who will do what. Four basic sets of information need to be included:

1. State a personal history of related work and experience (including formal resumes in an Appendix).
2. List and describe specific duties and responsibilities of each individual.
3. List benefits and other forms of compensation for each individual.
4. Identify other professional resources available to the business such as: accountant, lawyer, insurance broker, banker. Describe the relationship of each to the business. For example, an accountant available on a part-time hourly basis, as needed; initial agreement calls for services not to exceed a specified number of hours per month at certain rate per hour.

To make this section graphically clear, start with a simple organisational chart that lists specific tasks and show who (type of person is more important than individual name other than for principals) will do what indicated by arrows, work flow, and lines of responsibility and/or communications. Consider the following examples:

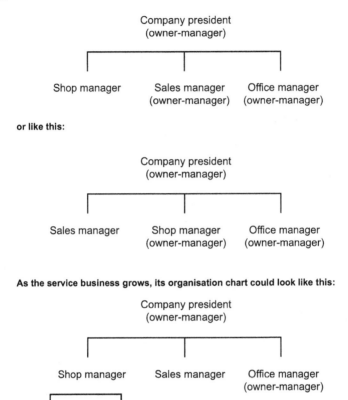

or like this:

As the service business grows, its organisation chart could look like this:

1.2.3 Human resources plan

Human Resources (HR) is concerned with how people are managed within organisations, focusing on policies and systems. HR departments in businesses are typically responsible for a number of activities including: recruitment, training and development, performance appraisal, and rewarding (i.e. managing pay and incentive systems). HR is also concerned with industrial relations.

In the HR plan, the following questions should be answered:

- What are your personnel needs now?
- What skills must each key person have?
- Are the people needed available? Name them and indicate full or part time and salary rates.

Detail a proposed work schedule by week and month for at least the first year. Calculate total salaries, wages, fringe benefits, and payroll taxes for each month of the first year.

If any gaps in personnel skills have been identified, state with a detailed plan how these will be overcome by training, purchase of outside services, or subcontracting. Check with the nearest employment service office for assistance.

1.2.4   The marketing plan

The marketing plan is the core of a business rationale. To develop a consistent sales growth, the management must become knowledgeable about the market. To demonstrate understanding, answer several basic questions:

Who is in your market?

Describe the profile of your typical customer by age, sex, income, number in a family, location, buying patterns, and the reason to buy the product.

Geographically describe the trading area, i.e. local, regional, national, and/or international. Economically describe the trading area, i.e. single family, average earnings, number of children.

How large is the market?

What is the estimated total number of units or value of the market? Is the market growing, steady, or decreasing? If the market is growing, what is the annual growth?

Who are your competitors?

No business operates in isolation. Get to know and respect the competition. Target the marketing plans. Identify direct competitors (both in terms of geography and product lines) and those who are similar or marginally comparative. Begin by listing names, addresses, and products or services. Detail briefly, but concisely, the following information concerning each of your competitors:

- Who are the nearest ones?
- How are their businesses similar or competitive to yours?
- Do you have a unique 'niche'? Describe it.
- How will your product be better or more saleable than your competition?
- Are their businesses growing? Stable or declining? Why?
- What can be learned from observing their operations or talking to their present or former clients?

Remember, your business can become more profitable by adopting good competitive practices and by avoiding their errors.

To help evaluate how successful your product will be, consider the following list of standard characteristics and/or others and make a candid evaluation of your competitive edge. On a scale of '0' (theirs is best) to '10' (yours is best) indicate the potential for your success and a total score. A total score of less than 6 indicates that you might reconsider the viability of your venture or think about how you can improve it. Over 8 points indicates a clear competitive edge, e.g. feature, durability, price, versatility, performance, appearance, or packaging.

What percentage of the market will you penetrate?

Estimate the market in total units (or value). Estimate your planned volume and the amount your volume will add to the total market.

What pricing and sales terms are you planning?

The primary consideration in pricing a product is the value that it represents to the customer. If, on the previous checklist of features, your product is truly ahead of the competition, you can command a premium price. On the other hand, if it is a 'me too' product, you may have to 'buy' a share of the market to get your foothold and then try to move price later. This is always risky and difficult. One rule will always hold: ultimately, the market will set the price. If your selling price does not exceed your costs and expenses by the margin necessary to keep your business healthy, you will fail. Know your competitors' pricing policies. Is there discounting? Special sales? Price leaders?

What is your sales plan?

Describe how you will sell and distribute what you sell. Common practices are: Distributors are a convenient way to get wide distribution if you can find the right match for your product(s).

Direct Sales is often the most common approach to start. You know your product and how it should be positioned.

What is your advertising plan?

Each product will need its own advertising strategy as part of the total business-marketing plan. Before developing an advertising campaign for your business plan, take time to review a few basic assumptions. By definition, advertising is any form of paid, nonpersonal promotion that communicates with a large number of potential customers at the same time. The purpose of advertising is to inform, persuade, and remind customers about your company's products. Every advertising activity should have specific goals. Some common examples are:

- To bring in sales orders or contracts
- To promote special events such as sales, business openings, new products
- To bring in requests for estimates or for a sales representative to call
- To use special media at the outset may be a goal to establish before start-up and to get potential customer 'feedback'.

1.2.5 Operational plan

Operations management is concerned with overseeing, designing, and controlling the process of production and redesigning business operations in the production of goods or services. It involves the responsibility of ensuring that business operations are efficient and effective in terms of meeting customer requirements. It is concerned with managing production. The purpose of this section is to summarise from previous sections the various operations of your business and to link them to the finance section of your business plan.

1.2.6 Supply chain plan

Supply chain management is the management of the flow of goods. Supply chain management is a cross-functional approach that includes managing the movement of raw materials into the business, some aspects of the internal processing of materials into finished goods, and the movement of finished goods out of the business to the consumer. It includes the movement and storage of raw materials, work-in-process inventory, and finished goods from point of origin to point of consumption.

1.2.7 Financial plan

The most critical section of your Business Plan Document is the Financial Plan. In formulating this part of the planning document, you will establish vital schedules that will guide the financial health of your business through the challenges of the first year and beyond.

To be able to build the Financial Plan, some basic knowledge of accounting is essential to the productive management of your business. Conviction that your product is of good quality is vitally important to success and the product may indeed be very good, but the business will fail if you do not make a profit. If the finances in the business are not controlled, you are not in a good position to assure the profitability.

Many small businesses will use the cash method of accounting with a system of record keeping in which is recorded all receipts and all expenditures, backed up by a few forms of original entry (invoices, receipts, cash tickets). A larger business will need a more sophisticated system. Computerised accounting systems are commonly available and range from very complex to quite simple, depending on needs.

What is your banking plan?

What will be the location and type of bank accounts opened for the business? A word of caution for new business ventures: keep business accounts separate from personal or family accounts. These vital records will be necessary for future tax and accounting purposes. Describe your banking plan.

How is your credit rating?

There may be several partial answers to this question. All will be of importance to the future of the business. What is your personal history of paying debts? Just to be safe, purchase a copy of your personal credit record from the 'Credit Reporting Agency' for a small fee and make sure that it is accurate.

To establish a credit rating, it is necessary to secure credit with a number of businesses and to use it. Your rating will be based upon your record for paying for goods and services based upon the agreed terms. If your prior credit rating is poor, discuss with your lawyer, accountant, and banker options for improvement before seeking and being refused business credit.

The Financial Plan includes the following:

1. **Financial Planning Assumptions**: These are short statements of the conditions under which you plan to operate and will cover market health, date of start-up, sales build-up, gross profit margin, equipment, furniture and fixtures required, and payroll in addition to other key expenses that will impact the financial plan.

2. **Operational Plan Profit and Loss Projection**: This is also referred to as the profit and loss statement, prepared for the first year, and broken into 12 individual months. The profit and loss statement is valuable as a planning tool and as a key management tool to help control operations to reach business goals. It enables the owner/manager to develop a 'preview' of the amount of profit, or loss, generated each month, and for the business year – based on reasonable predictions of monthly levels of sales, costs, and expenses. The owner/manager can compare the year's expected profits or losses against the profit goals and needs established for the business. A completed Profit and Loss statement allows the owner/manager to compare actual figures with the monthly projections and to take steps to correct any problems. The Profit and Loss statement is prepared in the following manner and should become your first year's Budget. Refer to Table 1.1. Create a wide sheet of analysis paper with a 3 in wide column at the extreme left and 13 narrow columns across the page. Write at the top of the first page the planned name of your business. On the second line of the heading, write 'Profit and Loss Projection.' On the third line, write 'First Year.'

   Note the headings on the table and copy them onto your 13-column sheet. If start-up is indefinite, just write 'Month #1', 'Month #2', etc. Column 13 should be headed 'Total Year'.

   In the wide, unnumbered column on the left of your 13-column sheet, copy the headings from the similar area on Table 1.1. Then follow the example and list all of the other components of your income, cost, and expense structure. You may add or delete specific lines of expense to suit your business plan. Guard against consolidating too many types of expense under one account lest you lose control of the components. Table 1.1 provides ample detail for most small businesses.

   In the small column just to the left of the first monthly column, you will want to note which of the items is to be estimated on a monthly (M) or a yearly (Y) basis. Items such as sales, cost of sales, and variable expenses will be estimated monthly based on planned volume and seasonal or other estimated fluctuations. Fixed expenses can usually be estimated on a yearly basis and divided by 12 to arrive at even monthly values. The 'M' and 'Y' designations will be used later to distinguish between variable and fixed expenses.

   Depreciation allowances for fixed assets such as production equipment, office furniture and machines, vehicles, etc. will be calculated from the source of funds schedule.

Revenue (sales)

List the departments within the business.

- In the 'Estimate' columns, enter a reasonable projection of monthly sales for each department of the business. Include cash and on-account sales.
- In the 'Actual' columns, enter the actual sales for the month as they become available.
- Exclude from the Revenue section any revenue that is not strictly related to the business.

Table 1.1 **Operating plan forecast (profit and loss projection)**

	—Month 1—	—Month 2.12—	—Totals—
	% Estimate actual	% Estimate actual	% Estimate actual
Revenue (sales)			
Total revenue (sales)			
Cost of sales			
Total cost of sales			
Gross profit			
Expenses:			
Salary expense: Sales people, office and other			
Payroll expenses (taxes, etc.)			
Outside services			
Supplies (office and operating)			
Repairs and maintenance			
Advertising			
Car, delivery, and travel			
Accounting and legal			
Rent			
Telephone			
Utilities			
Insurance			
Taxes (real estate, etc.)			
Interest			
Depreciation			
Other expenses (specify each)			
Miscellaneous (unspecified)			
Total expenses			
Net profit			

Cost of sales

*   Cite costs by department of the business, as above.
*   In the 'Estimate' columns, enter the cost of sales estimated for each month for each department. For product inventory, calculate the cost of the goods sold for each department (beginning inventory plus purchases and transportation costs during the month, minus the inventory).
*   Enter 'Actual' costs when known each month.
*   Gross Profits – Subtract the total cost of sales from the total revenue.

Expenses

*   Salary Expenses – Base pay plus overtime.
*   Payroll Expenses – Include paid vacations, sick leave, health insurance, unemployment insurance, social security taxes.

- Outside Services – Include costs of subcontracts, overflow work farmed out, special or one-time services.
- Supplies – Services and items purchased for use in the business, not for resale.
- Repairs and Maintenance – Regular maintenance and repair including periodic large expenditures such as painting or decorating.
- Advertising – Include not sure here and yellow pages expenses.
- Car, Delivery, and Travel – Include charges if personal car used in business, including parking, tolls, buying trips, etc.
- Accounting and Legal – Outside professional services.
- Rent – List only real estate used in the business.
- Telephone and fax – Self-explanatory.
- Utilities – Water, electricity, etc.
- Insurance – Fire or liability on property or products, workmen's compensation.
- Taxes (real estate, etc.) – Inventory, sales, excise tax, others.
- Interest: Self-explanatory.
- Depreciation – Amortisation of capital assets.
- Other Expenses (specify each) – Such as tools, leased equipment.
- Miscellaneous (unspecified) – Small expenditures without separate accounts.

Net Profit = gross profit – total expenses

3. **Source of Funds Schedule:** This shows the source(s) of your funds to capitalise the business and how they will be distributed among your fixed assets and working capital. To create this schedule, you will need to create a list of all of the assets that you intend to use in your business, how much investment each will require, and the source of funds to capitalise them. A sample of such a list is shown below:

Asset	Cost	Source of funds
Cash	$2500	Personal savings
Accounts receivable	3000	From profits
Inventory	2000	Vendor credit
Pickup truck	5000	Currently owned
Packaging machine	10,000	Installment purchase
Office desk and chair	300	Currently owned
Calculator	75	Personal cash
Electric typewriter	500	Personal savings

Before you leave your source of funds schedule, indicate the number of months (years × 12) of useful life for depreciable fixed assets. For more information on allowances for depreciation, you can get assistance from your local Internal Revenue Service office. Divide the cost of each fixed asset item by the number of months over which it will be depreciated. You will need this data to enter as monthly depreciation on your profit and loss projection. All of the data on the source of funds schedule will be needed to create the balance sheet.

4. **Pro Forma Balance Sheet:** 'Pro forma' refers to the fact that the balance sheet is before the fact, not actual. This form displays assets, liabilities, and equity of the business. This will indicate how much investment will be required by the business and how much of it will be used as working capital in its operation.

Table 1.2 is a Balance Sheet form. This form varies considerably, and you may find it prudent to ask your banker for the form that the bank uses for small businesses. It will make

Table 1.2 Company name balance sheet as of (current date)

Assets		Liabilities	
Current assets		*Current liabilities*	
Cash	$____	Accounts payable	$____
Accounts receivable $ less allowance for		Short-term notes	$____
doubtful accounts $		Current portion of long-term notes	$____
Net realisable value	$____	Interest payable	$____
Inventory	$____	Taxes payable	$____
Temporary investments	$____	Accrued payroll	$____
Prepaid expenses	$____	Total current liabilities	
Total current assets	$____	*Long-term liabilities*	
Long-term investments		Notes payable	$____
(Detailed list)	$____	Total long-term liabilities	$____
Total investments	$____	*Total liabilities*	$____
Fixed assets		**Equity**	
Land	$____	Total Owner's equity (proprietorship)	$____
Buildings: $____ at cost, less accumulated			Or
depreciation of $		(Name's) equity	$____
Net book value	$____	(Name's) equity	$____
Equipment: $____ at cost, less accumulated		(partnership)	
depreciation of $		Total partners' equity	$____
Net book value	$____		Or
Furniture/Fixtures: $____ at cost, less		Shareholders' equity (corporation)	
accumulated depreciation of $		Capital stock	$____
Net book value	$____	Capital paid-in in excess of par	$____

Autos/Trucks: $ _____ at cost, less accumu-
lated depreciation of $
Net book value $ _____
Total net fixed assets $ _____

Other assets
(Detailed list)
Total other assets $ _____
Total assets $ _____

Retained earnings $ _____
Total shareholders' equity $ _____
Total liabilities and equity $ _____

Reconcilement of equity
As of (current date)
Equity at beginning of period $ _____
Plus: Net income (or minus: Net Loss) after taxes $ _____
Plus: Additional capital contributions (investments by
owner(s) or stock purchases by shareholders) $ _____
Less: Total deductions (withdrawals by owner(s)
or dividends to shareholders) $ _____
Equality as shown on current Balance Sheet $ _____

it easier for them to evaluate the health of your business. You can use Table 1.2 to get started and transfer the data to a preferred form later. The following is information that describes line by line how to develop the Balance Sheet.

Even though you may plan to stage the purchase of some assets through the year, for the purposes of this pro forma balance sheet, assume that all assets will be provided at the start up.

The following text covers the essential elements of a Balance Sheet. Figures used to compile the Balance Sheet are taken from the previous and current Balance Sheet as well as the current Income Statement (or Profit and Loss statement). The report is usually attached to the Balance Sheet.

a. **Heading** – The legal name of the business, the type of statement, and the day, month, and year must be shown at the top of the report.

b. **Assets** – Anything of value that is owned or legally due the business. Total assets include all net realisable and net book (also net carrying) values. Net realisable and net book values are amounts derived by subtracting any estimated allowances for doubtful accounts, depreciation, and reductions of future service – such as amortisation of a premium during the term of an insurance policy – from the acquisition price of assets.

c. **Current Assets –** Cash and resources that can be converted into cash within 12 months of the date of the Balance Sheet (or during one established cycle of operations). Besides cash (money on hand and demand deposits in the bank, e.g. checking accounts and regular savings accounts), resources include:
   - *Accounts Receivable* **–** The amounts due from customers in payment for merchandise or services.
   - *Inventory* **–** Includes raw materials on hand, work in process, and all finished goods either manufactured or purchased for resale.
   - *Temporary Investments* – Interest- or dividend-yielding holdings expected to be converted into cash within a year. Also called marketable securities or short-term investments, they include stocks and bonds, certificates of deposit, and time-deposit savings accounts. List on the Balance Sheet at either their cost or market value, whichever is less.
   - *Prepaid Expenses* – Goods, benefits, or services a business buys or rents in advance of use. Examples are office supplies, insurance protection, and floor space.

d. **Long-Term Investments** – Also called long-term assets. They are holdings the business intends to keep for at least a year and that typically yield interest or dividends. Included are stocks, bonds, and savings accounts earmarked for special purposes.

e. **Fixed Assets** – Fixed assets, frequently called plant and equipment, are the resources a business owns or acquires for use in operations and does not intend for resale. Land is listed at its original purchase price, with no allowance for appreciation or depreciation. Other fixed assets are listed at cost minus depreciation. Fixed assets may be leased. Depending on the leasing arrangement, both the value and the liability of the leased property may need to be listed on the Balance Sheet.

f. **Other Assets** – Resources not listed with any of the above assets. Examples include tangibles such as outdated equipment saleable to the scrap yard and intangibles such as trademarks.

g. **Liabilities** – All monetary obligations of a business and all claims creditors have on its assets.

h. **Current Liabilities** – All debts and obligations payable within 12 months or within one cycle of operations. Typically they are:
   - Accounts Payable – Amounts owed to suppliers for goods and services purchased in connection with business operations.
   - Short-Term Notes – The balance of principal due to pay off short-term debt for borrowed funds.
   - Current Portion of Long-Term Notes – Current amount due of total balance on notes the terms of which exceed 12 months.
   - Interest Payable – Any accrued fees due for use of both short- and long-term borrowed capital and credit extended to the business.
   - Taxes Payable – Amounts estimated by an accountant to have been incurred during the accounting period.
   - Accrued Payroll – Salaries and wages currently owned.

i. **Long-Term Liabilities** – Notes, contract payments, or mortgage payments due over a period exceeding 12 months or one cycle of operations. They are listed by outstanding balance, less the current portion due.

j. **Equity** – Also called net worth. Equity is the claim of the owner(s) on the assets of the business. In a proprietorship or partnership, equity is each owner's original investment plus any earnings after withdrawals. In a corporation, the owners are the shareholders. The corporation's equity is the sum of contributions plus earnings retained after paying dividends.

k. **Total Liabilities and Equity** – The sum of these two amounts must always match that for Total Assets.

l. **Reconcilement of Equity** – Used for proprietorships and partnerships, this report reconciles the equity shown on the current Balance Sheet. It records equity at the beginning of the accounting period and details additions to or subtractions from this amount made during the period. Typically, additions and subtractions are net income or loss and owner contributions and/or deductions.

For corporations, the same type of report is called the Statement of Retained Earnings. It lists increases or decreases in the accumulated net income since the beginning of the current period.

5. **Cash-flow projection** – This will forecast the flow of cash into and out of your business through the year. It helps you plan for staged purchasing, high-volume months, and slow periods.

An important subsidiary schedule to your financial plan is a monthly cash-flow projection. Prudent business management practice is to keep no more cash in the business than is needed to operate it and to protect it from catastrophe. In most small businesses, the problem is rarely one of having too much cash. A cash-flow projection is made to advise management of the amount of cash that is going to be absorbed by the operation of the business and compares it against the amount that will be available.

See Table 1.3. Your projection should be prepared on 13-column analysis paper to allow for a 12-month projection. Table 1.4 represents a line-by-line description and explanation of the components of the cash-flow projection, which provides a step by-step method of preparation.

Table 1.3 Monthly cash flow projection

Year _____

	Pre-start-up position		Months 1 thru 12		Annual	
	Estimate	Actual	Estimate	Actual	Estimate	Actual
1. Cash on hand (beginning of month)						
2. Cash receipts						
a. Cash sales						
b. Collections from credit accounts						
c. Loan or other cash injection (specify)						
3. Total cash receipts (2a + 2b + 2c = 3)						
4. Total cash available (before cash out) (1 + 3)						
5. Cash paid out						
a. Purchases (merchandise)						
b. Gross wages (excludes)						
c. Payroll expenses (taxes)						
d. Outside services						
e. Supplies (office and operating)						
f. Repairs and maintenance						
g. Advertising						
h. Car, delivery, and travel						
i. Accounting and legal						
j. Rent						
k. Telephone						
l. Utilities						
m. Insurance						
n. Taxes (real estate, etc.)						

o. Interest
p. Other expenses (specify)
q. Miscellaneous (unspecified)
r. Subtotal
s. Loan principal payment
t. Capital purchases (specify)
u. Other start-up costs
v. Reserve and/or escrow (specify)
w. Owner's withdrawal

6. **Total cash paid out** (total 5a through 5w)

7. **Cash position** (end of month) (4 minus 6)

Essential operating data
(Non-cash flow information)
a. Sales volume (dollars)
b. Accounts receivable (E)
c. Bad debt (end of month)
d. Inventory on hand (end of month)
e. Accounts payable (end of month)
f. Depreciation

Table 1.4 Components of a cash flow projection

1.	**Cash on hand** (Beginning of month)	Cash on hand same as (7). Cash Position Previous Month.
2.	**Cash receipts**	
	a. Cash sales	All cash sales. Omit credit sales unless cash is actually received.
	b. Collections from credit accounts	Amount to be expected from all credit accounts.
	c. Loan or other cash injection (specify)	Indicate here all cash injections not shown in 2(a) or 2(b) above. See 'A' of 'analysis'.
3.	**Total cash receipts** (2a + 2b + 2c = 3)	Self-explanatory.
4.	**Total cash available** (before cash out) (1 + 3)	Self-explanatory.
5.	**Cash paid out**	
	a. Purchases (merchandise)	Merchandise for resale or for use in product (paid for in current month).
	b. Gross wages (excludes)	Base pay plus overtime (if any).
	c. Payroll expenses (taxes)	Include paid vacations, paid sick leave, health insurance, unemployment insurance, etc. (this might be 10–45% of 5(b)).
	d. Outside services	This could include outside labor and/or material for specialised or overflow work, including subcontracting.
	e. Supplies (office and operating)	Items purchased for use in the business (not for resale).
	f. Repairs and maintenance	Include periodic large expenditures such as painting or decorating.
	g. Advertising	This amount should be adequate to maintain sales volume-include telephone book yellow page cost.
	h. Car, delivery, and travel	If personal car is used, charge in the column-include parking.
	i. Accounting and legal	Outside services, including, for example, bookkeeping.
	j. Rent	Real estate only (see 5(p) for other rentals).
	k. Telephone	Self-explanatory.
	l. Utilities	Water, heat, light, and/or power.
	m. Insurance	Coverages on business property and products, e.g. fire, liability, also workman's compensation, fidelity, etc. Exclude 'executive' life (include in 5w).
	n. Taxes (real estate, etc.)	Plus inventory tax, sales tax, and excise tax, if applicable.
	o. Interest	Remember to add interest on loan as it is injected (see 2(c) above).
	p. Other expenses (specify)	Unexpected expenditures may be included here as a safety factor. Equipment expensed during the month should be included here (noncapital equipment). When equipment is rented or leased, record payments here.

q. Miscellaneous (unspecified) — Small expenditures for which separate accounts would not be practical.

r. Subtotal — This subtotal indicates cash out for operating costs.

s. Loan principal payment — Include payment on all loans, including vehicle and equipment purchases on time payment.

t. Capital purchases (specify) — Nonexpensed (depreciable) expenditures such as equipment, building, vehicle purchases, and leasehold improvements.

u. Other start-up costs — Expenses incurred prior to first month projection and paid for after the 'start-up' position.

v. Reserve and/or escrow (specify) — Example: Insurance, tax, or equipment escrow to reduce impact of large periodic payments.

w. Owner's withdrawal — Should include payment for such things as owner's income tax, social security, health insurance, 'executive' life insurance premiums, etc.

6. **Total cash paid out** (total 5a thru 5w) — Self-explanatory.

7. **Cash position** (end of month) (4 minus 6) — Enter this amount in (1) cash on hand following month-See 'A' of 'analysis'.

Essential operating data (Noncash flow information) — This is basic information necessary for proper planning and for proper cash flow projection. In conjunction with this data, the cash flow can be evolved and shown in the above form.

a. Sales volume (dollars) — This is a very important figure and should be estimated carefully, taking into account size of facility and employee output as well as realistic anticipated sales (actual sales performed – not orders received).

b. Accounts receivable (E) — Previous unpaid credit sales plus current month's credit sales, less amounts received current month (deduct 'C' below).

c. Bad debt (end of month) — Bad debts should be subtracted from (B) in the month anticipated.

d. Inventory on hand (end of month) — Last month's inventory plus merchandise received and/or manufactured current month minus amount sold current month.

e. Accounts payable (end of month) — Previous month's payable plus current month's payable minus amount paid during month.

f. Depreciation — Established by your accountant, or value of all your equipment divided by useful life (in months) as allowed by the Internal Revenue Service.

1.3   Conclusion

Ideally, a Business Plan should be completed before the business is operational, but if the business is already a going concern, having a formal Plan will aid growth and profitability and give direction. Drawing up a Business Plan takes a great deal of time and effort, but it will result in a set of management tools to guide you in your venture. Once the processing plant is in production, you will be inundated by the details, problems, and challenges. A Business Plan should be a living document, referred to regularly and often revised. Begin a planning cycle that expands this first-year plan into one that spans three or five years out. Update it at regular intervals.

Acknowledgements

The information in this chapter was based on information taken from: The Business Plan for Home based Business (1983). Business development Publication MP15. US Small business administration, office of business development.

Food regulations, standards, and labelling

2

2.1 Introduction

Managers/operators of food businesses are obliged to insure that all stages of production, processing, and distribution of food under their control satisfy the relevant food safety, hygiene, and processing requirements laid down by the relevant authorities in the areas in which they are producing and marketing their products.

Food business operators carrying out any stage of production, processing, and distribution of food must comply with the general hygiene standards, and, in most countries, must put in place, implement, and maintain a permanent procedure or procedures based on the hazard analysis and critical control point (HACCP) principles. In addition, the food business operators must provide the competent authority with evidence of their compliance with the requirement to have procedures based on the HACCP principles and insure that any documents describing the procedures developed in accordance with this are up-to-date at all times and should retain any other documents and records for an appropriate period.

The personal hygiene of all workers who are exposed to any part of food processing is also regulated. These regulations will cover issues like the good personal hygiene of personnel and the type of protective clothing that may be required. Anybody employed in a food business and who is likely to encounter food must immediately report any illness or symptoms of illnesses to the food business operator, who must act accordingly. It is important that all food workers are trained, supervised, and instructed in food hygiene matters.

The building of food processing factories and the standards to which these facilities have to comply are specified by regulations and legislation. See Figure 2.1 for an overview of some of the legislative requirements for a food business.

2.2 Codex Alimentarius

The Codex Alimentarius (Latin for 'Book of Food') is a collection of internationally recognised guidelines, standards, and codes of practice and other recommendations relating to foods, food production, and food safety. The texts are developed and regularly updated by the Codex Alimentarius Commission. The Codex Alimentarius Commission was established in 1961 by the Food and Agriculture Organization of the United Nations (FAO) and was joined by the World Health Organization (WHO) in 1962. The first session was held in Rome in October 1963. The Commission's main goals are to protect the health of consumers and insure fair practices in the international food trade. The Codex Alimentarius is recognized by the World Trade Organization as an international reference point for the resolution of disputes concerning food safety and consumer protection.

A Complete Course in Canning and Related Processes. http://dx.doi.org/10.1016/B978-0-85709-677-7.00002-5

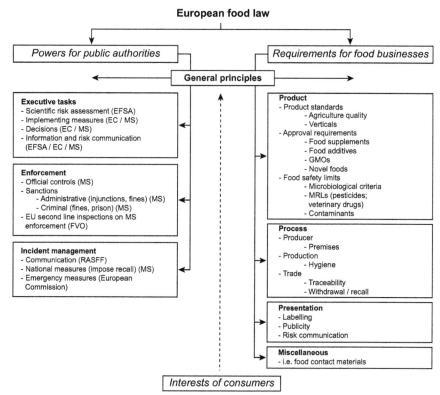

Figure 2.1 European food law: an overview of the powers for public authorities and requirements for food businesses (Holah & Lelieveld, 2011).

The Codex Alimentarius covers all foods (raw, semiprocessed, and processed). It contains general standards covering matters such as food labelling, food hygiene, food additives, and pesticide residues. It contains standards for specific foods. It also contains guidelines for the management of official import and export inspection and certification systems for foods.

The Codex Alimentarius programme has been actively supported by a large number of countries. Work has been carried on concurrently in developing standards for many food products; a substantial number of completed standards have now been submitted to participating countries for their acceptance or rejection. A country may accept a Codex standard fully with no reservations, or it may accept it with changes, or it may conclude not to accept it.

2.3 Food and drug administration in the United States of America

The Food and Drug Administration (FDA or USFDA) is an agency of the United States Department of Health and Human Services. It is responsible for protecting and promoting public health through the regulation and supervision of food safety, tobacco products, dietary supplements, prescription and over-the-counter pharmaceutical drugs (medications), vaccines, biopharmaceuticals, blood transfusions, medical devices, electromagnetic radiation-emitting devices (ERED), cosmetics, and veterinary products in the US.

US Federal Regulations require commercial processors of shelf-stable acidified foods and low-acid canned foods in a hermetically sealed container to be sold in the United States to register each establishment and file scheduled processes with the Food and Drug Administration for each product, product style, container size, and type and processing method. Their website contains detailed information useful to manufacturers of these types of products as well as instructions for establishment registration and process filing.

The Code of Federal Regulations (CFR), which govern the commercial processing of foods and food-related items, are Parts one through 199 of Title 21. USDA/FSIS regulations are published in Title 9. Part 318.300 contains the regulations for canned meat products, and Part 381.300 contains the regulations for canned poultry products. Certain processed fruits and vegetables must also meet the US Standard for grades and these requirements are published in Part 52 of Title 7 of the CFR (7CFR52) and are administered by the Agricultural Marketing Service/ US Department of Agriculture (USDA).

The sections of greatest interest to a commercial canning operation are

- Food Labelling (21CFR101-FDA; 7CFR317-Meat and 7CFR381.1-Poultry)
- Current Good Manufacturing Practice in Manufacturing, Processing, Packing, or Holding Human Food
- Food (21CFR110)
- Thermally Processed Low-Acid Foods Packaged in Hermetically Sealed Containers (21CFR113)
- Canned Vegetables (Standards; 21CFR155)
- Processed Fruits and Vegetables, Processed Products Thereof, and Certain Other Processed Food
- Products (USDA/Agriculture Marketing Service; US Standards for Grades; 7CFR52)
- Mandatory Meat Inspection (9CFR301-335) and/or Mandatory Poultry Products Inspection
- (9CFR381)
- Draft Guidance for Industry: Acidified Foods (21 CFR part 108.25 and 114)

Both Title 21, Code of Federal Regulations (CFR), and the Federal Register may be obtained from the Superintendent of Documents, Government Printing Office, Washington, DC 20204. Inquiries regarding specific subjects in the CFR may be directed to the Food and Drug Administration, Industry Guidance Branch, 200C Street, SW, Washington, DC 20204. These documents are readily available and provide excellent general guidance to the canning industry even if the US is not the place of manufacture or the destination of the final product.

Title 21, Code of Federal Regulations is updated on April 1st of each year. The annual edition contains three volumes applicable to FDA regulations for foods.

A description of each volume is as follows:

- Part 1 to 99. General regulations for the enforcement of the Federal Food, Drug, and Cosmetic Act, the Fair Packaging and Labelling Act, and Colour Additives.

- Part 100-169. Food standards, good manufacturing practices for food, low-acid canned foods, acidified foods, and food labelling.
- Part 170-199. Food additives.

Some of the details of Title 21 are summarised below:

2.3.1   Prohibited acts (violations)

The Federal Food, Drug, and Cosmetic Act prohibits distribution in the United States, or importation, of articles that are adulterated or misbranded. The term 'adulterated' includes products that are defective, unsafe, filthy, or produced under unsanitary conditions (Secs. 402, 501, 601). 'Misbranded' includes statements, designs, or pictures in labelling that are false or misleading and failure to provide required information in labelling (Secs. 403, 502, 602). Further explanations are provided later in this chapter. Detailed definitions of adulteration and misbranding are in the law itself, and hundreds of court decisions have interpreted them.

The law also prohibits the distribution of any article required to be approved by FDA, if such approval has not been given, the refusal to provide required reports, and the refusal to allow inspection of regulated facilities pursuant to Sec. 704.

2.3.2   Food adulteration

Food adulteration provisions of the Food, Drug, and Cosmetic Act govern three types of adulteration: economic, aesthetic, and hazardous.

Economic adulteration is the oldest form of consumer fraud. Watered milk, for example, is not a health problem, but it is an economic problem and is prohibited by law.

Aesthetic adulteration involves more analysis. Under the FD&C Act, food that contains any filthy or decomposed matter, or that is 'otherwise unfit for food', is adulterated and therefore illegal. This is true, even if the food is absolutely sterile and presents no health problem. These aesthetic provisions in the law protect the consumer's sensitivity and health.

Any food held under unsanitary conditions that may present a danger to health is adulterated and illegal. The FDA has taken most of its enforcement action under the 'unsanitary conditions' provision of the food law, the backbone of food and drug protection in the United States.

Regulation of added poisonous or deleterious substances has been an important issue in US food safety in recent years. Food additives include basic components of food and other items, such as preservatives, emulsifiers, fillers, colours, etc. Indirect food additives include things that are exposed to food such as the container, processing machinery, pesticide residues, animal drugs, etc. US laws treat these different substances on an individual basis.

A food safety law should also include a way to allow some food ingredients that cannot be banned because of their importance to the food supply, tolerances for these ingredients in food can be set. For example, if a tolerance of 0% for aflatoxins in peanuts were set, most of the peanut supply would be considered unsuitable for consumption. The FDA, therefore, sets a tolerance for aflatoxins in peanuts. A food law that does not provide for exemptions, as well as rules, will not be functional.

2.3.3   Enforcement

Enforcement is the key to making laws work. Four major enforcement methods exist under US food law.

1. Inspection of food plants and warehouses for unsanitary conditions, good manufacturing practices, and generally recognised principles of hygiene and sanitation are the first method. In some countries (but not in the United States), authorities can also look at records kept by the food manufacturer.
2. Analysis of products including checks for filth, microbial contamination, decomposition, toxic substances, etc., may be performed without access to records or inspection of the factory.
3. After violation has been found, formal legal action can be brought, or informal enforcement methods used. Three methods of court enforcement action are possible: seizure of the product, issuance of an injunction against the illegal activity, or criminal prosecution by fine or jail sentence.
4. Informal enforcement action is often used, instead of formal action, because it is more efficient and less expensive. Letters to violating companies, recalls, and publicity, the most potent weapon of all, are informal enforcement tools.

2.3.4   Food misbranding

Food labelling involves two elements: economics and safety. Today's food standards exist almost solely for economic reason, i.e. to define food in a way that meets the economic expectations of producers and consumers. Likewise, net contents and the name and address of the manufacturer are on the label for economic reasons. The rationale for a statement of ingredients on the label is partly economic, so that the consumer will understand what he is getting, and partly safety, in the event the consumer has a known allergic reaction to an ingredient in that food product. Nutritional information is on the label, partly for economic reasons, to show the value of the food, and partly for health reasons.

2.3.5   Premarket testing and approvals

The Federal Food, Drug, and Cosmetic Act and the Public Health Service Act require manufacturers of certain consumer products to establish, before marketing, that such products meet the safety and effectiveness requirements of the law and are properly labelled. New drugs, biological drugs, and certain devices (including their labelling) must be approved for safety and effectiveness. Substances added to food must be 'generally recognised as safe,' 'prior sanctioned,' or approved by specific FDA regulations based on scientific data. Samples of colour additives and insulin drugs must be tested and certified by the FDA laboratories. Residues of pesticides in food commodities must not exceed safe tolerances set by the Environmental Protection Agency and enforced by FDA. Such premarketing clearances are based on scientific data provided by manufacturers, subject to review and acceptance by government scientists for scope and adequacy. Submission of false data to secure approvals is a criminal violation of the laws, which prohibit giving false information to the government.

FDA regulations prescribe the type and extent of premarket testing that must be conducted, depending on the legal requirements applicable to the particular product and on the technology available to fulfil those requirements. Testing may include physical and chemical studies, clinical laboratory studies, animal tests, or clinical trials in humans.

The importance of toxicological and other data derived from such investigations demands that they be conducted according to scientifically sound protocols and procedures. FDA regulations prescribing good laboratory practices for nonclinical research are published in 21 CFR 58.

2.3.6 Imports

Although legal requirements that must be met are the same for imported and domestic products, enforcement procedures are necessarily different.

Imported products regulated by the Food and Drug Administration are subject to inspection at the time of entry through US Customs. Shipments found not to comply with the laws and regulations are subject to detention. They must be brought into compliance, destroyed, or re-exported.

At the discretion of the Food and Drug Administration, an importer may be permitted to try to bring an illegal importation into compliance with the law before a final decision is made as to whether it may be admitted. Any sorting, reprocessing, or relabelling must be supervised by an FDA investigator at the expense of the importer.

Both foreign shippers and importers in the United States should realise that conditional release of an illegal importation to bring it into compliance is not a right but a privilege. Abuse of the privilege, such as repeated shipments of the same illegal article, may result in denial of the privilege in the case of subsequent importations.

2.3.7 Exports

Many US producers are interested in exporting some of their products. If the item is intended for export only, meets the specifications of the foreign purchaser, is not in conflict with the laws of the country to which it is to be shipped, and is properly labelled, it is exempt from the adulteration and misbranding provisions of the Act (Sec. 801(d)).

2.3.8 Low-acid canned foods registration and process filing

All commercial processors of heat processed 'low-acid' canned foods and acidified foods are required to register their establishments and file processing information for all such products with the Food and Drug Administration, on appropriate forms. Registration and process filing is required for both US establishments and those in other countries which export such foods to the United States (21 CFR 108.25 and 108.35).

Registration forms, process filing forms, and instructions are obtainable from the Food and Drug Administration, LACF Registration Coordinator, 200C Street, SW, Washington, DC 20204. New establishments must register within 10 days from the date the plant begins processing. Processes must be filed no later than 60 days after registration and before packing a new product. Modifications and changes in previously filed processes are made through an amended submission. Foreign firms must register and file processing information before shipping any low-acid canned foods and acidified foods to the United States. Failure to comply may result in legal proceedings against the US firms or their products, or the detention of shipments from foreign firms.

Full text of the low-acid canned food and acidified food regulations may be found in the Code of Federal Regulations, Title 21, Parts 108, 113, and 114. The regulations provide a comprehensive guide to requirements for the processing of low-acid canned foods and acidified foods.

Additional information is included in the section on 'Canned Foods' in this chapter.

2.3.9   Colour additives

The Federal Food, Drug, and Cosmetic Act provides that foods, drugs, cosmetics, and some medical devices are adulterated if they contain colour additives that have not been proven safe to the satisfaction of the Food and Drug Administration for that particular use. A colour additive is a dye, pigment, or other substance, whether synthetic or derived from a vegetable, animal, mineral, or other source, which imparts a colour when added or applied to a food, drug, cosmetic, or the human body, as described in the Act (Sec. 201(t)).

Regulations (21 CFR 73, 74, and 81) list the approved colour additives and conditions under which they may be safely used, including the amounts that may be used when limitations are necessary. Separate lists are provided for colour additives for use in or on foods, drugs, medical devices, and cosmetics. Some colours may appear on more than one list.

Testing and certification by the Food and Drug Administration of each batch of colour is required before that batch can be used, unless the colour additive is specifically exempted by regulation.

Before using colour additives in foods, drugs, devices, or cosmetics, regulations should be reviewed to ascertain which colours have been listed for various uses. Colour additive labels, which are required to contain sufficient information to assure safe use, should be read. 'For food use only', directions for use where tolerances are imposed, or warnings against use, such as 'Do not use in products used in the area of the eye', are examples of information available.

Manufacturers of certifiable colours may address requests for certification of batches of such colours to the Food and Drug Administration, Office of Cosmetics and Colour, 200C Street, SW, Washington, DC 20204. Certification is not limited to colours made by US manufacturers. Requests will be accepted from foreign manufacturers, if signed by both such manufacturers and their agents residing in the United States. Certification of a colour by an official agency of a foreign country cannot, under the provisions of the law, be accepted as a substitute for certification by the Food and Drug Administration.

2.3.10 Scientific functions: methods of analysis

Modern scientific methods are required to enforce laws such as the Federal Food, Drug, and Cosmetic Act. Insuring the wholesomeness of foods and the safety and efficacy of drugs would be impractical without reliable methods of laboratory analysis to determine if products are up to standard. Food and drug scientists in both government and industry must know the normal composition of products to distinguish them from those that are defective. They also investigate the toxicity of ingredients, study the causes of food poisoning, and test the potency of vitamins and thousands of drugs; these investigations also cover the adequacy of controls over processing, packaging, and storage practices. Such research requires experts in many fields but especially chemistry, microbiology, microanalysis, pharmacology, and both human and veterinary medicine. Any action taken by the FDA must be based on scientific facts which can be supported in court. The principal authority relied upon for laboratory methods is 'Official Methods of Analysis of The Association of Official Analytical Chemists (AOAC)'. This compendium of tested analytical methods, published since 1895, is the leading internationally recognised guide to analytical procedures for law enforcement.

2.4   Principal requirements of food law

Section 201(f) of the US Federal Food, Drug, and Cosmetic Act defines food as follows:

The term 'food' means: (1) articles used for food or drink for man or other animals; (2) chewing gum; and (3) articles used for components of any such article.

A synopsis of the principal requirements of the Act relating to foods follows. The numbers in parentheses are the pertinent sections of the statute itself, or sections in the Code of Federal Regulations (CFR).

2.4.1   Health safeguards

A food is illegal (adulterated) if it bears or contains an added poisonous or deleterious (harmful) substance which may render it injurious to health (402(a)(1)).

A food is illegal if it bears or contains a naturally occurring poisonous or deleterious substance which ordinarily renders it injurious to health (402(a)(1)).

Food additives (201(s)) must be determined to be safe by FDA before they may be used in a food or become a part of a food because of processing, packaging, transporting, or holding the food (409).

Raw agricultural products are illegal if they contain residues of pesticides not authorised by, or in excess of, tolerances established by regulations of the Environmental Protection Agency (408).

A food is illegal if it is prepared, packed, or held under unsanitary conditions, whereby it may have been rendered injurious to health (202(a)(4)).

Food containers must be free from any poisonous or deleterious substance which may cause the contents to be injurious to health (402(a)(6)). Some packaging materials, for example, plastic or vinyl containers, may be 'food additives' subject to regulations (409).

Only those colours found safe by the Food and Drug Administration may be added to food (706). A food is illegal if it bears or contains an unsafe colour(s) (402(c)). Unless exempt by regulation, colours for use in food must be from batches tested and certified by the Food and Drug Administration (706(c)).

A food is illegal if any part of it is filthy, putrid, or decomposed (402(a)(3)).

A food is illegal if it is prepared, packed, or held under unsanitary conditions, whereby it may become contaminated with filth (402(a)(4)).

A food is illegal if it is the product of a diseased animal or one that has died otherwise than by slaughter (402(a)(5)).

2.4.2 Economic safeguards

Damage or inferiority in food must not be concealed in any manner (402(b)(3)). Example: artificial colouring or flavouring added to a food to make it appear a better value than it is, as in the case of yellow colouring used to make a food appear to contain more eggs than it actually contains.

Food labels or labelling (circulars, etc.) must not be false or misleading in any particular (403(a)). Labelling is misleading, not only if it contains false or misleading statements but also if it fails to reveal material facts (201(n)).

A food must not be sold under the name of another food (403(b)). Example: canned bonito labelled as tuna fish.

A substance recognised as being a valuable constituent of a food must not be omitted or subtracted in whole or in part, nor may any substance be substituted for the food in whole or in part (402(b)(l) and (2)). Example: an article labelled 'milk' or 'whole milk' from which part of the butterfat has been skimmed.

Food containers must not be so made, formed, or filled as to be misleading (403(d)). Example: a closed package filled to less than its capacity. A food for which a standard of fill of container has been prescribed (402) must comply with the fill requirements, and if the fill falls below that specified, its label must bear a statement that it falls below such standards (403(h)(2)).

2.4.3 Required label statements

The law states that required label information must be conspicuously displayed and in terms that the ordinary consumer is likely to read and understand under ordinary conditions of purchase and use (403(f)). Details concerning label requirements are listed later in this chapter.

2.4.4 Foods for special dietary uses

Section 403(j) of the Food, Drug, and Cosmetic Act classes a food as misbranded: 'If it purports to be, or is represented for special dietary uses, unless its label bears such information concerning its vitamin, mineral, and other dietary properties as the Secretary determines to be, and by regulations prescribes as, necessary in order fully to inform purchasers as to its value for such uses'.

Section 411(c)(3) of the Act defines 'special dietary use' as a particular use for which a food purports, or is represented to be used, including, but not limited to, the following:

1. Supplying a special dietary need that exists because of a physical, physiological, pathological, or other condition, including, but not limited to, the conditions of disease, convalescence, pregnancy, lactation, infancy, allergic hypersensitivity to food, underweight, or the need to control the intake of sodium.
2. Supplying a vitamin, mineral, or other ingredient for use by humans to supplement the diet by increasing the total dietary intake.
3. Supplying a special dietary need because of being a food for use as the sole item of the diet.

Regulations under this section of the Act (21 CFR 105) prescribe appropriate information and statements which must be given on the labels of foods in this class.

Importers and foreign shippers should consult the detailed regulations under sections 403(j), 411, and 412 before importing foods represented by labelling or otherwise as special dietary foods.

When special dietary foods are labelled with claims of disease prevention, treatment, mitigation, cure, or diagnosis, they must comply with the drug provisions of the Act. Regulations regarding foods for special dietary use can be found in 21 CFR Part 105.

2.4.5 Infant foods

The Infant Formula Act of 1980 (21 USC 412) amended the Federal Food, Drug, and Cosmetic Act to establish nutrient requirements for infant formulas and to give FDA authority to establish requirements for quality control, record keeping, reporting, and recall procedures. The Act also extends the FDA's factory inspection authority to permit access to manufacturers' records and test results necessary to determine compliance.

The Act specifies that an infant formula is adulterated: (1) if it fails to provide nutrients as required, (2) if it fails to meet the nutrient quality factors required by regulation, or (3) if processing does not comply with appropriate quality control procedures or record retention requirements as prescribed by regulation. Formulas for infants with inborn errors of metabolism, or other unusual medical or dietary problems, are exempted.

Under authority of the Act, FDA has promulgated regulations which specify the quality control and recall procedures for infant formulas. Current labelling regulations (21 CFR 107.10) specify the kind of label information, which must be present on these foods, in addition to the mandatory label information required on all packaged foods. These regulations, however, predate the Infant Formula Act and are being revised. In addition, FDA regulations on foods for infants with special medical or dietary needs may be found in 21 CFR 105.65. Inquiries should be directed to the Food and

Drug Administration, Division of Regulatory Guidance (HFS-150), 200C Street, SW, Washington, DC 20204.

2.4.5.1 Sanitation requirements

One of the basic purposes of the Food, Drug, and Cosmetic Act is protection of the public from products that may be deleterious, that are unclean or decomposed, or have been exposed to unsanitary conditions that may contaminate the product with filth or may render it injurious to health.

Sanitation provisions of the Food, Drug, and Cosmetic Act go further than to prohibit trade in products that are carriers of disease. The law also requires that foods be produced in sanitary facilities; it prohibits distribution of foods which may contain repulsive or offensive matter considered as filth, regardless of whether such objectionable substances can be detected in the laboratory. Filth includes contaminants such as rat, mouse, and other animal hairs and excreta, whole insects, insect parts and excreta, parasitic worms, pollution from the excrement of man and animals, as well as other extraneous materials that, because of their repulsiveness, would not knowingly be eaten or used. The presence of such filth renders foods adulterated, whether or not harm to health can be shown.

The law thus requires that food be protected from contamination at all stages of production. Such protection includes extermination and exclusion of rodents, inspection and sorting of raw materials to eliminate the insect-infested and decomposed portions, quick handling and proper storage to prevent insect development or contamination, the use of clean equipment, control of possible sources of sewage pollution, and supervision of personnel who prepare foods so that acts of misconduct may not defile the products they handle.

Foods that are free from contamination when they are shipped sometimes become contaminated en route and must be detained or seized. This emphasises the importance of insisting on proper storage conditions in vessels, railroad cars, or other conveyances. Although the shipper may be blameless, the law requires action against illegal merchandise, no matter where it may have become illegal. All shippers should pack their products to protect them against spoilage or contamination en route and should urge carriers to protect the merchandise by maintaining sanitary conditions and segregating food from other cargo which might contaminate it. For example, vessels transporting foods may also carry ore concentrates and poisonous insecticides. Improper cargo handling or disasters at sea have resulted in shipments becoming seriously contaminated, with detentions required.

When import shipments become contaminated after Customs entry and landing (for example, in truck accidents, fire, barge sinkings, etc.), legal actions are not taken under the import provisions of the law but by seizure proceedings in a federal district court, the same as domestic interstate shipments (304).

Fumigation of commodities already infested with insects will not result in a legal product, because dead insects or evidence of past insect activity are objectionable. Fumigation may be employed, when necessary, to prevent infestation, but care is required to prevent buildup of non-permitted chemical residues from fumigation.

2.5   Current good manufacturing practice regulations

To explain what is needed to maintain sanitary conditions in food establishments, FDA has published a set of Current Good Manufacturing Practice (GMP) Regulations. These tell what kinds of buildings, facilities, equipment, and maintenance are needed and the errors to avoid to insure sanitation. They also deal with such matters as building design and construction, lighting, ventilation, toilet and washing facilities, cleaning of equipment, materials handling, and vermin control. Food firms may obtain copies of these regulations by writing to the Food and Drug Administration.

Many food materials are intended for further processing and manufacture into finished foods; such processing in no way relieves the raw materials from the requirements of cleanliness and freedom from deleterious impurities.

In the 1970s, the following regulations were published by the FDA:

- 21 CFR Part 108, 'Emergency Permit Control';
- 21 CFR Part 110, 'Current Good Manufacturing Practice in Manufacturing, Processing, Packing and Holding Human Food (Commonly referred to as the "Umbrella Regulation")';
- 21 CFR Part 113, 'Thermally Processed Low-Acid Food Packaged in Hermetically Sealed Containers';
- 21 CFR Part 114, 'Acidified Foods'.

Part 113, Thermally Processed Low-Acid Foods Packed in Hermetically Sealed Containers, is the most extensive and detailed GMP regulation. It delineates equipment and its hook-up, operating procedures required to assure safety of the foods, records to be maintained, and the training required for plant personnel supervising retort operations and container closure operations. The regulation was first proposed to FDA by industry through the National Canners Association (now the National Food Processors Association) in November 1971. It was published as a final order on January 26, 1973.

The legal basis for 21 CFR Part 113 and other GMPs is Section 402(a) four of the US Food, Drug, and Cosmetic Act, which states: 'A food shall be deemed to be adulterated if it has been prepared, packed or held under unsanitary conditions, whereby it may have become contaminated with filth, or whereby it may have been rendered injurious to health'.

Microbiological spoilage due to any cause is, of course, of concern to industry and to FDA, whether or not it has public health significance. The provisions of Part 113 are intended to render a food 'commercially sterile', that is, free of all viable micro-organisms capable of reproducing in the food under normal nonrefrigerated conditions of storage and distribution. The thermal processes required to achieve commercial sterility are more than adequate to inactivate the spores of *Clostridium botulinum*, the dreaded organism in the canning industry. This organism, widely prevalent in nature, particularly in soils, grows only under anaerobic conditions. Although the organism is ingested regularly in fresh foods without any ill effects, when the organism is presented an anaerobic environment in the sealed container, it can multiply and produce its deadly toxin, if the pH of the product is higher than 4.6. There are some exceptions to this valid generalisation.

Part 113 contains both mandatory provisions and advisory suggestions, all of which are recognised as current good manufacturing practices. Meticulously followed, these practices greatly improve the safety of low-acid foods. In one respect, the regulation is unique; it recognises the importance of people and requires that all retort operations and all closure operations and inspections be performed under the supervision of personnel who have successfully completed a training course conducted by a school approved by the Commissioner of Food and Drugs. These courses are known as 'Better Process Control Plan' schools.

The Current GMP Regulations for low-acid canned foods (Part 113) first became effective in March 1973 and were last revised effective May 15, 1979.

The Emergency Permit Control Regulation for thermal processed low-acid canned foods packaged in hermetically sealed containers (part 108.35) requires all commercial processors of low-acid canned foods to register their establishments with FDA and requires processors to file scheduled processes (processing times and temperatures, etc.) for each low-acid canned food and/or container size. Plant registration and process filing is required for both US establishments and those in other countries which export low-acid canned foods to the US.

Another regulation, the Current GMP Regulations for acidified food (part 114), became effective in May 1979. The GMPs for acidified foods outline the manufacturing, processing, and packing procedures for acidified food processors, which must be followed to insure production of a safe product.

The Emergency Permit Control regulation for acidified foods (part 108.25) became effective in July 1979. Under this regulation, all food processors of acidified foods must register their processing establishment with FDA; processors must also file processing information (pH, water activity, etc.) for each acidified product. Both food-processing establishments in the US, and those in other countries which ship to the US, must register and file process information with FDA.

Because improperly processed low-acid foods or acidified foods present life-threatening hazards, registration of those processors who manufacture, process, or pack low-acid canned foods or acidified foods is necessary to permit adequate monitoring of compliance with regulations and to provide for immediate application of emergency permit control should an emergency situation arise.

To assist FDA in monitoring compliance with the regulations, filed scheduled processes are reviewed but are not approved by FDA. It is the responsibility of the processor to ascertain the adequacy of any process before it is used.

2.5.1   Process deviations

A process deviation is defined as a change in any critical condition of the scheduled process which reduces the sterilising value of the process or which raises a question regarding the public health safety and/or commercial sterility of the product lot.

Examples of changes in critical conditions are reductions in initial temperature, in time and/or temperature of processing, or not meeting the critical factors associated with the process such as headspace, consistency, and drained weight.

Examples of items which could raise questions regarding the sterility of the lot are

1. Faulty record keeping on the written and recorded processing record.
2. Failure to vent or improper venting procedures.
3. Failure to use heat-sensitive indicators or other effective means on containers, retort baskets, trucks, cars, or crates, or unretorted product, to prevent a retort bypass.
4. Not having the scheduled processes appropriately established.
5. Not adhering to the scheduled process.
6. Operating with improper equipment, such as improper mercury-in-glass thermometers.

Any lot of product which has a process deviation must either be destroyed, fully reprocessed, or the product set aside and evaluated by a competent processing authority, using procedures recognised as adequate to detect any potential public health hazard.

In all cases of process deviations, records must be kept, regardless of the degree of seriousness of the deviation. Full records must be kept, documenting the conditions of reprocessing and/or other measures taken to detect and eliminate potential public health hazards. Should the decision be made to destroy any portion on all of the questioned lot, records should document the destruction of the lot.

2.5.2   Foods not covered under the low-acid canned foods regulations (21 CFR 108.35 and 113)

Because the following foods are not considered low-acid foods, processing plants are not required to register and file processing information for these products.

1. Acid foods (natural or normal pH 4.6 or below).
2. Alcoholic beverages.
3. Fermented foods.
4. Foods processed under the continuous inspection of the meat and poultry inspection programme of the USDA under the Federal Meat Inspection Act and the Poultry Products Inspection Act.
5. Foods with water activity of 0.85 or below.
6. Foods which are not thermally processed.
7. Foods which are not packaged in hermetically sealed containers.
8. Foods stored, distributed, and retailed under refrigerated conditions.
9. Tomatoes and tomato products having a finished equilibrium pH less than 4.7.

2.5.3   Foods not covered under the acidified foods regulations (21 CFR 108.25 and 114)

Because the following foods are not considered acidified foods, processing plants do not have to register and file processing information for these products.

1. Acid foods (natural or normal pH 4.6 or below).
2. Acid foods (including such foods as standardised and nonstandardised food dressings and condiment sauces) that contain small amounts of low-acid food(s) and have a resultant finished equilibrium pH that does not significantly differ from that of the predominant acid or acid food.
3. Alcoholic beverages.

4. Carbonated beverages, jams, jellies, or preserves.
5. Fermented foods.
5. Foods with water activity of 0.85 or below.
6. Foods stored, distributed, and retailed under refrigerated conditions.

In addition, figs, pears, or pineapples, which are not acidified or are acidified and have a finished equilibrium pH above 4.6, are considered low-acid foods; if they are thermally processed and packaged in hermetically sealed containers, they are subject to part 113. If figs, pears, or pineapples are acidified to a pH of 4.6 or below, they are subject to Part 108.25 and 114, if the normal pH prior to processing is greater than 4.6.

2.5.4   The relationship between pH, water activity, and low-acid or acidified foods regulations

The following table (Table 2.1) illustrates the relationships between pH, water activity, and current low-acid or acidified food regulations. The table indicates whether a product is a low-acid or an acidified food. A product that is either a low-acid food or an acidified food must meet the registration and process filing requirements of FDA. The finished equilibrium pH and water activity (a_w) of the product must be determined to use the table. The relationship between pH and water activity defines a product as low acid or acidified. For example, if a food product has a pH less than 4.6 and a water activity of 0.85 or below, it would not be covered by the low-acid regulations or the acidified regulations. Thus, filing scheduled process information for that product would not be required.

Table 2.1   Relationship between pH, water activity, and acidified food regulations

		Registration and processing required	
pH (final equilibrium)	Water activity (a_w)	Low acid[a] acidified (not naturally) (21 CFR 108.35/113)	Acidified (not naturally)[b](21 CFR 108.35/114)
<4.5	≤0.85	no	no
<4.5	>0.95	no	yes
4.6	≤0.85	no	no
4.6	>0.85	no	yes
≥4.7	≤0.85	no	no
≥4.7	>0.85	yes	no

< = less than; > = greater than; ≤ = less than or equal to; ≥ = greater than or equal to.
[a]Low-Acid (21 CFR Part 108.35 and 113) A 'yes' under this column defines the product as low-acid and requires registration of the processing plant and filing scheduled process information for that low-acid product. A 'no' in this column means that registration of the processing plant or filing scheduled process information is not required.
[b]Acidified (21 CFR Part 108.25 and 114) A 'yes' under this column defines the product as acidified and that registration of the processing plant and filing scheduled process information is required. A 'no' under this column means neither registration nor filing scheduled process information is required because the product is not acidified.

2.5.5   Important information for processing plants in other countries

Imported products, such as low-acid foods, are subject to inspection at the time of entry through US Customs. Shipments found not to comply with laws and regulations may be detained at the port of entry. They must be either brought into compliance with laws and regulations, destroyed, or re-exported.

2.5.5.1   Reasons why import products may be detained

Low-acid or acidified foods may be detained at the port of entry for the following reasons:

1. Analytical results indicating product adulteration;
2. Labelling violations;
3. Firm is not registered;
4. Firm has not filed a process filing form indicating process information;
5. Lack of container code;
6. Scheduled process information is questionable and/or incomplete.

2.5.5.2   Information to be included in invoices or shipment manifest

To avoid delays in clearing low-acid or acidified products for entry into the United States, the shipping invoice must include the following:

1. Name and address of registered processing establishment where low-acid or acidified product is processed, manufactured, or packed.
2. Name of low-acid or acidified product, style, and packing medium.

   For example: Tuna in Brine.
   Dill Pickle (whole).
   Endives (whole) in Brine.

3. FCE # – Food Canning Establishment Number
4. Container Dimensions, listed as follows:

   Two dimensions – Diameter × Height.
   For example: 211 × 400 for a 2 11/16 × 4 in container.
   Three dimensions – Length × Width × Height (or thickness).
   For example: 405 × 211 × 014 for a 45/16 in × 2 11/, 6 in × 14/16 in container.

2.5.5.3   Deviations from GMP regulations

Low-acid canned food regulations require that products be prepared and packed in containers in such a way that they can be adequately processed, that the heating process be designed by persons with expert knowledge of thermal processing requirements, and that the heating process achieve commercial sterility. They also require proper design, controls, and instruments for all of the common retorting systems being used and operating practices that will provide the necessary safety. Records must be kept of all coding, processing, and container closure inspections so that management can review the records to assure commercial sterility prior to release for shipment.

If a processor finds from the records that a retort load is under-processed, he must either reprocess the load or set it aside for evaluation. If, after evaluation, the set-aside load does not present a health problem, it may be shipped. If it cannot be shown that there is no public health concern, then the product must be destroyed. Container closures must be inspected to assure that containers are properly sealed.

Supervisors of heat processing operators and container inspectors must receive instruction in a school approved by FDA. FDA has collaborated with other countries and has included foreign food safety personnel in training courses.

The most frequently reported deviations from GMPs are (1) inadequate, incomplete records that are not properly signed or inspected, (2) registration and process filing forms that are incomplete and sometimes nonexistent, (3) process deviations that have not been adequately evaluated to determine whether a public health hazard exists, (4) faulty temperature recording devices, (5) adequacy not verified, (6) improper venting of the retort, (7) initial temperature not controlled, (8) inadequate retorts, (9) reference thermometers not in compliance, and (10) critical factors not controlled.

In order for their products to be accepted for importation into the United States, foreign low-acid canned food producers must register with the FDA. Registration requirements include the name of the product, the manufacturer's name and address, and even the can size; the processing method used must be on file. Products must not be under-processed or abnormal nor should they contain more than the allowable amount of filth. In some cases, decomposition and contamination have been problems with imports.

2.5.5.4 Tolerances for filth

Many inquiries are received by the Food and Drug Administration concerning permitted variations from absolute cleanliness or soundness in foods. The act does not explicitly provide for 'tolerances' for filth or decomposition in foods. It states that a food is adulterated if it consists in whole or in part of a filthy, putrid, or decomposed substance.

This does not mean that a food must be condemned because of the presence of foreign matter in amounts below the irreducible minimum after all possible precautions have been taken. FDA recognises that it is not possible to grow, harvest, and process crops that are totally free of natural defects. The alternative, to increase the use of chemicals to control insects, rodents, and other sources of contamination, is not acceptable because of potential health hazards from chemical residues. To resolve this problem, FDA has published a list of 'defect action levels' stating amounts of contamination which will subject food to enforcement action. Copies may be obtained by request from the nearest FDA office.

The Food Defect Action Levels are established at levels that post no hazard to health and may be changed from time to time. Any products which might be harmful to consumers are subject to regulatory action, whether or not they exceed the 'defect' levels. In addition, manufacturing processes that do not conform to FDA's Current Good

Manufacturing Practice Regulations will result in regulatory action by FDA whether or not the product exceeds the defect level. These levels are not averages – actually, the average levels are much lower and FDA continues to lower these action levels as industry performance improves. The mixing of food to dilute contamination is prohibited and renders a product illegal, regardless of any defect level in the final product.

2.5.5.5   Generally recognised as safe substances

A number of substances are classified by the US Food and Drug Administration as 'generally recognised as safe', or GRAS. That group of substances is exempted from the requirements of proving that they are safe. Those materials are classified as GRAS on the basis of prior scientific evaluation or from experience based on long-time usage in food.

Typical GRAS substances include common salt (sodium chloride); common sugar (sucrose); ascorbic acid (vitamin C); fruit and beverage acids, like citric, malic, and phosphoric; caramel; benzoic acid and sodium benzoate; emulsifiers, such as fatty acid monoglycerides and diglycerides; potassium iodide; glycerol; numerous flavouring materials; and many others.

Criteria for eligibility for classification of substances in the GRAS list is included in 21 CFR 170.30.

The 'GRAS List' of approximately 670 substances is contained in Title 21, CFR Part 182. This part of the CFR, which includes the GRAS List, as well as food additive regulations, can be purchased from the Superintendent of Documents, US Government Printing Office, Washington, DC 20402.

2.5.5.6   Food additives

'Food Additives' are substances which may, by their intended uses, become components of food, either directly or indirectly, or may otherwise affect the characteristics of the food. The term 'specifically' includes any substance intended for use in producing, manufacturing, packing, processing, preparing, treating, packaging, transporting, or holding food and any source of radiation intended for any such use (409).

However, the law excludes from the definition of a 'food additive'

1. Substances generally recognised as safe by qualified experts;
2. Substances used in accordance with a previous approval ('prior sanction') under either the Federal Food, Drug, and Cosmetic Act, the Poultry Products Inspection Act (21 USC 451), or the Meat Inspection Act;
3. Pesticide chemicals in or on raw agricultural products;
4. A colour additive;
5. A new animal drug (but 3, 4, and 5 are subject to similar safety requirements of other sections of the law).

Manufacturers or importers, not certain whether chemicals or other ingredients used in their foods are subject to the safety clearance requirements of the Food Additives Amendment, may seek an opinion from the Food and Drug Administration. If premarket approval is required, this may mean that studies, including animal feeding tests, will have to be carried out in accordance with recognised scientific procedures and the results submitted to the Food and Drug Administration for evaluation. General principles for evaluating the safety of food additives are published in 21 CFR 170.20. Detailed instructions for preparing a food additive petition are in 21 CFR 171. The 'Delaney Clause' in the law (409(c)(3)) provides that no food additive may be found safe if it produces cancer when ingested by man or animals or if it is shown by other appropriate tests to be a cancer-producing agent, except that such ingredient may be used in animal feeds, if it causes no harm to the animal and provided there are no residues of the ingredient in the meat or other edible products reaching the consumer. This later provision is primarily applicable to veterinary drugs added to animal feed.

If the Food and Drug Administration concludes from evidence submitted that an additive will be safe, a regulation permitting its use will be issued. This regulation may specify the amount of the substance which may be present in or on foods, foods in which it is permitted, manner of use, and any special labelling required.

A substance cleared under the Food Additive Regulations is still subject to all the general requirements of the Food, Drug, and Cosmetic Act. The Saccharin Study and Labelling Act, passed November 23, 1977, prohibited for 18 months any new regulations restricting or banning the sale of saccharin or products containing it. Congress has extended this legislation three times, most recently through 1985. It requires further scientific evaluation of the carcinogenic potential of saccharin and a label warning: 'Use of this product may be hazardous to your health. This product contains saccharin, which has been determined to cause cancer in laboratory animals'.

Artificially sweetened products are required to be labelled as Special Dietary Foods (21 CFR 105.66). The foods permitted to contain such sweeteners and the amounts are specified in the Food Additive Regulations (21 CFR 180.37).

2.5.5.7 Housewares

Manufacturers of food-contact articles for use in the home or in food service establishments should make sure that nothing from the articles imparts flavour, colour, odour, toxicity, or other undesirable characteristics to food, thereby rendering the food adulterated.

Although food-packaging materials are subject to regulation as 'food additives', ordinary housewares are not. Such housewares include dishes, flatware, beverage glasses, mugs, cooking utensils, cutlery, and electrical appliances. This means that manufacturers do not have to submit data to the FDA showing that the materials used are safe nor are they required to pre-clear their housewares with FDA. However, housewares are not exempt from the general safety provisions of the Federal Food, Drug, and Cosmetic Act. Regulatory actions have been taken against cookware and ceramic dinnerware containing lead or cadmium.

2.6   Food standards

Food standards are a necessity to both consumers and the food industry; they maintain the nutritional values and the general quality of a large part of the national food supply. Without standards, different foods would have the same names, and the same foods different names, both situations confusing and misleading consumers and creating unfair competition.

Section 401 of the Federal Food, Drug, and Cosmetic Act requires that whenever such action will promote honesty and fair dealing in the interest of consumers, regulations shall be promulgated fixing and establishing for any food, under its common or usual name so far as practicable, a reasonable definition and standard of identity, a reasonable standard of quality, and/or reasonable standards of container fill. However, no definition and standard of identity, or standard of quality, may be established for fresh or dried fruits, fresh or dried vegetables, or butter, except that definitions and standards of identity may be established for avocados, cantaloupes, citrus fruits, and melons.

2.6.1 Standards of identity

The Standards of Identity define what a given food product is, its name, and the ingredients which must be used, or may be used, and which ones must be declared on the label. Standards of quality are minimum standards only and establish specifications for quality requirements. Fill-of-container standards define how full the container must be and how this is measured. FDA standards are based on the assumption that the food is properly prepared from clean, sound materials. Standards do not usually relate to such factors as deleterious impurities, filth, and decomposition, but there are exceptions. For example, the standards for whole egg and yolk products and for egg white products require these products be pasteuris or otherwise treated to destroy all viable *Salmonella* micro-organisms. Some standards for foods set nutritional requirements, such as those for enriched bread, or nonfat milk with added vitamins A and D, etc. A food which is represented, or purports to be a food for which a standard of identity has been promulgated, must comply with the specifications of the standard in every respect.

2.6.2 Imitation foods

A food that imitates another food is misbranded unless its label bears, in type of uniform size and prominence, the word 'imitation' and immediately thereafter the name of the food imitated (403(c)). Under the law, a food is an imitation if it is a substitute for, and resembles another food, but is nutritionally inferior to that food. Nevertheless, if the food is not nutritionally inferior to the food it imitates, it is permitted to be labelled descriptively according to the regulations in 21 CFR 101.3(e) and need not bear the term 'imitation'.

2.6.3 Standards of quality

Established under the Food, Drug, and Cosmetic Act, the Standards of Quality should not be confused with Standards for Grades that are published by the USDA for meat and other agricultural products and the US Department of the Commerce for fishery products. A standard of quality under the Food, Drug, and Cosmetic Act is a minimum standard only. If a food for which a standard of quality or fill of container has been promulgated falls below such standard, it must bear, in a prescribed size and style of type, label statements showing it to be substandard in quality, or in fill of container; for example, 'Below Standard in Quality, Good Food – Not High Grade'.

The USDA grades established are usually designated 'Grade A' or 'Fancy', 'Grade B' or 'Choice', or 'Grade C' or 'Standard'. The US Department of the Interior grades for fishery products are usually designated 'Grade A', 'Grade B', or 'Grade C'. These grade designations are not required by the Food, Drug, and Cosmetic Act to be stated on the labels, but, if stated, the product must comply with specifications for the declared grade.

2.6.4   Fill-of-container standards

Fill-of-container standards established for certain foods designate the quantity in terms of the solid or liquid components, or both. The existing fill-of-container standards for canned fruits and vegetables may be grouped as follows: (1) those that require the maximum practicable quantity of solid food that can be sealed in the container and processed by heat without crushing or breaking such component (limited to canned peaches, pears, apricots, and cherries); (2) those requiring a minimum quantity of solid food in the container after processing (the quantity is commonly expressed either as a minimum drained weight for a given container size or as a percentage of the water capacity of the container); (3) those requiring that the food, including both solid and liquid packing, shall occupy not less than 90% of the total capacity of the container; (4) those requiring both a minimum drained weight and the 90% minimum fill; and (5) those requiring a minimum volume of the solid component irrespective of the quantity of liquid (canned green peas and canned field peas). Fill-of-container standards specifying minimum net weight or minimum drained weight have been promulgated for certain fish products.

2.7   Composition and labelling guidelines

FDA food standards govern both labelling and composition and should be consulted for detailed specifications. The standards are published in the annual editions of the Code of Federal Regulations, Title 21, Parts 103–169. The following sections provide pertinent information on requirements for specific groups of foods.

2.7.1   Canned fruits and fruit juices

Standards of identity, quality, and fill of container have been promulgated for a number of canned fruits and fruit juices. The specific standards should be consulted by anyone intending to ship canned fruit to the United States.

Labels on canned fruits and fruit juices which meet the minimum quality standards, and canned fruits and fruit juices for which no standards of quality have been promulgated, need not make reference to quality, but, if they do, the product must correspond to the usual understanding of the labelled grade. Particular care must be taken to use the terms 'Fancy' or 'Grade A' only on those products meeting the specifications established for such grades by the USDA.

2.7.1.1 Fill-of-container standards – fruits and juices

Fill-of-container standards have been promulgated for a number of canned fruit and fruit juices, but in packing any other canned fruit, the container must be well filled with the fruit and with only enough packing medium added to fill the interstices; otherwise the container may be deceptive and prohibited by the Act. In judging the finished product, due allowance is made for natural shrinkage in processing.

The standards of fill for canned peaches, pears, apricots, and cherries require that fill of the solid food component be the maximum practicable quantity, which can be sealed in the container and processed by heat without crushing or breaking such component.

Standards for canned fruit cocktail, grapefruit, and plum specify minimum drained weights for the solid food component, expressed as a percentage of the water capacity of the container. These drained-weight requirements are as follows: fruit cocktail – 65%, grapefruit – 50%, whole plums – 50%, and plum halves – 55%. An additional 90% fill requirement, based on the total capacity of the container for the solid food and the liquid packing medium, has been established for plums. This 90% fill requirement also applies to applesauce in metal containers (85% fill for applesauce in glass), crushed pineapple, and pineapple juice.

Minimum drained weight requirements for canned pineapple provide that 'crushed pineapple' may be labelled as 'heavy pack' or 'solid pack'; the standard of quality requires a minimum drained weight for crushed pineapple of 63% of the net weight.

All fruit used for canning or juice should be mature and sound, that is, free from insect infestation, mouldiness, or other forms of decomposition.

2.7.2 Canned vegetables

Canned vegetable products must be prepared from sound, wholesome raw materials, free from decomposition. Definitions and standards of identity for a wide variety of canned vegetables provide that 'the food is processed by heat, in an appropriate manner, before or after being sealed in a container so as to prevent spoilage'. The importance of adequate heat processing of canned vegetables, particularly the nonacid types, is emphasised by the special regulations for low-acid canned foods.

Cans of food which have become swollen, or otherwise abnormal, are adulterated and should be destroyed.

Standards of quality have been promulgated for many vegetables. These are minimum standards only and establish specifications for quality factors, such as tenderness, colour, and freedom from defects. If the food does not meet these standards, it must be labelled in bold type 'Below Standard in Quality', followed by the statement 'Good Food-Not High Grade', or a statement showing in what respect the product fails to meet the standard, such as 'excessively broken', or 'excessive peel' (21 CFR 130.14). The standards of quality for canned vegetables supplement the identity standards. In other words, the identity standards define what the product is and set floors and ceilings on important ingredients, whereas the quality standards provide for special labelling for any product which does not meet the usual expectations of the consumer but is nevertheless a wholesome food.

2.7.2.1   Fill of container

If a fill-of-container standard has not been promulgated for a canned vegetable, the container must nevertheless be well filled with the vegetable, with only enough packing medium to fill the interstices. In the case of canned tomatoes, no added water is needed or permitted.

The standards of fill for canned tomatoes and canned corn require that the solid food and the liquid packing medium occupy not less than 90% of the total capacity of the container. The standard for canned corn also specifies a minimum drained weight of 61% of the water capacity of the container for the corn ingredient, whereas the standard of quality for canned tomatoes specifies a minimum drained weight of 50% of the water capacity of the container for the solid tomato ingredient. Canned mushrooms must meet minimum drained weight requirements, stated in ounces, for various can sizes; canned green peas and field peas must comply with volumetric fill requirements which state that if the peas and liquid packing medium are poured from the container and returned thereto, the levelled peas (irrespective of the liquid) completely fill the container.

2.7.3   Tomato products

Shippers of tomato products (canned tomatoes, tomato juice, paste, puree, and catsup) should consult the standards of identity for these items (21 CFR 155.190-4). Attention is called to the salt-free tomato solids requirements for puree and paste and to the fact that neither artificial colour, nor preservatives, are permitted in any of these products. Tomato juice is unconcentrated; tomato puree must contain not less than 8%, and tomato paste not less than 24%, salt-free tomato solids.

These tomato products are occasionally contaminated with rot because of failure to remove decayed tomatoes from raw material entering the cannery. Flies and worms are also filth contaminants of tomato products. The preparation of a clean tomato product requires proper washing, sorting, and trimming of the tomatoes and frequent cleaning of the cannery equipment, including tables, utensils, vats, and pipelines.

In judging whether tomato products have been properly prepared to eliminate rot and decay, the Food and Drug Administration uses the Howard Mould Count test; it refuses admission to import shipments and takes action against domestic shipments if mould filaments are present in excess of amounts stated in the Food Defect Action Levels. Methods of testing tomato products are given in the Official Methods of the AOAC.

2.7.4   Fruit jams (preserves), jellies, fruit butters, marmalades

The standards (21 CFR 150) require that these products be prepared by mixing not less than 45 parts by weight of certain specified fruits (or fruit juice in the case of jelly), and 47 parts by weight of other designated fruits, to each 55 parts by weight of sugar or other optional nutritive carbohydrate sweetening ingredient. Only sufficient pectin may be added to jams and jellies to compensate for a deficiency, if any, of the natural pectin content of the particular fruit. The standards also require that for both jams (preserves) and jellies, the finished product must be concentrated to not less than 65% soluble solids.

Standards of identity have also been established for artificially sweetened jams and jellies; for these products, the fruit ingredient must be not less than 55% by weight of the finished food product.

Fruit butters are defined as smooth, semisolid foods made from not less than five parts by weight of fruit ingredient to each two parts by weight of sweetening ingredient. As is the case with jams and jellies, only sufficient pectin may be added to compensate for a deficiency, if any, of the natural pectin content of the particular fruit. The fruit butter standard requires that the finished product must be concentrated to not less than 43% soluble solids.

There is no formal standard of identity for citrus marmalade. However, FDA expects a product labelled sweet orange marmalade to be prepared by mixing at least 30 lbs (13.6 kg) of fruit (peel and juice) to each 70 lbs (31.8 kg) sweetening ingredients. Sour or bitter (Seville) orange marmalade, lemon marmalade, and lime marmalade would be prepared by mixing at least 25 lbs (11.4 kg) of fruit (peel and juice) to each 75 lbs (34.1 kg) of sweetening ingredient. The amount of peel should not be in excess of the amounts normally associated with fruit. The product should be concentrated to not less than 65% soluble solids. Such products should not be regarded as misbranded.

Jams, jellies, and similar fruit products should, of course, be prepared only from sound fruit. Decayed or decomposed fruits and insect-contaminated fruits should be sorted out and discarded.

2.7.5   Fishery products

Fish and shellfish are imported into the US in the fresh, frozen, pickled, canned, salted, dried, or smoked conditions. The raw materials are, by nature, extremely perishable and must be handled rapidly under adequate refrigeration, if decomposition is to be avoided. Conditions of handling are also likely to result in contamination, unless particular safeguards are imposed.

Sometimes preservatives are to prevent or retard spoilage in fish. Any preservative used must be one accepted by FDA as safe and must be declared on the label.

Chemical contamination of lakes, rivers, and the oceans has been found to concentrate in certain species of fish. Excessive residues of pesticides, mercury, and other heavy metals are prohibited.

2.7.5.1   Names for seafoods

To prevent substitution of one kind of seafood for another, and subsequent deception of the consumer, it is essential that labels bear names which accurately identify the products designated. Words like 'fish', 'shellfish', or 'mollusk' are not sufficient; the name of the specific seafood must be used. Many fish, crustaceans, and mollusks have well established common or usual names throughout the United States (for example, 'pollock', 'cod', 'shrimp', and 'oyster'). These may not be replaced with other names, even though the other names may be used in some areas or countries. Neither may they be replaced with coined names, even though the coined names may be considered more attractive or to have greater sales appeal.

A more difficult problem is that of deciding what constitutes a proper designation for a seafood which has not previously been marketed in the United States and thus which has not acquired an established common or usual name in this country. In selecting an appropriate name for such a product, full consideration must be given to its proper biological classification and to avoiding a designation which duplicates or may be confused with the common or usual name in this country of some other species.

Labels on frozen fishery products, for which no standards of quality have been promulgated, need not refer to quality, but, if they do, the product must correspond to the usual understanding of the labelled grade. Particular care must be taken not to use the term 'Grade A' on products that do not meet the established understanding for these terms in the United States.

2.7.5.2 Canned fish

Canned fish generally are 'low-acid canned foods' and packers are therefore subject to the registration requirements outlined earlier in this chapter. Failure to declare the presence of added salt or the kinds of oil used as the packing medium in canned fish has resulted in the detention of fish products. If permitted, artificial colours or chemical preservatives are used; their presence must be conspicuously declared in the labelling. Artificial colouring is not permitted if it conceals damage or inferiority or if it makes the product appear better or of greater value than it is.

The packing of canned fish and fish products with excessive amounts of packing medium has resulted in many detentions. If fish is in a packing medium, such as anchovies in oil, the container should be as full as possible of fish with the minimum amount of oil. The fact that oil may be equal in value, or even more expensive than the fish, does not affect this principle. Canned lobster paste and similar products have been encountered which were deceptively packaged because of excessive headspace, i.e. excessive space between the lid of the can and the surface of the food in the can.

Canned Pacific Salmon is required to comply with standards of identity and fill of container. The standards establish the species and names required on labels and the permitted styles of pack (21 CFR 161.170).

Anchovies. Products represented as anchovies should consist of fish of the family *Engraulidae*. Other small fish, such as small herring and herring-like fish, which may superficially resemble anchovies, cannot properly be labelled as anchovies. The product should be prepared from sound raw material and salting or curing should be conducted in such a manner that spoilage does not occur.

Sardines. The term 'sardines' is permitted in the labelling of the canned products prepared from small-size clupeoid fish. The sea herring (*Clupea harengus*), the

European pilchards (*Sardina pilchardus* or *Clupea pilchardus*), and the brisling, or sprat (*Clupea sprattus*), are commonly packed in small-size cans and labelled as sardines. The terms 'brisling sardines' and 'sild sardines' are permissible in the labelling of canned small brisling and herring, respectively. Large-size herring cannot be labelled sardines. These canned products must be free from all forms of decomposition, such as 'feedy', 'belly-blown' fish and must be adequately processed to prevent active spoilage by micro-organisms. Fish are called 'feedy fish' when their stomachs are filled with feed at the time the fish are taken from the water. Such fish deteriorate rapidly until the viscera and thin belly wall disintegrate, producing a characteristic ragged appearance called 'belly-blown'.

Tuna. A standard of identity defines the species of fish that may be canned under the name 'tuna' (21 CFR 161.190(a)); there is also a standard of fill of container for canned tuna (21 CFR 161.190(c)). The standards provide for various styles of pack, including solid pack, chunk or chunk style, flakes, and grated tuna. Provision is also made for various types of packing media, certain specified seasonings and flavourings, colour designations, and methods for determining fill of containers. The standard of fill of container for canned tuna specifies minimum values for weights of the pressed cake of canned tuna, depending on the form of the tuna ingredient and the can size.

The canned fish *Sarda chilensis*, commonly known as the bonito or bonita, may not be labelled as tuna, because it is not a true tuna, but must be labelled as bonito or bonita. The fish *Seriola dorsalis*, commonly known as 'yellowtail', must be labelled as yellowtail and may not be designated as tuna.

2.7.5.3 Shellfish certification and standards for canned oysters

Because raw shellfish – clams, mussels, and oysters – may transmit intestinal diseases, such as typhoid fever, or be carriers of natural or chemical toxins, it is most important that they be obtained in a sanitary manner.

Shellfish must comply with the general requirements of state health agencies cooperating in the National Shellfish Sanitation Program (NSSP), administered by FDA, and in the Interstate Shellfish Sanitation Conference (ISSC).

Standards of identity are set for raw oysters, Pacific oysters, and canned oysters. These standards define the sizes of oysters and prescribe the methods of washing and draining to prevent adulteration with excess water. A fill-of-container standard for canned oysters fixes the drained weight at not less than 59% of water capacity of the container (21 CFR 161.130–161.145).

2.7.5.4 Rock lobster, spiny lobster, sea crayfish

The sea crayfish, *Palinurus vulgaris*, is frequently imported into the United States in the form of frozen tails, frozen cooked meat, or canned meat. By long usage, the terms 'Rock Lobster' and 'Spiny Lobster' have been established as common or usual names for these products. No objection has been offered to either of these terms, providing the modifying words 'rock' or 'spiny' are used in direct connection with the word 'lobster' in type of equal size and prominence.

In examination of imports, decomposition has sometimes been detected in all three forms of the product. In the canned product, decomposition resulted from the packing of decomposed raw material and from active bacterial spoilage. In the frozen cooked products, detentions have been necessary, also because of the presence of micro-organisms indicative of pollution with human or animal filth, as well as of decomposition.

2.7.5.5 Shrimp

Standards set minimum requirements for canned wet and dry pack shrimp and frozen raw breaded shrimp (21 CFR 161.173). This product must also comply with the regulations for low-acid canned foods.

2.7.6 Meat and meat food products

Meat or meat products derived from cattle, sheep, swine, goats, and horses are subject to the provisions of the Wholesome Meat Act enforced by the Food Safety and Inspection Service of the USDA, as well as certain provisions of the FDC Act. Foreign meat products must originate in countries whose meat inspection programmes have been approved. Each shipment must be properly certified by the foreign country and upon inspection by the federal meat inspectors found to be completely sound, wholesome, and fit for human food before entry into the United States. Requests for inspection should be made to the Food Safety and Inspection Service, USDA, Washington, DC 20250. The imported product, after admission into the United States, becomes a domestic article subject not only to the Federal Meat Inspection Act but also to the Federal Food, Drug, and Cosmetic Act to the extent the provisions of the Meat Inspection Act do not apply. Wild game, however, is subject to the requirements of the Federal Food, Drug, and Cosmetic Act and its regulations.

2.7.7 Poultry and poultry products

Poultry and poultry products offered for import into the US are subject to the Wholesome Poultry Act, also enforced by the Food Safety and Inspection Service, to which inquiries concerning such products should be addressed. The term 'poultry' means any live or slaughtered domesticated bird (chickens, turkeys, ducks, geese, or guineas), and the term 'poultry product' means any poultry which has been slaughtered for human food, from which the blood, feathers, feet, head, and viscera have been removed in accordance with rules and regulations promulgated by the Secretary of Agriculture, or any edible part of poultry, or, unless exempted by the Secretary, any human food product consisting of any edible part of poultry, separately, or in combination with other ingredients.

Poultry and poultry products are also subject to the Federal Food, Drug, and Cosmetic Act to the extent that the provisions of the Poultry Products Inspection Act do not apply.

Soups generally are under the jurisdiction of the USDA; those containing less than 3% raw meat or 2% cooked meat, poultry, or both, as flavouring ingredients, are subject to regulation by FDA.

2.7.8   Nuts and nut products

Nuts are adulterated if they are insect infested or insect damaged, mouldy, rancid, or dirty. Empty or worthless unshelled nuts should be removed by careful hand sorting or by machinery. Care should be taken to eliminate infested, dirty, mouldy, or rancid nuts from the shipment. Conditions which may cause nuts to be refused admission for import into the US are described below.

*   Insect infestation – Nuts are insect infested if they contain live or dead insects, whether larvae, pupae, or adults, or if they show definite evidence of insect feeding or cutting, or if insects' excreta pellets are present.
*   Dirt – Nut meats may become dirty because of lack of cleanliness in cracking, sorting, and packaging.
*   Rancidity – Nuts in this class have an abnormal flavour characterised by rancidity. Rancid nuts are frequently soft and have a yellow, dark, or oily appearance.
*   Extraneous material – Stems, shells, stones, and excreta should not be present. Defect action levels have been established for tree nuts. Deliberate mixing of good and bad lots to result in defects under these levels is prohibited, even though the percentage of defects in the mixed lots is less than the defect action level.
*   Aflatoxin – Aflatoxins are a group of chemically related substances produced naturally by the growth of certain common moulds. Aflatoxins, especially aflatoxin B1, are highly toxic, causing acute liver damage in exposed animals; Aflatoxin B1 also exhibits highly potent cancer-producing properties in certain species of experimental animals. Studies of certain population groups reveal that the consumption of aflatoxin-containing foods is associated with liver cancer in humans. The presence of aflatoxin in nuts and other products at levels above established action levels is a significant public health problem and is a basis for seizing or refusing imports of products containing it.
*   Bitter almonds – Because of their toxicity, bitter almonds may not be marketed in the United States for unrestricted use. Shipments of sweet almonds may not contain more than 5% of bitter almonds. Almond paste and pastes made from other kernels should contain less than 25 parts per million hydrocyanic acid (HCN) naturally occurring in the kernels. Nuts and nut meats must be prepared and stored under sanitary conditions to prevent contamination by insects, rodents, or other animals. Nuts imported for pressing of food oil must be just as clean and sound as nuts intended to be eaten as such or to be used in manufactured foods.

2.7.8.1   Standards for nut products

Mixed tree nuts, shelled nuts, and peanut butter are subject to FDA standards (21 CFR 164). The standards establish such factors as the proportions of various kinds of nuts and the label designations for 'mixed nuts', the fill of container for shelled nuts, and the ingredients and labelling for peanut butter. All packers and shippers of nut products should be aware of the requirements of these standards.

2.7.9   Edible oils

Olive oil is the edible oil expressed from the sound, mature fruit of the olive tree. Refined or extracted oil is not entitled to the unqualified name 'olive oil'. Other vegetable oils should be labelled by their common or usual name, such as cottonseed, sunflower, peanut, or sesame. Mixtures of edible oils should be labelled to show all the oils present and names should be listed on the labels in the descending order of predominance in the product. The terms 'vegetable oil' or 'shortening' are not permitted to be used in food labelling without disclosing the source of each oil or fat used in the product (21 CFR 101.4(a)(14)). Pictures, designs, or statements on the labelling must not be misleading concerning the kind or amount of oils present or as to their origin.

Cod liver oil is a drug, as well as a food, because it is recognised in the United States Pharmacopeia (USP). Its value as a food, whether intended for human or animal use, depends mainly on its vitamin D content. Articles offered for entry as cod liver oil must comply with the identity standard prescribed by the USP and conform to other specifications set forth in that official compendium.

2.8   Colour additives

The Federal Food, Drug, and Cosmetic Act provides that foods, drugs, devices, and cosmetics are adulterated if they contain colour additives which have not been proved safe to the satisfaction of the Food and Drug Administration for the particular use. A colour additive is a dye, pigment, or other substance, whether synthetic or derived from a vegetable, animal, mineral, or other source, which imparts a colour when added or applied to a food, drug, cosmetic, or the human body.

Regulations are issued listing permitted colour additives and the conditions under which they may be safely used, including the amounts that may be used when limitations are necessary. Separate lists are provided for colour additives for use in or on food, drugs, devices, and cosmetics, though some colours may appear on more than one list.

Testing and certification by the Food and Drug Administration of each batch of colour is required before use in foods, drugs, or cosmetics, unless certification of the colour additive is specifically exempted by regulation.

Regulations should be checked to ascertain which colours have been listed for various uses as colour additives in foods, drugs, devices, or cosmetics. Information on labels, which are required to contain sufficient information to assure safe use, and observance of any limitations placed on use, such as 'for food use only', directions for use where tolerances are imposed, or warnings against use, such as 'Do not use in products used in the area of the eye' should be used.

Manufacturers of certifiable colours may address requests for certification of batches of such colours to the Food and Drug Administration. Certification is not limited to colours made by United States manufacturers; requests will be accepted from foreign manufacturers, if signed by both such manufacturers and their agents residing in the United States. Certification of a colour by an official agency of a foreign country cannot, under the provisions of the law, be accepted as a substitute for certification by the Food and Drug Administration.

For copies of regulations governing the listing, certification, and use of colours in foods, drugs, and cosmetics shipped in interstate commerce, or offered for entry into the United States, or answers to questions concerning them, contact the Food and Drug Administration, 200C St., SW, Washington, DC 20204.

2.9   Suggestions to foreign exporters and United States importers to expedite entries

Packaged foods, drugs, and cosmetics must be labelled. Unlabelled goods must necessarily be detained because of lack of the mandatory label information. This causes delays that might not otherwise occur.

Products known to be illegal should not be imported on the theory that they can be brought into compliance with the law upon arrival. Conditional releases for that purpose are discretionary under the Food, Drug, and Cosmetic Act and, if the privilege is granted, delays and added expense to importers inevitably occur before final release or denial of release. Correct at the source obviously objectionable conditions responsible for detentions, such as infestation of crude drugs, foods, etc. with insects and other vermin.

US Customs entries should be made promptly upon arrival of shipments at the port of entry. The Food and Drug Administration cannot ordinarily act upon an importation until entry is actually filed and the Customs Bureau has notified the Administration's local office. Entry documents should be 'flagged' for the attention of the FDA representative at the appropriate office.

In carrying on business transactions with the importing trade of the United States, foreign shippers, as well as the importing purchasers, should bear in mind they have a definite obligation to comply fully with United States laws affecting these commodities. Sustained effort should be made to bring about fundamental correction of any existing conditions which adversely affect purity, strength, or quality of such articles.

Such corrective measures might well be undertaken by individual producers and associations of producers and possibly by foreign government agencies in the countries of production. By such procedures, eradication of causes of deterioration and contamination in a number of commodities has already been successfully accomplished in various parts of the world, with a resulting reduction in the number of adverse actions by the Food and Drug Administration when shipments were offered for entry into the United States. Moreover, individuals and associations and various foreign government agencies have, in a number of instances, instituted systems for sampling and examination of specific lots intended for shipment to the United States for predetermining whether specific lots so intended meet the requirements of the US Laws.

2.9.1   Retail food protection

FDA relies mainly on state and local authorities to insure the safety of food passing through retail channels, such as restaurants, cafeterias, supermarkets, snack bars, and vending machines. Given the size and diversity of the retail food industry, FDA directs its efforts to promote effective state and local regulation. It assists and supports the development of model food protection laws and uniform standards by providing technical assistance, exchanging information, conducting training programmes for regulatory personnel, and, upon request, evaluating state programmes.

Preventing the mishandling of food after it reaches the retailer is the primary concern of this joint federal/state effort. Food-borne disease has numerous causes but results most often from improper holding and storage temperatures, inadequate cooking, poor hygienic practices by food handlers, and lack of cleanliness in facilities. Inquiries should be addressed to Food Protection Branch (HFS-627), Food and Drug Administration, 200C Street, SW, Washington, DC 20204.

2.9.2   Interstate travel sanitation

Sanitation requirements relating to passengers and crew on interstate carriers are contained in the regulations for Control of Communicable Diseases (21 CFR Part 1240) and Interstate Conveyance Sanitation (21 CFR Part 1250). These regulations contain specific requirements for equipment and operations for handling food, water, and waste, both on conveyances (aircraft, buses, railroads, and vessels) and those located elsewhere, i.e., support facilities such as caterers and commissaries, watering points, and waste servicing areas. The regulations also specify requirements for reviewing plans and inspecting construction of equipment, conveyances, and support facilities. Carriers are required to use equipment and support facilities which have been approved by the Agency.

For further information write to: Food and Drug Administration, Division of Regulatory Guidance, 200 C Street, SW, Washington, DC 20204.

2.10   Food plant inspection

Food plant inspection, as historically practised in the United States, has been based on the observation of operating conditions on the day of inspection. Many corrective actions were achieved through this approach. This type of inspection undoubtedly will be continued because it is quite applicable to certain segments of the food industry; the inspection of warehouses is a good example.

This type of inspection provides little information on the manner in which the operation is conducted when an inspector is not present at the plant. It also does not tell much about procedures employed by management to monitor and control operations on a continuing basis.

To determine whether a plant is doing a good job consistently, it is necessary to audit production logs, charts, and written quality control records covering a long period. On the day of an inspection, for example, observing that adequate times and temperatures are being used in the processing of a food does not always mean that this is an everyday procedure. If clear instructions are posted and detailed records exist showing evidence of a properly functioning quality assurance system, then there is reasonable assurance of an acceptable product every day. If quality control records are not up-to-date, or are nonexistent, then it is probable that some of the finished product is substandard or even unsafe.

2.11 US food standards

As US standards are promulgated, they are published, first as proposals to allow for comments from interested parties and then published as a final regulation in the Federal Register of the United States. They are later compiled in an annual edition of the CFR, Title 21. Copies of individual standards may be obtained from the appropriate US government agency.

Proposed standards offered by the food industry, the government, or some interested person are formally published in the Federal Register to obtain comments and suggestions from all interested parties, including consumers.

After evaluation of the comments and other pertinent facts involved, an order is published which adopts, modifies, or rejects the proposal.

An order or a 'rule' may be stayed upon the basis of an objection made by an adversely affected party who requests a public hearing, and in addition, there are provisions in the law for carrying an appeal to a Federal Circuit Court of Appeals.

FDA food standards of identity are mandatory or regulatory. They set requirements which products must meet if they move in interstate commerce. They protect against deception, because they define what a food product must consist of to be legally labelled 'mayonnaise', for example.

USDA grade standards for food are voluntary. Federal law does not require that a food processor or distributor use the grade standards. These standards are widely used, however, as an aid in wholesale trading, because the quality of a product affects its price. The grade (quality level) is often shown on food products in retail stores so consumers may choose the grade that best fits their needs.

Food standards established by the federal government usually fall into two general classes: voluntary or mandatory. A third class consists of standards recommended for adoption by state and local governments.

Below is a listing of the principal kinds of federal standards for food.

2.11.1 Voluntary standards

2.11.1.1 USDA grade standards

Under authority of the Agricultural Marketing Act of 1946 and related statutes, USDA has issued grade standards for some 300 food and farm products.

Food products for which grade standards have been established are: beef, veal and calf, lamb and mutton, poultry, including turkey, chicken, duck, goose, guinea, and squab, eggs, manufactured dairy products, including butter, cheddar cheese, and instant nonfat dry milk, fresh fruits, vegetables, and nuts, canned, frozen, and dried fruits and vegetables and related products, such as preserves, and rice, dry beans, and peas. US grade standards are also available for grains but not for the food products, such as flour or cereal, into which grain is processed.

USDA provides official grading services, often in cooperation with state departments of agriculture, for a fee to packers, processors, distributors, or others who wish official certification of the grade of a product. Grade standards are also often used by packers and processors as a quality control tool.

Federal law does not require use of US grade standards or official grading services. Official grading is required under some state and local ordinances and some industry marketing programmes.

Products which have been officially graded may carry the USDA grade name or grade shield, such as the familiar purple 'USDA Choice' shield seen on cuts of beef, or the 'US Grade A' on cartons of eggs. Grade labelling, however, is not required by federal law, even though a product has been officially graded. On the other hand, a packer or processor may not label his product with an official grade name, such as Grade A (even without the 'US' prefix), unless it actually measures up to the federal standard for that grade. Mislabelling of this sort would be deemed a violation of the US Federal Food, Drug, and Cosmetic Act.

2.11.1.2   National marine fisheries service grade standards

The US Department of Commerce's National Marine Fisheries Service provides grade standards and grading services for fishery products similar to those provided by USDA for other foods. To date, 15 US grade standards have been developed for frozen processed fishery products, covering such products as semiprocessed raw whole fish, fish blocks, and cut fish portions, steaks and fillets, breaded raw and precooked fish portions and stocks, raw headless and breaded shrimp, and raw and fried scallops. Such products, when produced and graded under the US Department of Commerce inspection programme, may carry the USDA 'Federally Inspected' mark and/or the US grade shield. However, as under the USDA grading programmes, grade labelling is not required by federal law, even though products are officially inspected and graded.

2.11.2   Mandatory standards

2.11.2.1   Standards of composition and identity

USDA has established minimum content requirements for federally inspected meat and poultry products (usually canned or frozen) under the Federal Meat Inspection Act and the Poultry Products Inspection Act and these regulations are covered in Title nine CFR Part 319.

To be labelled with a particular name – such as 'Beef Stew' – a federally inspected meat or poultry product must meet specified content requirements. These requirements assure the consumer that he's getting what the label says he's getting. They do not, however, prevent different companies from making distinctive recipes. The USDA minimum content requirement for products listed in Title nine CFR Part 319 are 'Chile con Carne' – 40% of fresh meat weight; 'Chile con Carne with Beans' – 25% of fresh meat weight; 'Hash' – 35% of cooked meat weight;

'Corned Beef Hash' – not less than 35% of cooked meat weight of fresh beef; 'Cured Beef' or 'Canned Corned Beef', for meat stew – 25% fresh meat of species named on label; 'Tamales' – 25% fresh meat; 'Spaghetti with Meat and Sauce' and similar products – not less than 12% of fresh meat weight; 'Spaghetti sauce with Meat' – 6% of fresh meat; 'Tripe with Milk' – 65% tripe; 'Beans with Frankfurters in Sauce', 'Sauerkraut with Wieners and Juice', and similar products – not less than 20% by weight of the cooked meat product; 'Lima Beans with Ham in Sauce', 'Beans with Bacon', and similar products – 12% of cooked product; 'Chow Mein Vegetables with Meat' – 12% of fresh meat; 'Pork or Beef with Barbecue Sauce' – 50% of cooked meat; 'Beef with Gravy' shall not contain from meat with more than 30% trimable fat; 'Deviled Ham', tongue, or similar products – not more than 35% fat; and 'Ham or Tongue Spread' – not less than 50% of the meat ingredient named. In addition, other guidelines in the above-mentioned standards include setting specific and optional ingredients. Interested parties should consult the CFR for complete details.

USDA has also established complete standards of identity for pizza, oleomargarine, lard, mixed-fat shortening, and meat extracts.

2.11.2.2   US food and drug administration standards

The Federal Food, Drug, and Cosmetic Act provides for three kinds of mandatory standards for products being shipped across state lines: standards of identity, standards of minimum quality, and standards of fill of container. All these standards are administered by the Food and Drug Administration of the US Department of Health and Human Services. The law sets forth penalties for noncompliance.

Standards of Identity – FDA standards of identity (like USDA's) establish what a given food product is – that is, what a food must be to be labelled 'preserves'. FDA standards of identity also provide for use of optional ingredients, in addition to the mandatory ingredients that identify the product. Standards of identity have eliminated from the market such things as 'raspberry spread' – a product made from a little fruit and a lot of water, pectin, sugar, artificial colouring and flavouring, and a few grass seeds to suggest a fruit preserve; such a product may not be labelled as 'preserves'.

FDA has standards of identity for a large number of food products (excluding meat and poultry products, which are covered by USDA).

Types of products for which standards of identity have been formulated by FDA include: cacao product, cereal flour and related products, macaroni and noodle products, bakery products, milk and cream products, cheese and cheese products, frozen desserts, food flavouring, dressings for food, canned fruits and fruit juices, fruit butters, jellies and preserves, nonalcoholic beverages, canned and frozen shellfish, eggs, and egg products, oleomargarine or margarine, nut products, canned vegetables, and tomato products.

Standards of Minimum Quality – FDA standards of quality have been set for a number of canned fruits and vegetables to supplement standards of identity. These are minimum standards for such factors as tenderness, colour, and freedom from defects. They are regulatory, as opposed to USDA grand standards of quality, which are for voluntary use.

If a food does not meet FDA quality standards, it must be labelled 'Below Standard in Quality Food – Not High Grade' or, words may be substituted for the second part of that statement to show in what respect the product is substandard. The label could read, 'Below Standard in Quality Food, Excessively Broken', or 'Below Standard in Quality Food, Excessive Peel'. The consumer seldom, if ever, sees a product with a substandard label.

When USDA grade standards are developed for a product for which FDA has a minimum standard of quality, requirements for the lowest grade level USDA sets are at least as high as the FDA minimum. USDA grade standards for canned tomatoes, for example, are US Grades A, B, and C. Grade C is comparable to FDA's minimum standard of quality.

Standards of Fill of Container – These standards tell the packer how full a container must be to avoid deception. They prevent the selling of air or water in place of food.

2.12   A general guide to canned food

The packaging and labelling of food is subject to regulation in most jurisdictions to promote food safety by informing the consumer to make the right choices and to prevent false advertising. Food labelling is an area of much debate and interest. It is very important to understand any claims that may be made, any allergens that may be present (even if in trace amounts), and the target market. In most countries, it is the responsibility of the manufacturer or importer of a food to insure that the products comply with current food labelling. Below is a summary of a manual available from the US FDA that addresses frequently asked questions on food labelling. These are good guidelines for labelling of any food, but the actual requirements for specific countries or areas must be followed.

2.12.1   Summary of frequently asked questions and answers from the FDA

2.12.1.1   General food labelling requirements

1. Where should label statements be placed on food containers and packages?

   There are two ways to label food packages and containers:

   a. Place all required label statements on the front label panel (the principal display panel), or
   b. Place certain specified label statements on the principal display panel, and other labelling on the information panel (the label panel immediately to the right of the principal display panel, as seen by the consumer facing the product).
2. What are the principal display panel and the alternate principal display panel areas of the label?

   The principal display panel, or PDP, is that portion of the package label that is most likely to be seen by the consumer at the time of purchase. Many food containers are designed with two or more different surfaces that are suitable for display as the PDP.

3. What label statements must appear on the principal display panel?

Place the statement of identity, or name of the food, and the net quantity statement, or amount of product, on the PDP and on alternate PDP.

4. Which label panel is the information panel?

The information panel is the label panel immediately to the right of the PDP, as displayed to the consumer. If this panel is not usable due to package design and construction, i.e. folded flaps, then the information panel is the next label panel immediately to the right.

5. What is information panel labelling?

The phrase 'information panel labelling' refers to the label statements that are required to be placed together, without any intervening material, on the information panel, if such labelling does not appear on the PDP. These label statements include the name and address of the manufacturer, packer or distributor, the ingredient list, and nutrition labelling (when present).

6. What type size, prominence, and clarity is required?

For information panel labelling, use a print or type size that is prominent, conspicuous, and easy to read. The lettering must conform to specific size requirements and must contrast sufficiently with the background to be easy to read. Do not crowd required labelling with artwork or nonrequired labelling.

7. What is the prohibition against intervening material?

Nonessential, intervening material is not permitted to be placed between the required labelling on the information panel.

8. What name and address must be listed on the label?

Name and address of the manufacturer, packer, or distributor (unless the name given is the actual manufacturer, it must be accompanied by a qualifying phrase which states the firm's relation to the product, e.g. 'manufactured for' or 'distributed by').

2.12.1.2   Name of food

1. What is the name of the food statement called and where must it be placed?

The statement of identity is the name of the food. It must appear on the front label or principal display panel, as well as any alternate PDP.

2. Should the statement of identity stand out?

Use prominent print or type for the statement of identity. It shall be in bold type. The type size must be reasonably related to the most prominent printed matter on the front panel and should be one of the most important features on the PDP. The size of the lettering may be specified.

3. What name should be as the statement of identity?

The common or usual name of the food, if the food has one, should be used as the statement of identity. If there is none, an appropriate descriptive name, that is not misleading, should be used.

4. How should the statement of identity be placed on the label?

Place the statement of identity in lines generally parallel to the base of the package.

5. When are fanciful names permitted as the statement of identity?

When the nature of the food is obvious, a fanciful name commonly used and understood by the public may be used.

6. Is it necessary to use the common or usual name instead of a new name?

The common or usual name must be used for a food, if it has one. It would be considered misleading to label a food that has an established name with a new name.

7. Should modified statements of identity be used for sliced and unsliced versions of a food?

Labels must describe the form of the food in the package if the food is sold in different optional forms, such as sliced and unsliced, whole or halves, etc.

8. What food must be labelled as an 'imitation'?

A new food that resembles a traditional food and is a substitute for the traditional food must be labelled as an imitation, if the new food contains less protein or a lesser amount of any essential vitamin or mineral.

9. What type size and degree of prominence are required for the word 'imitation' in the product name?

Use the same type size and prominence for the word 'imitation' as is used for the name of the product imitated (see preceding example).

2.12.1.3   Net quantity of contents statements

1. What is the net quantity of contents?

The net quantity of contents (net quantity statement) is the statement on the label which provides the amount of food in the container or package.

2. Where is the net quantity statement placed on the label?

The net quantity statement (net quantity of contents) is placed as a distinct item on the bottom 30% of the principal display panel, in lines generally parallel with the base of the container.

3. Should the net quantity of contents be stated in both grams and ounces?

In the US, printed food labels must show the net contents in both metric (grams, kilograms, millilitres, litres) and US Customary System (ounces, pounds, fluid ounces) terms.

The metric statement may be placed either before or after the US Customary statement, or above or below it.

4. Why is it necessary to calculate the area of the principal display panel?

The area of the principal display panel (calculated in square inches or square centimetres) determines the minimum type size that is permitted for the net quantity statement (see next question).

5. What is the minimum type size?

In the US, for the net quantity statements, the minimum type size is the smallest type size that is permitted based on the space available for labelling on the principal display panel. Determine the height of type of printing by measuring the height of the lower case letter 'o' or its equivalent when mixed upper and lower case letters are used or the height of the upper case letters when only upper case letters are used.

6. What are the conspicuousness and prominence requirements for net quantity statements?

Choose a print style that is prominent, conspicuous, and easy to read. The letters must not be more than three times as high as they are wide, and lettering must contrast sufficiently with the background to be easy to read. Do not crowd the net quantity statement with artwork or other labelling (minimum separation requirements are specified in the regulation).

7. What is included in the net quantity of contents statement?

Only the quantity of food in the container or package is stated in the net quantity statement. Do not include the weight of the container or wrappers and packing materials. To determine the net weight, subtract the average weight of the empty container, lid, and any wrappers and packing materials from the average weight of the container when filled with food.

8. Is water or other packing medium included in determining the net quantity of contents in a food container?

The water or other liquid added to food in a container is usually included in the net quantity declared on a label. In some cases in which the packing medium is normally discarded, the drained weight is given, e.g. olives and mushrooms.

9. What is the net quantity of contents for a pressurised can?

The net quantity is the weight of volume of the product that will be delivered from the pressurised container together with the weight or volume of the propellant.

2.12.1.4 Ingredient lists

1. What is the ingredient list?

The ingredient list on a food label is the listing of each ingredient in descending order of predominance.

2. What is meant by the requirement to list ingredients in descending order of predominance?

Descending order of predominance means that the ingredients are listed in order of predominance by weight, that is, the ingredient that weighs the most is listed first, and the ingredient that weighs the least is listed last.

3. Where is the ingredient list placed on the label?

The ingredient list is placed on the same label panel as the name and address of the manufacturer, packer, or distributor. This may be either the information panel or the principal display panel. It may be before or after the nutrition label and the name and address of the manufacturer, packer, or distributor.

4. What type size is required for ingredient lists?

The type size is specified and must be easy to read.

5. Should water be listed as an ingredient?

Water added in making a food is considered an ingredient. The added water must be identified and listed in its descending order of predominance by weight.

6. Should the common or usual name always be used for ingredients?

Always list the common or usual name for ingredients unless there is a regulation that provides for a different term. For instance, use the term 'sugar' instead of the scientific name 'sucrose'.

7. Is it necessary to declare trace ingredients?

It depends on whether the trace ingredient is present in a significant quantity and has a function in the finished food. If a substance is an incidental additive and has no function or technical effect in the finished product, then it need not be declared on the label. An incidental additive is usually present because it is an ingredient of another ingredient.

8. What foods may list alternative fat and oil ingredients?

Listing alternative fat and oil ingredients ('and/or') is permitted only in the case of foods that contain relatively small quantities of added fat or oil ingredients (foods in which added fats or oils are not the predominant ingredient) and only if the manufacturer is unable to predict which fat or oil ingredient will be used.

9. What ingredient listing is necessary for chemical preservatives?

When an approved chemical preservative is added to a food, the ingredient list must include both the common or usual name of the preservative and the function of the preservative by including terms such as 'preservative', 'to retard spoilage', 'a mould inhibitor', 'to help protect flavour', or 'to promote colour retention'.

10. How are spices, natural flavours, or artificial flavours declared in ingredient lists?

These may be declared in ingredient lists by using either specific common or usual names or by using the declarations 'spices', 'flavour' or 'natural flavour', or 'artificial flavour'.

11. What listing is used for a spice that is also a colouring?

Spices, such as paprika, turmeric, saffron, and others that are also colourings must be declared, either by the term 'spice and colouring' or by the actual (common or usual) names, such as 'paprika'.

12. What ingredient listing is used for vegetable powder?

Vegetable powders must be declared by common or usual name, such as 'celery powder'.

13. What ingredient listing is used for artificial colours?

It depends whether the artificial food colour is a certified colour.

Certified colours: List by specific or abbreviated name such as 'FD&C Red No. 40' or 'Red 40'.

Noncertified colours: List as 'artificial colour', 'artificial colouring', or by their specific common or usual names, such as 'caramel colouring' and 'beet juice'.

2.12.1.5 Nutrition labelling

The nutrition label is required on most food packages with some exemptions. The legislation is very clear as to what may and may not be claimed, how the nutritional facts must be presented, and what claims may be made. A claim on a food product that directly, or by implication, characterises the level of a nutrient in the food (e.g. 'low fat' or 'high in oat bran'). Nutrient content claims are also known as 'descriptors'.

2.12.1.6 Miscellaneous

1. Are mail order sales covered by the food labelling laws?

The same labelling laws apply to all categories of retail sale, including mail orders. Foods sold by mail order must be fully labelled.

2. In the US, are foreign language labels permitted?

All required label statements must appear in both English and the foreign language, if any representations appear in a foreign language.

3. On labels that have two languages, may nutrition information be provided in one 'bilingual' 'Nutrition Facts' label?

When nutrition labelling must be presented in a second language, the nutrition information may be presented in separate nutrition labels for each language or in one label with the second language, translating all required information, following that in English. Numeric characters that are identical in both languages need not be repeated.

4. Where should the country of origin be declared on labels of food imported into US?

The law does not specifically require that the country of origin statement be placed on the principal display panel but requires that it be conspicuous. If a domestic firm's name and address is declared as the firm responsible for distributing the product, then the country of origin statement must appear in close proximity to the name and address and be at least comparable in size of lettering.

5. Which foods require warning statements?

Warnings are required on self-pressurised food containers and on some protein-based weight reduction products and dietary supplements. A warning is also required on products containing saccharin. Place the warning statements on the PDP or on the information panel, if there is one.

6. Is it permissible to use stickers to make changes in labelling?

Correcting label mistakes in any manner is acceptable, if the final label is correct and complies with all regulations at the time of retail sale. The stickers should not cover other mandatory labelling and should adhere tightly.

7. Are there restrictions on label artwork?

Do not use artwork that hides or detracts from the prominence and visibility of required label statements or that misrepresents the food.

Acknowledgements

The author acknowledges with thanks the contributions made to earlier editions of this chapter, 'Food Laws, Regulations and Standards', by Dr. Merton V. Smith, 11, Dr. Melvin R. Johnston, James R. Taylor, Jr., R. E. Newberry, Joseph Pendergast, Robert Lake, and Gerad McCowin, all of the US Food and Drug Administration.

The labelling questions and answers were summarised from an FDA manual available from the Division of Programs and Enforcement Policy (HFS-155), Office of Food Labeling, CFSAN-FDA, 200C Street, SW, Washington, DC 20204, Tel-202-205-5229.

The United States Government Manual; Federal Food, Drug, and Cosmetic Act, as amended, and related laws; Federal Register, Code of Federal Regulations and Requirements of Laws and Regulations enforced by the U.S. Food and Drug Administration are for sale by the Superintendent of Documents, U.S. Government Printing Office, Washington, D.C. 20402, www.fda.gov/food/.

References

Codex Alimentarius, www.codexalimentarius.org.
Holah, J., & Lelieveld, H. L. M. (2011). *Hygienic design of food factories*. Woodhead Publishing Ltd.

Kosher and halal food regulations

3

3.1 Introduction

Around the world, both kosher and halal are major considerations in the purchase of products by people in many countries.

Kosher foods are those that conform to the regulations of kashrut (Jewish dietary law). The term kosher is an English one. Some of the reasons for food not being kosher include the presence of ingredients derived from non-kosher animals or from kosher animals that were not slaughtered in a ritually proper manner, a mixture of meat and milk, wine, and grape juice (or their derivatives) produced without supervision, and the use of non-kosher cooking utensils and machinery (Regenstein and Regenstein, 1990a).

Halal foods are foods that Muslims are allowed to eat or drink under Islamic Law that specifies what foods are allowed and how that food must be prepared (halal means proper and permitted). The most well-known example of non-halal (haram) food is pork. Pork is the only meat that cannot be eaten by Muslims at all (due to cultural and religious reasons); meat other than pork can also be haram as a result of its source, the cause of the animal's death, and how it was processed. Halal food may not contain any alcohol or traces thereof (Chaudry, 1992).

Not all consumers buy kosher or halal products for religious reasons. In the US, almost half of grocery goods carry a kosher symbol. Of the consumers who buy kosher products, it is reported that some eat kosher for health and safety reasons (there is a perception that products that carry a kosher or halal mark are of higher quality), some are vegetarians or who simply like the taste, and others do so in an effort to maintain a halal diet. Kosher and halal are an important part of the food industry, and one needs to understand the implications of participation.

3.2 Importance of the process

To help the food processor consider whether to undertake kosher or halal certification, a broad definition of the scope of these regulations is offered. Because only a broad discussion of major points to be considered follows, close work with the religious supervisors who apply the details to a particular plant and process is mandatory (Regenstein and Regenstein, 1979 and 1988).

Both halal (Muslim) and kosher (Jewish) regulations control many aspects of food production. To succeed as a kosher or halal processor, a company must be willing to make the necessary changes in their processes and purchasing of ingredients to meet the requirements of the supervising agency representing the religious community.

A Complete Course in Canning and Related Processes. http://dx.doi.org/10.1016/B978-0-85709-677-7.00003-7

3.3   Major concerns

3.3.1   Allowed animals

Both kosher and halal regulations consider certain animals acceptable for food, whereas others are not permitted. For kosher, the mammals must be ruminants, i.e. animals with split hoofs and that chew their cud. For poultry, most land-based domestic birds are acceptable, and for fish they must have fins and scales. These scales must be removable without tearing the skin. All animals must be specifically ritually slaughtered. Animals from the allowed category must also be inspected for many different abnormalities and injuries that render them unfit.

Halal specifically prohibits the use of anything from a pig. This prohibition extends beyond just food use into all situations in which products from pigs might be involved. Other animals, particularly carnivores or carrion, are also prohibited, along with animals with specific defects.

3.3.2   Prohibition of blood

Blood, as the life fluid, is prohibited by both groups. Kosher ritual slaughter is designed to maximize blood removal. Once the blood is drained out, the meat must be processed to remove and/or open major arteries and veins to remove any congealed blood and then the meat is subject to a ritual salting ('koshering') to further remove blood.

Halal ritual slaughter is very similar to kosher slaughter. A key additional component is that the name of Allah must be invoked over every animal, preferably at the time of slaughter. Hunting is permitted with carefully defined limitations.

3.3.3   No alcohol

Muslims do not permit the use of alcohol (ethanol, ethyl alcohol). This is particularly important with flavors and other extractives that normally use alcohol as a carrier.

3.3.4   Separation of milk and meat

The kosher dietary laws require that meat and milk be kept completely separate. Thus, all kosher equipment, utensils, and meals must be divided into one of three categories: meat, dairy, or parev (pareve or parev, meaning neutral). Once products are produced in one of these categories, they are usually marked so that consumers can tell in which category the product fits. These markings of kosher foods provide information useful to other consumer groups.

3.4   Equipment kosherisation or preparing for halal production

In switching from non-supervised production to either kosher or halal production, equipment must be well cleaned and then ritually prepared. In the case of kosher, the transition between the different categories of meat, dairy, and parev also requires equipment kosherisation (Regenstein and Regenstein, 1990b). (Note: going from parev to dairy or to meat does not require any special changeover, but once the production is dairy, it does require a rabbi to supervise changeover back to parev.)

The process of equipment kosherisation is complex and is a major undertaking in operating a plant that switches between kosher and non-kosher, or between parev and dairy. (Most meat operations are distinctly separate and simply stay as kosher meat all the time, but many of the same rules would apply.)

1. It must be determined whether the system is operating hot (involving cooking). For kosher purposes, temperature of 43–46 °C (110 °F–115 °F) is used. (Note: Some rabbis use different temperatures, i.e. higher ones.) If no heat is used, then changeover is fairly simple. If cooking is involved, then equipment may have to sit idle for 24 h and then require very hot water (from 82 °C to boiling, depending on the rabbi) to cover all internal and external surfaces. If recirculating steam or water is used, this may present particular problems, as the steam may be considered 'non-kosher' if it was used to heat non-kosher foods and may be considered dairy if it was used to heat kosher dairy products. One solution to this problem has been to add a strong 'bittering' agent to the steam so that it is 'inedible' and, therefore, does not absorb the characteristics of the materials it is heating. Food-approved ingredients exist for this purpose. In most cases, these additives (materials) have been tested so that they will not negatively impact either the products or sensitive inline instrumentation.
2. The materials used to make the equipment must be determined. Certain materials, i.e. those religiously defined as non-absorbing, are more easily cleaned than those that absorb foods (e.g. stainless steel) but can be induced to desorb them.
3. Finally, there is a category of materials which can absorb, but not desorb (e.g. earthenware), and which therefore cannot be koshered.

With respect to baked goods, i.e. the use of ovens, the temperature for 'koshering' is much higher and often involves bringing the ovens to a red-hot or white-hot temperature.

As should be obvious from the short discussion above, there are many rules in this area, and the details of koshering a particular operation need to be worked out with the supervising rabbi.

Halal equipment preparation also involves a careful cleaning of equipment, including scrubbing with steel wool, etc., when deemed necessary, followed by hot water or higher heat.

3.4.1 Passover

In addition to the general rules that apply all year around, there are specific additional (not instead of) rules that are applied to Passover, the Jewish holiday in the spring, commemorating the Exodus from Egypt. During this period, special prepared and supervised unleavened bread (matzos) is used. This already baked bread is then ground into either a meal or flour. All other 'baked' products use this matzos meal or flour as a starting material. All other grains, legumes, and similar products and most of their derivatives (e.g. even 'pure' chemicals made from these ingredients) are not permitted (e.g. products from wheat, oats, rye, barley, split soybean, corn, rice, and peanuts). All equipment must be specially cleaned for production of foods for this holiday. The absence of all of these plant kingdom materials in Passover products can be used by people with various allergies or intolerances to obtain products that have been specifically prepared for this holiday that meet their special dietary needs. On the other hand, the additional stringencies of this holiday make it particularly difficult for many food companies to produce food products for this eight-day period.

3.5   Who prepares the product?

The cooking and preparing of many kosher products requires the input of a Sabbath-observing Jew. In many cases, this simply means that the Jew lights the pilot light. In other cases, it may mean that the Jew adds the coagulating enzymes to the cheese vat. It also means that an all Jewish crew is needed to press grape juice. If it is then pasteurised (i.e. cooked), it can be handled with minimal special requirements.

3.6   Examples of a few special issues of concern

3.6.1   Coatings on cans and plastics

The packaging industry uses many ingredients in preparing materials for the food industry. All of those that can be expected to come in contact with foods needing regulatory approval. Some of these ingredients can be non-kosher and non-halal. In previous years, some of the steel used for steel drums, pails, and the regular food could have been coated with lard after annealing. This is objectionable to the groups identi-fied in this chapter. Most steel manufacturers are aware of the problem and have used other coating materials in recent years (Regenstein and Regenstein, 1991).

Plastics often have chemicals included during their manufacture that are meant to interact with the product to serve as 'releasing' agents and for other functional pur-poses. Some of these compounds could have been derived from animal stearates. It should be ensured that these stearates are from plant or microbial sources, rather than animal sources.

3.6.2   Flavour

A few flavour are derived directly from non-kosher animals: castoreum, ambergesis, and musk. These are inherently not kosher. The Muslim community also does not use these materials. For halal products, no alcohol can be used as a carrier, even with natural permitted flavour.

3.6.3   Biotechnology

Muslim and Jewish religious leaders are continually debating the implications of biotechnology on their religious food requirements (Chaudry and Regenstein, 1994). Currently, Jewish religious leaders have accepted the addition of one or a few genes, using current technologies, from an animal into a plant or microbe as producing a food material the origin of which is now religiously equated as coming from the plant or microbe. The Muslim community has accepted the process with respect to acceptable animals, e.g. chymosin (rennet) from a calf.

3.6.4 Salt tablets

Some companies use salt tablets, which may be compounded with lactose. This would make the product a dairy product.

3.6.5 Lactic acid

This technology indicates the potential complexities that may arise. For many years, this product was derived from plants or made synthetically. It appears on some lists of 'kosher foods' as, therefore, always being kosher. However, a product derived from whey has been produced that makes a parev product into a dairy product.

3.6.6 Insects

The presence of visible insects in food products is objectionable, because these are prohibited animals. Therefore, many fruits and vegetables must be inspected to assure that they are insect-free.

3.6.7 Kosher and halal gelatine

The Jewish community has varying standards with respect to gelatine. The religious laws in this area are quite complex and there is disagreement about the standards that should apply. The mainstream Orthodox community does not accept any beef or pork gelatine, except for a beef gelatine made solely from the hides or bones of kosher-killed animals.

The Muslim community totally rejects pork gelatine and is also concerned with beef gelatine produced from non-halal (or kosher) slaughtered animals. A solution to this problem would be to develop both kosher and halal gelatines from fish skins. This is currently being done. For the food processor, the fish gelatine has an interesting additional property – the melting temperature can be selected over a much wider range (e.g. probably from about) 12 to $-3\,°C$ ($10\,°F$–$27\,°F$), compared to the restricted range for beef and pork gelatine, which is approximately -3 to $-1\,°C$ ($27\,°F$–$30\,°F$).

3.7 Kosher and halal supervision agencies

Many agencies are available for assisting with kosher supervision, along with individual rabbis. As indicated above, different standards apply. The more stringent the rules followed by the rabbi, the larger the audience for the product. The choice of how strict a route is taken is a commercial choice a company must make for retail products. However, for an ingredient manufacturer, it is recommended that the product must be able to satisfy most rabbis or rabbinical supervision agencies, or else the cost (which can be significant) will not be recovered. Thus, the standard referred to as 'mainstream Orthodox' is a minimal requirement for ingredients.

Similarly, there are agencies that can supervise the production of halal food. Inspectors and supervisors require training not only in religious teachings but also in production quality and control, product flow systems, cleaning, and the proper use of the production equipment. Training must include a working knowledge of ingredients and potential suppliers.

The decision as to which kosher or halal supervision agency to go to should be based on all the normal business decisions: price, quality of service, which includes such issues as the ability to provide supervisory personnel when needed, turn-around time for documentation, confidentiality, and ability to understand the operation. Most of these agencies have a trademarked symbol that companies can put on their products that tells the consumer who is providing the supervision. This can be important, because the sophisticated consumer will make judgments as to which agencies he/she will accept. The generic 'K' may be used by any rabbi or supervision agency. Many consumers will not use such products, unless they first check on who is supervising the product. By failing to indicate the supervision agency used, many companies decrease the value of their investment in kosher food production.

Acknowledgements

This chapter was revised from information in the 13th Ed compiled by Joe M. Regenstein Ph.D., Cornell Kosher Food Initiative, Department of Food Science, Cornell University, Ithaca, NY, e-mail: jmr9@cornell.edu.

References

Chaudry, M. M. (1992). Islamic food laws: philosophical basis and practical implications. *Food Technology*, *46*(10), 92.

Chaudry, M. M., & Regenstein, J. M. (1994). Implications of biotechnology and genetic engineering for kosher and halal foods. *Trends in Food Science & Technology*, *5*(5), 165.

Regenstein, J. M., & Regenstein, C. E. (1979). An introduction to the kosher (Dietary) laws for food scientists and food processors. *Food Technology*, *33*(1), 89.

Regenstein, J. M., & Regenstein, C. E. (1988). The kosher dietary laws and their implementation in the food industry. *Food Technology*, *42*(6), 86.

Regenstein, J. M., & Regenstein, C. E. (1990a). Kosher certification of vinegar: a model for industry/rabbinical cooperation. *Food Technology*, *44*(7), 90.

Regenstein, J. M., & Regenstein, C. E. (1990b). Looking in on kosher supervision of the food industry. *Judaism*, *39*(4), 408.

Regenstein, J. M., & Regenstein, C. E. (1991). Current issues in kosher foods. *Trends in Food Science & Technology*, *2*(3), 50.

Part Two

Design and upkeep of canned food factories

Plant location and construction

4

4.1 Introduction

Many factors determine the suitability of location for a cannery. The more important are as follows:

- Availability of an ample supply of raw products of the desired quality at prices that are satisfactory to the growers
- An adequate supply of water of suitable quality
- The availability of adequate help during the canning season
- Availability of regular transportation at reasonable rates between the cannery and the primary markets
- Adequate facilities that comply with local laws for the disposal of plant waste.

The economies of production will determine in a very large measure the location of a factory, with reference to these important items: raw materials, water supply, labour, and shipping facilities. The order of the importance of these factors will not be the same in each line of canning but will be dependent upon the amount of labour required, the nature of the raw materials as to perishability, and the amount of waste generated.

In the case of fruits requiring much hand labour, or a factory operating over a considerable part of the year, it will be found more economical and satisfactory to haul the product by rail or truck rather than to transport the help. In the case of corn, of which the waste in husk and cob is such a large part of the whole, and in which machinery has displaced workers, the reverse may be true. Peas formerly belonged in the same class as corn, but because the vining is being done in the field or at viner stations, and only the shelled peas are trucked, the factory may be located at any good site. Choosing a location with reference to labour should go further than merely getting near, or in, a city or large town, but should be in vicinity of a good quality labour pool. Physical equipment may be purchased or built, but labour can be made efficient and satisfactory only through proper training. Moreover, efficiency cannot be developed with different personnel each year. Waste treatment and disposal are usually more expensive in cities or large towns than in areas that are more open.

There is no question as to the advantage of locating a factory near the fresh material. It is a necessity for the proper handling in the case of some products, such as asparagus and corn, and is desirable in all cases, as it permits leaving the product to mature to its best stage and lessens injury from handling and deterioration from changes after harvesting.

The building site should be one easily accessible for the receipt and shipment of raw and finished materials and for help. It should have ample space and be in a clean locality. There must be opportunity for the treatment and final disposal of waste,

A Complete Course in Canning and Related Processes. http://dx.doi.org/10.1016/B978-0-85709-677-7.00004-9

whether it be carried by the wash water or be bulky and need to be removed by truck. Air and light should be accessible to all sides, and the yards hard, with ample slope for drainage. A food factory should be the cleanest place in the community.

It does not follow that the same kind of location is necessary for all factories. Many meat-canning establishments are near a slaughterhouse, and usually the yards, with their attendant distinctive odors, cause conditions that should be sufficient to condemn the place as suitable for the packing of fruits and vegetables. The same conclusion might be made for a fish factory, and yet both could be thoroughly clean and suited to their particular lines of packing. Most seafood canneries are in isolated locations, but some operate near centers of population. A waterfront location near a residential area is not desirable, even though it may not be forbidden by zoning regulations. No matter how careful the management might be to keep the establishment clean and to eliminate any possible nuisances, the presence of the plant will be resented, the operations will encounter hostility, and the product will receive unfavourable publicity.

4.2   Factors influencing plant location

It is the purpose of this section to call attention to those factors that should be considered in the location of a new plant. The basic objective is to choose the one location that results in the best possible combination of individual factors affecting total production cost, thus assuring maximum profits from a company's operations.

A complete objective study of the plant location factors by a company's staff or by outside qualified consultants should be considered. Unfortunately, in many instances, no thorough analysis is made and plant location is determined on the basis of a few rapidly gleaned impressions during a hurried site tour or by likes and prejudices that normally should not play a part in the solution of the problem (OSHA, 1992).

The approach to this problem will depend upon one's starting position, that is, whether the inquiring individual or organisation is searching for a geographical location, a site within a location, or just a site within an identified location. In the later situation, there might be restrictions, such as existing facilities, or a large growing area of which to take advantage.

Ortiz (1992) identifies a two-step process in locating a food plant: location and site selection analysis. The location analysis deals with those factors that influence the geographical location, such as labour, energy, and distribution costs and would involve the following tasks: identifying the search area, identifying the selection criteria, collecting data and conducting analysis, and then evaluating and ranking possible locations.

A site selection analysis deals with evaluating variables relating to a specific community and then comparing communities. These variables may be local land availability, tax structures and tax incentives, and local infrastructure. The tasks involved in the site selection analysis would be: defining plant requirements, defining client's desires regarding the host community, collecting site data and analyzing with engineering and economic criteria, and then ranking sites.

4.3 Selecting a geographical area

A compilation of basic questions that management must answer and evaluate to select the best possible site for a new plant follows (Section 4.4). From this list should be selected all questions that refer to specific factors that contribute to the total cost of products and services of the plant being considered. The factors to which questions refer are not of equal importance; their relative importance will vary from plant to plant. Sometimes, the scale of importance can be established by evaluating the percentage of the total cost of a plant's products upon which each factor bears. This percentage can then be used as the 'weighted value' for each factor (Doe, 1989).

If this is impractical, it is sometimes expedient to prepare a simple listing, in order of their importance, of factors that affect the total cost of products and services of the plant being considered. If 30 items are involved, the most important is given a weighted value of 30, the next important a value of 29, etc., until the least important receives a value of 1. These figures can then be used as the 'weighted values'.

In some cases, it may be necessary to assign a 'weighted value' on an arbitrary basis of considered and experienced judgment in which it is agreed that a special factor warrants unusual value. For instance, there have been cases in which plant management has been restricted in its choice of a plant site due to the inability of obtaining competent supervisory personnel in areas at some distances from the home plant.

When the final list of factors has been established and weighted values agreed upon, the selection of general areas favourable to the new plant can proceed. Generally, there are two or three factors that automatically limit the areas where a plant can be built, such as water and energy supplies and labour costs.

Another method of defining general areas for plant location is to study raw material sources and market destinations. The objective is to develop factual data on incoming and outgoing freight to determine the areas that combine maximum economies for receipt of raw material and for distribution of manufactured product. Any area defined by this method should be carefully scrutinised to insure that other important factors do not suffer to a degree that would overbalance the freight consideration.

4.3.1 Search for a specific location

The next step is the detailed search for a specific geographical or site location within general areas already identified. Once a list of potentially suitable sites has been developed, field crews can be assigned to visit and inspect each of them. The objective of inspections should be to rate or grade each site as to the degree with which it meets requirements of factors that have been assigned 'weighted values'. The grading should be done numerically, i.e., '5' for excellent, '4' for good, '3' for fair, '2' for poor, and '1' for unsatisfactory.

The 'weighted value' for each factor is then multiplied by the 'grade'. The product represents a numerical value of the factor that combines its importance with the production process and its degree of excellence as related to a specific site. The sums of these products for the various sites can be used as a fair and reasonably accurate basis for the comparison of one site with another.

4.3.2   Value to management

The great value to management of a careful and scientific plant location study may not always be appreciated. If properly developed and interpreted, such a study will insure that the location selected will establish the lowest possible level of production cost. This, in turn, helps to establish the plant's competitive strength in relation to other competing plants seeking to command the same markets. Competitive strength contributes to steady growth and thus to an increasing share of the profits available in the industry. Location studies can easily be made by the company's own engineering staff if it is properly equipped to do so. Otherwise, competent outside engineering talent should be engaged.

4.4   Basic considerations in evaluating plant location

1. Production materials
   a. What are delivered prices of raw and partially manufactured materials at the site?
   b. Are these costs competitively favourable?
   c. Are materials available in sufficient quantities?
   d. What is their comparative quality?
   e. Are the sources of materials dependable over a long period?
   f. Are emergency sources available?
2. Marketing
   a. What is the cost of delivering finished goods to markets and customers?
   b. Are these costs competitive?
   c. Can the products be serviced quickly and easily?
   d. What percentage of the total market can be serviced directly from the plant?
   e. Can sales, sales promotion, and advertising be effectively conducted?
3. Transportation and distribution
   a. What transportation facilities are available, including railroads, highways, waterways, airways, postal service, etc.?
   b. How dependable are the transportation facilities? Are there seasonal fluctuations in availability?
   c. What is the quality of service and the ability to handle such tonnages that are likely to be required in the foreseeable future?
   d. What are the freight rates, switching rates, speed of movements, headway between departures, correlation of freight schedules with production and switching schedules?
   e. Are adequate storage, warehousing, and terminal facilities available together with grading and inspection of goods, etc.?
4. Labour
   a. Is an adequate labour supply in the vicinity already available or potential through movement of workers within range of the site?
   b. What percentage of the labour supply is male? Female?
   c. What is the available range of skills in the labour market?
   d. Are highly specialised skills, required by your production, available in sufficient quantity?
   e. What is the relative efficiency of labour, as established by the experience of others?
   f. How stable is local labour as indicated by turnover rates of other manufacturers?

g. What is the percentage of homeowners in the labour group?

h. How restrictive are shop rules governing workers in other plants?

i. What is the state of labour relations, as indicated by local experiences of other employers in strikes, mass picketing, civil disturbance, etc.?

j. What is the local unemployment experience?

k. Does any local industry dominate the entire local labour market?

l. How large is the area from which employees can be drawn and what is the employable population of this area?

m. What are local rates of pay for jobs comparable to those required in the new plant?

n. Do these rates represent any competitive advantage?

o. How stable are these rates?

5. Supervision

   a. Does the new location facilitate general management supervision from executive headquarters?

   b. Can adequate supervisory personnel be drawn from the area selected or induced to locate there?

6. Utilities, service, and fuels

   a. What sources and quantities of water are available?

   b. What is the cost of water?

   c. What reserves of water are available and how large are they?

   d. What is the chemical, physical, and microbiological quality of the water related to its suitability for process or steam?

   e. How can the problem of industrial waste be handled?

   f. Are local facilities available for sewage disposal?

   g. What are the power sources and related costs for present and future?

   h. Are power sources available in adequate quantity?

   i. How dependable are sources of power supply?

7. Bank facilities

   a. What is the number and size of banks in the locality under consideration?

   b. What is the supply of loanable capital?

   c. What is the maximum loan any one bank can make to one customer?

   d. Are banking facilities adequate for handling the payroll?

   e. What are the banks' correspondents and financial connections?

   f. What are the banks' attitudes toward industry?

8. Laws and taxes

   a. What are existing laws and practices with regard to:

   b. Workmen's compensation

   c. Unemployment compensation

   d. State and local building codes

   e. Safety and insurance regulations

   f. Water and air pollution and waste disposal

   g. What is the public attitude toward industry, as indicated by existing laws and regulations?

   h. Are existing laws well administrated and is this administration alert and progressive?

   i. What are local taxes for:

   j. Real property

   k. Classified property such as machinery, inventories of raw materials, and finished goods

   l. Deposits and investments

   m. Corporate and personal income

   n. Sales and use taxes

    o. Organisation or franchise tax

    p. Will an adjacent city annex the plant site and thus eliminate tax advantages?

9. Community attitude toward industry

    a. How sincere a welcome would be extended to your business by the community under consideration?

    b. Will the community attract your type of industry by offering tax or other inducements?

    c. Does the community show an interest by offering assistance in matters such as site location and selection, extension of utility lines, modification of obsolete restrictions, and the like?

10. Living conditions

    a. How do living costs in the locality compare with other sections of the country?

    b. Are housing facilities available for workers coming from other sections of the country? If not, does the community have adequate plans for relieving the housing shortage?

    c. Are community facilities, including schools, churches, hospitals, and recreational facilities, adequate?

11. Climate

    a. What are the range and mean weather conditions, including snowfall, rainfall, storms, humidity, etc.?

    b. How do local weather conditions affect plant costs, labour, production, transportation, and living conditions?

4.5 Site characteristics

Factors to be considered in the selection of a specific site within a desirable plant location follow (OSHA, 1993):

1. Are existing plants or buildings available in the proposed location? Could these be converted at a reasonable cost to meet new plant requirements and maintain optimum efficiency?

2. If new facilities are to be built, is there sufficient acreage available to permit an attractive setting for the plant, future expansion, and adequate parking facilities?

3. Is the contour of the site such that a minimum of grading or other treatment will be required?

4. Is the site elevation sufficient to avoid any possibility of flooding?

5. What are the subsoil characteristics? Will expensive foundation, piling, or drainage costs be encountered?

6. Is land cost in line with the inherent advantages of the specific site?

7. Can the site be purchased outright or only leased?

8. What is the cost of connecting utilities and access roads to the plant?

Note: See section on regulations regarding environmental issues of site preparation.

4.6 Building a canning plant

4.6.1 Construction factors

In building a modem cannery, many factors will play an important role in its operation and will be seriously felt if, by chance, they are overlooked. These might be listed as follows, and their order of importance would be somewhat changed according to the size of the operation:

1. Adequate water supply to meet current, as well as future needs.
2. Adequate sewage outlet to carry off water used in washing product, as well as cooling finished products, for current, but also for future increased, output. Few canners realise how many litres of water are used daily and what would happen if sewers were too small to accommodate wastewater if output should be considerably augmented. It does not require much of an operation to use two to four million litres of water daily. In calculating wastewater output, allowance must be made for increased load upon the sewer caused by expansion of other industries in the adjacent area or an increase in homes using the same sewage outlet. Independent sewage treatment facilities may have to be considered.
3. Adequate land of sufficient load-resisting capacity for immediate needs and expansion, as well as for storage and parking facilities for employees.
4. Adequate strength in the building to withstand overloading (e.g. heavy snow on roofs), adequate drainage outlets to take care of sudden excesses of rainfall, and adequate reinforcing to meet demands of heavy loads resulting from high stacking of finished goods and the shock of movement of heavily loaded lift-trucks across floors.
5. Ample height to permit free use of conveying systems, or double decking of machinery and equipment, if necessary.
6. Adequate ventilation to prevent condensation of steam from raining onto products and insure worker comfort and sanitation.
7. Heating facilities (where needed) to insure comfortable working space for employees and to prevent freezing of finished product stored in cannery warehouses.
8. Adequate room for oversize rail sidings to prevent costly standby time of workers waiting for switch engines to bring more cars.
9. Streamlined buildings to have as few interior posts as possible. Building partitions should be erected so they can be easily removed to prevent the building from becoming a 'special purpose' building, in the event that canning operations should be suspended. This would facilitate getting a construction loan or selling later, if necessary.
10. All inside concrete walls to be brushed with waterproof cement to fill all crevices and holes to prevent accumulation of dust or mildew spores. A plant built in an east–west direction, or vice versa, is much better for lighting.

Local construction codes should be examined to insure compliance with all regulations relating to fire, worker safety, etc.

Many of the preceding factors will require different treatment according to existing conditions in the area where the plant is built. For example, the problem of snow load on roofs is not a factor in warm climates. In the north western US, engineers allow for a possible load of 150 kg of snow per square metre and drainage pipe for water from roofs is calculated to provide an outflow resulting from a flash flood of 2.54 cm (1 in) of rain in 30 min.

Earth resistance is very important. High stacking of canned foods with lift-trucks increases static loads greatly per unit area. When soil is soft, it is often necessary to drive pilings to prevent yielding and shifting of floors. Wherever floors are laid over sewers, or places where trees have been removed, floors should be more heavily reinforced to prevent cracking because of low soil resistance.

Reinforcing of floors requires attention. It can be expected that a forklift truck will move a three-ton load on four points of contact (where wheels touch the floor, in an area of only about 0.1 m^2). This places a tremendous strain on floors. Moving loads create greater strain than static loads of the same weight per unit area.

Earthquake proofing is generally overlooked. Earthquake proofing a cannery consists mainly in tying the building together in such a way that when it is shaken, the whole building will move together as one unit. It is assumed, of course, that if the building is concrete, the walls, as well as floors, are steel reinforced.

Columns – It is not considered good practice to place column footings on top of floors, because this tends to contribute to cracking of concrete. Footings should be placed beneath floor levels; when concrete for floors is poured, a pressure strip should be placed around each column to isolate it from the main floor slab. This permits the floor to rise or recede without breaking concrete around the column.

To protect columns from lift-truck damage, the bottom of the column is encased with a steel pipe, which is then filled with concrete. Above the pipe, concrete should be rounded upward to eliminate a shelf on which dirt and rubbish can accumulate. If these pipe guards can be put around columns before the floor is laid, they can extend through the floor to the footing, which is a better way of handling this problem. Where columns protrude through the floor in processing or preparation areas and have no guards around them, concrete should be curved upward within their flanges to cause washing water to run away from them instead of pocketing against them and causing rust. Bases of windows should also be beveled to prevent water running under metal sashes and causing rust, as well as to prevent ledges from becoming resting places for cans and other debris. Wherever possible, floors should be constructed such that columns are at the high point of drainage.

To protect walls of the building against forklift-truck damage, especially where there are electrical switch boxes, control panels, etc., concrete curbs are built. Generally, they are about 13 cm (5 in) high and are beveled backward (concave profile) from the floor, at a rate of 2.54 cm (1 in) for each 7.6 cm (3 in) of rise to prevent lift-truck tires from being damaged in case the truck should bump the curb.

Where possible, double walls should be avoided to prevent giving rodents nesting places. The trend is to build plants with floors at ground level and to depress railroad sidings to permit driving lift-trucks directly into cars from cannery floor levels. This partially avoids the cost of heavy gravel fills and ramps and permits direct entry to the building at both the receiving and shipping areas by either lift-trucks or over-the-road trucks.

Industrial floors, especially those exposed to corrosive spillages, frequently receive less design study than any other part of the building process (Demetrakakes, 1994). The consequences of inattention can be serious. Production may be lost, the stability of structures compromised, and hazards to attendant personnel created. An easy and inexpensive method for reducing floor attack is to slope all floors and provide sufficient floor drains. Care must be exercised to avoid uneven areas that will hold liquids. Monolithic concrete has advantages in certain systems, whereas floors with sacrificial tile may be the most useful solution to spill damage in others. Metal decking with isolated skid-resistant protrusions is much more desirable than interlocking diamond or ring varieties. The latter retain solutions that enhance corrosion of surfaces.

Floors are made thick; usually 13 and 15 cm (5 and 6 in) in depth, and heavier reinforcing is used. In areas where preparation of raw product is to be done, 'dense mix' concrete is being used and often an extra bag of cement is added to each batch. In addition to these precautions, several coats of a silicate type filler are brushed onto the floor, which has the effect of 'case-hardening' the surface to a depth of 0.6 cm (0.25 in) and of making the surface more impenetrable by fruit and vegetable acids. A floor slope of between 1 and 2% toward floor drains is recommended to avoid standing water. Machinery should be leveled.

There is some difference of opinion whether floors should be made smooth for better sanitation or slightly rough to hold down accidents due to slipping on wet floors.

Current costs for floor drainage systems will prove a shock to anyone contemplating construction of a new canning plant. When a floor drain is constructed, it is necessary to build two walls and a bottom, or floor, in the drain. It requires two sets of concrete forms for each wall and one more for the bottom, for a total of five forms. At current costs for labour and materials, this rapidly runs into money, so the tendency is to place drains farther apart and increase the pitch in the floor. In that event, part of the handicap of distance is overcome by using drainpipes or troughs to convey water from equipment to floor drains.

Floor drains are made with round bottoms to get a better flow when a small amount of water is moved and to facilitate cleaning. Fruit and vegetable acids, after a while, penetrate these drain bottoms and create a very unsanitary condition in the concrete below. To offset this difficulty, many types of paint, including one with an asphalt base, have been used to put a protective film over the bottom of drains. Acid-resisting vitrified tile is split in half lengthwise, and these halves are imbedded in the concrete in the bottom of the drain. These provide maximum cleanliness and require practically no attention for years. It is necessary when laying this tile to 'point up' the joints with a special type of acid-resistant cement. If floor drains can be made wider and shallower, it permits easier cleaning, although it increases the cost of drain covers and tile somewhat. Drains are usually 15–20 cm (6–8 in) deep and 15–30 cm (6–12 in) wide (Imholte, 1984). To carry the weight of drain covers, it has been customary to lay at the top edge of the drain trough a shelf made of angle iron on which drain covers may lie. It is also customary to cover the edge of the drain at the floor level with angle iron, welding these two angle irons together for their entire length, not only to prevent seepage of raw product materials between them into the concrete of the floor but also to give additional strength. These welded angle-iron floor edgings and drain cover shelves are anchored securely into the floor with lags penetrating deeply into the concrete before it has solidified. Because of heavy shock stress from lift-trucks crossing these drain covers, it is necessary to provide strong edgings.

Drain covers are generally of sidewalk-type grating in sections no longer than 90 cm (36 in) for easy removal. They should have metal strips standing on the edge and should be forged together across the top side. This permits pieces of raw product to be washed into floor drains without danger of their being caught in meshes or lying on shoulders of floor drains.

Some canners put hose bibs into the ends of main drains, to provide an added flow of water, if necessary, where waste materials do not carry enough water to float out readily.

When floor drains are laid out, if it is desirable to keep vegetable waste separate from fruit waste, two master drains could be run out of the building far enough apart so that each could go into a separate compartment at one end of the waste removal screening pit. In this way, if it became possible to salvage some of the waste for stock feed or some other purpose, waste originating on one side of the cannery could be isolated from that originating on the opposite side.

4.6.2 Lighting

Good industrial lighting is a production tool that markedly affects employee accuracy, efficiency, and morale. It has been shown that good lighting that is properly designed and maintained results in more and better production through improved vision. Lighting systems that are tailored to a budget, rather than to the best-recommended lighting practices, often prove inadequate.

When lux levels exceed the recommended minimum levels shown in the following tables, better visual conditions are provided, more efficient use of plant equipment results, and manufacturing costs are reduced. When production quality standards are high or when a high degree of precision is required, lighting levels substantially higher than those recommended in Table 4.1 are suggested. The following are levels of illumination recommended by the Illuminating Engineering Society.

Table 4.1 Levels of illumination required in food-processing plants

Place in factory	Lux (lumen/ m^2) on tasks	Place in factory	Lux (lumen/ m^2) on tasks
Canning and preserving		**Candy making**	
Initial grading of raw materials	538	Box department	538
• Tomatoes	1076	Chocolate department	
• Colour grading	2153[a]	• Husking, winnowing, fat extraction, crushing, refining, feeding	538
Preparation			
Preliminary sorting			
• Apricots and peaches	538	Bean cleaning, sorting, packing	538
• Tomatoes	1076	Milling	1076
• Olives	1615	Cream making	
Cutting ad pitting	1076	• Mixing, cooking, moulding	538
Final sorting	1076	Gum drops and jellied forms	538
Canning		Hand decorating	1076
Continuous belt canning	1076	Hard candy	
Sink canning	1076	• Mixing, cooking, moulding	538
Hand packing	538	Die cutting and sorting	1076
• Olives	1076	Kiss making and wrapping	1076
Examination of canned samples	2153[b]		

Table 4.1 Levels of illumination required in food-processing plants – cont'd

Place in factory	Lux (lumen/ m^2) on tasks	Place in factory	Lux (lumen/ m^2) on tasks
		Bakeries	
Container handling		Mixing room	538
Inspection	2153[a]	Face of shelves	323
Can unscramblers	753	Inside of mixing bowls	538
Labelling and cartooning	323	Fermentation room	323
Dairy products: Fluid milk industry		Make-up room	
Boiler room	323	• Bread	323
Bottle storage	323	• Sweet yeast-raised products	538
Bottle sorting	538	Proofing room	323
Bottle washing	2153[b]	Oven room	323
Can washers	323	Filling and other ingredients	538
Cooling equipment	323	Decorating and icing	
Filling: Inspection	1076	• Mechanical	538
Gauges (on face)	538	• Hand	1076
Laboratories	1076	Scales and thermometers	538
Metre panels (on face)	538	Wrapping room	
Pasteurisers	323	**Offices**	323
Separators	323	Designing, detailed drafting	2153
Storage refrigerator Tanks and vats	323	Accounting, auditing, tabulating, computer operation, rough drafting	1615
• Light interiors	215	Regular office work, filing, index references	1076
• Dark interiors	1076	Reading or writing good quality material	753
Thermometer (on face)	538		
Weighing room	323	Reading printed material, interviewing, conferring, washrooms	323
• scales	753		
Meat packing			
Slaughtering	323	Corridors, stairways, elevators	215[c]
Cleaning, cutting, cooking, mincing	1076		

[a]Can be obtained with a combination of general lighting, plus specialised supplementary lighting. Care should be taken to keep within the recommended brightness ration. The seeing tasks generally involve the discrimination of fine detail for long periods and under consideration of poor contrast. The design and installation of the combination system must not only provide a sufficient amount of light but also the proper direction of the light, diffusion, colour, and eye protection. As far as possible, it should eliminate direct and reflected glare, as well as objectionable shadows.
[b]Special lighting such as that (1) the luminous area shall be large enough to cover the surface that is being inspected and (2) the brightness be within the limits necessary to obtain comfortable contrast conditions. This involves the use of sources of large area and relatively low brightness in which the source brightness is the principal factor rather than the foot-candles produced at a given point.
[c]Or not less than one-fifth levels in adjacent areas.

4.7   Layout of a canning plant

In planning a cannery, every effort should be made to locate boiler houses, transformers, pumping stations, etc., as close as possible to the point of use, to cut down transfer losses. In purchasing boilers, it is easy to be deceived into buying one with a lower cost per rated horsepower, whereas another boiler with a higher cost per rated horsepower will, because of higher output of steam per horsepower, be cheaper in the end. Canneries generally require a heavy amount of steam for a short period. Some boilers of the high-efficiency type can be run at a constant state of overload for a period of 30 days or more and in this way, although they cost slightly more per horsepower, savings can be made by buying a lower horsepower capacity.

When a new cannery is constructed, there is an opportunity to do a real job of condensation-damage prevention by installing a precast concrete roof or by running purlins and other lumber used in roofing through a solution of rot preventative. An undercoat of paint before this lumber is laid could also be put on, thereby giving the roof several extra years of life. In putting the finishing coat of paint on ceilings, the canner would do well to conduct extensive investigations into the matter of securing a mildew-proof paint in which the preventative will not dissolve and drop, through the dripping of condensation, into food that is in a state of preparation. Many manufacturers claim paints in this respect, but few such paints will actually pass the severe test they undergo in canneries.

In addition to choosing paint for its unique characteristics for the particular job, compliance with food contact and factory regulations is important. Volatile organic compounds (VOCs) in the paint may lead to air pollution. Solvent-rich paints are fast drying but are usually high in VOCs that lead to air pollution, as well as present a potential tainting problem with certain inventories. The fumes from the drying solvent may also present a health hazard to workers.

Some paints on the market contain a urethane base that cures by contact with moisture. There are positive and negative aspects to urethane paints; they dry fast but are more expensive.

No set rule can be laid down for sizes of different areas because of wide variation in space required for different types of operations. However, some so-called 'rule of thumb' bases may be applied in developing space areas. For instance, there should be receiving-room space sufficient to store at least double the amount of raw product that might be expected for the largest day's run. Ceilings in this area should be high enough to permit over-the-road trucks to back in, if necessary, without contacting any overhead conveying equipment.

Warehouse storage space should be sufficient, after allowing for aisles, to store 50% of the anticipated capacity of the plant. Obviously, with the possible exception of advertised goods, it is hardly likely that a canner would have 50% of the entire pack on hand at one time. However, loss of space through broken stacks, loss of space needed for labeling equipment, and often space needed for storage of emergency supplies of cans, ingredients, labels, etc. use up a considerable amount of warehouse and must be considered. Outlet doors from warehouse space could advantageously be high enough in size to permit over-the-road trucks to back directly into the warehouse for loading.

Firewalls between the warehouse and other areas will reduce insurance costs considerably; in some locales, they are required by ordinance.

Preparation and processing area must be planned to allow for the length of the longest lines of equipment in both these departments, and there should be ample aisle space at the beginning and end of these operations, as well as an aisle across, and between the preparation and the processing areas. These aisles must be wide enough to permit lift-trucks to maneuver in transporting raw or finished product and in placing or removing any of the cookers, coolers, preparation tables, or other equipment that may need to be removed from time to time. A great deal of labour cost can be saved if lift-trucks can be used for this purpose.

Loading platforms alongside buildings should be wide enough to permit two lift-trucks to pass with sufficient space for a pile of pallets at the loading edge, as well as room for a narrow curb to protect walls of the building, downspouts, etc. Platforms should be a width of 20 feet (6 m) or slightly more. In the wall of the platform, there should be imbedded, for the full length of the building, a conduit with outlets placed at convenient intervals to allow for lighting and other service needs.

Ceiling heights should permit free use of all types of conveying systems. The extra cost of 1 m (3–4 ft) more in height of walls and columns is not much compared with savings in costs if it becomes desirable to double-check some of the equipment. Cannery warehouses usually have 6 m (20 ft) clearance beneath roof trusses, and many have as much as 7.3 m (24 ft) clearance. There are some advantages in these heights beside those in the storage department. If cannery walls are high enough, glass block windows can be placed just beneath the roof the full length of the processing and preparation departments for natural lighting purposes. Where sidings are depressed, it is possible, with windows at this height, to run conveyors out to box piles, waste equipment, etc. without conflicting with legal clearance above railroad cars standing on sidings. Where windows are placed lower than roof level, steam tends to pocket in the area over the windows against the roof, causing disintegration of roof lumber. Where windows are at roof level, steam escaping through either monitors or fan-driven ventilators will suck in fresh air, much of which will follow the airflow against the ceiling and cut down steam saturation, as well as ceiling condensation. However, because of the many problems generated by windows, most modern food plants have eliminated them (Imholte, 1984).

There is some difference of opinion in the use of monitors in roofs for escaping steam, especially above processing areas, or of fan-operated roof ventilators. It is felt that in the preparation area, power-driven fans could be used for ventilation, but in areas where open-type cookers and exhaust boxes are being used, steam is being released into the air so fast that an adequate power-driven fan system might be used.

Worker comfort is another advantage of high ceilings. Where ceilings are 6 m or higher, air in the upper 3 m of elevation will be heavy with warmth and humidity, whereas the lower layer of about 3 m will be relatively cool and comfortable.

Waste treatment and disposal are complex and expensive and must follow the laws on water and air pollution. It is advisable to pump waste to screens located somewhat above ground to create a 'pressure head' on the sewer to insure adequate outflow of water that has passed through the screens.

Offices and other facilities: Allowance of space and plans will also have to be worked out for main offices, cafeteria, men's and women's toilets, and locker rooms. Offices should be grouped together, i.e. management, payroll, quality control office, and labouratory; warehouse office, label department, mechanical department, and other service-type employees. The first-aid room can be included with either of these.

Other rooms: The canner will want to provide for are: mechanical supervisor's office, machine shop, parts storage room, electrical equipment repair room, fireproof room for grease and oils, rodent and insect-proof room for storage of cans, ingredients, labels, pastes, glues, and other supplies.

Water: The water supply must be sufficient to comply with fire protection requirements in most cities, as well as for canning operations. Well-point water or city/municipal water can be used. Plants located where city water is not available necessitate the installation of adequate water treatment facilities to make available water of the quality needed to satisfy food-processing requirements.

4.8   Government regulations relating to building construction

Before any major commercial undertaking is begun, it is extremely important that compliance with government regulations be considered. Awareness of regulations is essential when building a food-processing or canning factory. There are various different departments that deal with air, water, and ground pollution problems; energy conservation in structures; workers' health and safety issues; as well as specific bodies that control some products like fish and meat.

There are many different agencies and regulatory bodies to consult and include in the planning. One will define the construction of meat plants and possibly another will outline the construction of other food plants. Some regulatory bodies regulate all food-processing facilities, whereas one agency may be concerned with safety, another may be concerned with energy conservation in building construction, and still others may be concerned with cleaning and sanitation.

All regulations related to the canning industry are not covered in this chapter, nor is there contained a definitive description of such regulations. This is merely an attempt to call attention to certain regulations and what each covers. A prospective business venture should be in contact with respective trade associations and local and state or regional government bodies to learn necessary regulations. It is important to get advice on which regulations pertain to the type of factory you wish to build in your specific area.

4.9   Environmental considerations

After a site has been selected to build a plant, some environmental questions should be considered. The following is taken from a questionnaire, the New York State Environmental Quality Review (SEQR, 1987), and is identified as a good checklist for new site development, regardless of the identified geographical location that is chosen.

Project impacts and their magnitude

General information

When answering the following questions, the applicant should be guided by the question: Have responses and determinations been reasonable? The applicant is not expected to be an expert on environmental matters. If the answer is yes to any of the following questions, then further investigation is encouraged. The examples provided are to assist the reviewer by showing types of impacts new construction may create. The examples are generally applicable for most situations. The impacts of each project, on each site, in each locality, will vary, and therefore, the examples are illustrative, not exhaustive, and are just a guide. The number of examples per question does not indicate the importance of each question; in identifying impacts, consider long-term, short-term, and cumulative effects.

Instructions

Answer yes if there will be any impact; 'maybe' answers should be considered as yes answers. If there is any doubt about size of the impact, then consider the impact as potentially large. If a potentially large impact can be avoided by change(s) in the project to a small or moderate impact, it should be considered.

Impact on land

1. Will Proposed Action result in a physical change to the project site?   NO   YES

Examples that would apply

- Any construction on slopes of 15% or greater (15 ft rise per 100 ft of length or 0.3 m/30 m) or where the general slopes in the project area exceed 10%.
- Construction on land where the depth to the water table is less than 0.9 m (3 ft).
- Construction of paved parking area for 1000 or more vehicles.
- Construction on land where bedrock is exposed or generally within 0.9 m (3 feet) of existing ground surface.
- Construction that will continue for more than one year or involve more than one phase or stage.
- Excavation for mining purposes that would remove more than 907,200 kg (1000 tons) of natural material (i.e. rock or soil) per year.
- Construction or expansion of a sanitary landfill.
- Construction in a designated floodway.
- Identify other impacts that may occur.
  2. Will there be an effect to any unique or unusual landforms found on the site? (i.e. cliffs, dunes, geological formations, etc.)   NO   YES
- Identify other impacts that may occur.

Impact on water

3. Will Proposed Action affect any water body designated as protected?   NO   YES

Examples that would apply

- Developable area of site contains a protected water body.
- Dredging more than 76 m^3 (100 cubic yards) of material from channel of a protected stream.
- Extension of utility distribution facilities through a protected water body.
- Construction in a designated freshwater or tidal wetland.
- Identify other impacts that may occur.

   4. Will Proposed Action affect any nonprotected existing or new body of water?   NO   YES

Examples that would apply

- A 10% increase or decrease in the surface area of any body of water or more than a 4 ha (0.04 km^2 or 10 acre) increase or decrease.
- Construction of a body of water that exceeds 10 acres (0.04 km^2 or 4 ha) of surface area.
- Identify other impacts that may occur.

   5. Will Proposed Action affect surface or ground water quality or quantity?   NO   YES

Examples that would apply

- Proposed Action will require a discharge permit.
- Proposed Action requires use of a source of water that does not have approval to serve proposed (project) action.
- Proposed Action requires water supply from wells with greater than 170 L (45 gallons) per minute pumping capacity.
- Construction or operation causing any contamination of a water supply system.
- Proposed Action will adversely affect ground water.
- Liquid effluent will be conveyed off the site to facilities that presently do not exist or have inadequate capacity.
- Proposed Action would use water in excess of 76 kL (20,000 gallons) per day.
- Proposed Action will likely cause siltation or other discharge into an existing body of water to the extent that there will be an obvious visual contrast to natural conditions.
- Proposed Action will require the storage of petroleum or chemical products greater than 4.2 m^3 (1–100 gallons).
- Proposed Action will allow residential uses in areas without water and/or sewer services.
- Proposed Action locates commercial and/or industrial uses that may require new or expansion of existing waste treatment and/or storage facilities.
- Identify other impacts that may occur.

   6. Will Proposed Action alter drainage flow or patterns or surface water runoff?   NO   YES

Examples that would apply

- Proposed Action would change flood water flows.
- Proposed Action may cause substantial erosion.
- Proposed Action is incompatible with existing drainage patterns.
- Proposed Action will allow development in a designated floodway.
- Identify other impacts that may occur.

Impact on air

   7. Will Proposed Action affect air quality?   NO   YES

Examples that would apply

- Proposed Action will induce 1000 or more vehicle trips in any given hour.
- Proposed Action will result in the incineration of more than 1 ton of refuse per hour.
- Emission rate of total contaminants will exceed 2.3 kg (5 lbs) per hour or a heat source producing more than 10 million kJ (10 million BTU) per hour.
- Proposed Action will allow an increase in the amount of land committed to industrial use.
- Proposed Action will allow an increase in the density of industrial development within existing industrial areas.
- Identify other impacts that may occur.

Impact on plants and animals

8. Will Proposed Action affect any threatened or endangered species?   NO   YES

Examples that would apply

- Reduction of one or more species listed on the governmental list, using the site, over or near site, or found on the site.
- Removal of any portion of a critical or significant wildlife habitat.
- Application of pesticide or herbicide more than twice a year, other than for agricultural purposes.
- Identify other impacts that may occur.

  9. Will   Proposed   Action   substantially   affect   nonthreatened   or   nonendangered species?   NO   YES

Examples that would apply

- Proposed Action would substantially interfere with any resident or migratory fish, shellfish, or wildlife species.
- Proposed Action requires the removal of more than 4 ha (10 acres or 0.04 km^2) of mature forest (over 100 years of age) or other locally important vegetation.

Impact on agricultural land resources

10. Will the Proposed Action affect agricultural land resources?   NO   YES

Examples that would apply

- The Proposed Action would sever, cross, or limit access to agricultural land (includes cropland, hayfields, pasture, vineyard, orchard, etc.)
- Construction activity would excavate or compact the soil profile of agricultural land.
- The Proposed Action would irreversibly convert more than 4 ha (10 acres) of agricultural land or, if located in an Agricultural District, more than 1 ha (2.5 acres) of agricultural land.
- The Proposed Action would disrupt or prevent installation of agricultural land management systems (e.g. subsurface drain lines, outlet ditches, strip cropping) or create a need for such measures (e.g. cause a farm field to drain poorly due to increased runoff).
- Identify other impacts that may occur.

Impact on aesthetic resources

11. Will Proposed Action affect aesthetic resources?   NO   YES

Examples that would apply

- Proposed land uses, or project components obviously different from, or in sharp contrast to, current surrounding land use patterns, whether manmade or natural.
- Proposed land uses, or project components visible to users of aesthetic resources that will eliminate or significantly reduce their enjoyment of the aesthetic qualities of that resource.
- Project components that will result in the elimination or significant screening of scenic views known to be important to the area.
- Identify other impacts that may occur.

Impact on historic and archaeological resources

12. Will Proposed Action impact any site or structure of historic, prehistoric or paleontological importance?   NO   YES

Examples that would apply

- Proposed Action occurring wholly or partially within or substantially contiguous to any facility or site listed on the State or National Register of Historic Places.
- Any impact to an archaeological site or fossil bed located within the project site.
- Proposed Action will occur in an area designated as sensitive for archaeological sites on the Site Inventory.
  - Identify other impacts that may occur.

Impact on open space and recreation

13. Will Proposed Action affect the quantity or quality of existing or future open spaces or recreational opportunities?   NO   YES

Examples that would apply

- The permanent foreclosure of a future recreational opportunity.
- A major reduction of an open space important to the community.
- Identify other impacts that may occur.

Impact on transportation

14. Will there be an effect to existing transportation systems?   NO   YES

Examples that would apply

- Alteration of present patterns of movement of people and/or goods.
- Proposed Action will result in major traffic problems.
- Identify other impacts that may occur.

Impact on energy

15. Will Proposed Action affect the community's sources of fuel or energy supply? NO   YES

Examples that would apply

- Proposed Action will cause a greater than 5% increase in the use of any form of energy in the municipality.
- Proposed Action will require the creation or extension of an energy transmission or supply system to serve a major commercial or industrial use.
- Identify other impacts that may occur.

Noise and odor impacts

16. Will there be objectionable odours, noise, or vibration because of the Proposed Action?   NO   YES

Examples that would apply

- Blasting within 460 m (1500 ft) of a hospital, school, or other sensitive facility.
- Odours will occur routinely (more than 1 h per day).
- Proposed Action will produce operating noise exceeding the local ambient noise levels for noise outside of structures.
- Proposed Action will remove natural barriers that would act as a noise screen.
- Identify other impacts that may occur.

Impact on public health

17. Will Proposed Action affect public health and safety?   NO   YES

Examples that would apply

- Proposed Action may cause a risk of explosion or release of hazardous substances (i.e. oil, pesticides, chemicals, radiation, etc.) in the event of accident or upset conditions, or there may be a chronic low-level discharge or emission.
- Proposed Action may result in the burial of 'hazardous wastes' in any form (i.e. toxic, poisonous, highly reactive, radioactive, irritating, infectious, etc.).
- Storage facilities for one million or more gallons (3785 m^3) of liquefied natural gas or other flammable liquids.
- Proposed Action may result in the excavation or other disturbance within 610 m (2000 ft) of a site used for the disposal of solid or hazardous waste.
- Identify other impacts that may occur.

Impact on growth and character of community or neighborhood

18. Will Proposed Action affect the character of the existing community?   NO   YES

Examples that would apply

- The permanent population of the city, town, or village in which the project is located is likely to grow by more than 5%.
- The municipal budget for capital expenditures or operating services will increase by more than 5% per year because of this project.
- Proposed Action will conflict with officially adopted zoning plans.
- Proposed Action will cause a change in the density of land use.
- Proposed Action will replace or eliminate existing facilities, structures, or areas of historic importance to the community.
- Development will create a demand for additional community services (e.g. schools, police and fire, etc.).
- Proposed Action will set an important precedent for future projects.
- Proposed Action will create or eliminate employment.
- Identify other impacts that may occur.
19. Is there, or is there likely to be, public controversy related to potential adverse environmental impacts? NO YES

4.10 Energy conservation considerations

Energy conservation is an important consideration, regardless of geographical location, in the decision to construct a new canning plant. These guidelines are intended to provide cost-effective ideas for saving energy without compromising the building's intended functional use or internal environmental conditions; they cover lighting, building envelope, electrical power and distribution, heating ventilation and air-conditioning systems and equipment, service water heating systems, and energy management. The building envelope refers to the elements of a building that enclose conditioned spaces through which thermal energy may be transferred to or from the exterior or to or from unconditioned spaces. Cost savings from energy conservation can come from building design and construction, equipment needs, and operations. Cost effectiveness of conservation methods will largely dictate their implementation.

Identification of energy needs. Before developing a plan for energy conservation in building construction, a clear picture of major energy requirements must be developed; that is, what may contribute to the largest annual energy costs, e.g. lighting, cooling, refrigeration, heating, ventilation, or equipment operation. If the major energy demand is electrical, then peak power demands and peak time rates should be studied. Consideration of peak demands may reduce the requirements for oversizing of an energy system.

A general strategy might be to address the building's energy needs in the following sequence: minimise impact of the building's functional requirements, minimise loads, and improve the efficiency of the distribution system. Energy efficiency is fundamental to running a cost-effective processing plant.

Lighting design. The lighting system should be designed to provide a productive, safe, and satisfactory visual environment for the intended use of any given space. Lighting is a major energy end-use and may be a major contributor to internal demands by increasing cooling loads and possibly decreasing heating loads. Some suggestions for reducing lighting energy demands are: use task lighting, whenever possible, to accommodate the need for higher lighting levels due to task visual difficulty, changing requirements, or individual visual differences (poor and aging eyesight); grouping similar activities requiring higher luminance or special lighting, if possible, for reducing a general higher level of lighting; the use of lighting controls that maintain proper lighting levels when and where needed but also allow reductions in lighting when tasks are completed, are less critical, or the space is not fully occupied; consider a complete lighting system construction that minimises light loss due to dirt collection; investigate the use of dimmers to reduce energy consumption when the system is new and capable of providing more light than necessary; use more efficient lamps and ballasts.

4.11   Safety considerations

All countries have prescribed Occupational Safety and Health (OSH) legislation that give prescribed guidelines, which must be followed to insure employee safety.

Employers should maintain a program, which provides systematic policies, procedures, and practices that protect their employees from job-related safety and health hazards. Such a program should include provisions for the systematic identification, evaluation, and prevention or control of general workplace hazards, specific job hazards, and potential hazards that might occur from foreseeable conditions. The program should not comply only with OSH standards but address all potential hazards.

The extent to which the program is described in writing is less important than how effective it is in practice. As the size or complexity of an activity increases, the need for written guidance increases to insure clear communication of policies and priorities, as well as consistent and fair application of rules.

There may be considered four main elements of an effective safety program.

1. *Management commitment and employee involvement.* The elements of management commitment and employee involvement are crucial and form the core of any occupational safety and health program. Management's commitment provides the motivating force and the resources for organising and controlling activities within an organization. Employee involvement provides the means by which workers develop and/or express their own commitment to workplace safety.
2. *Worksite analysis.* A realistic analysis of the workplace consists of worksite examinations to identify existing hazards and situations in which changes might occur, creating new hazards. Unawareness of a hazard resulting from lack of site examination is an indication that existing policies or practices are ineffective.
3. *Hazard prevention control.* When feasible, workplace hazards should be prevented by design of the job site. When it is not feasible to eliminate such hazards, they must be controlled to prevent unsafe exposure, and furthermore, they must be controlled immediately after recognition.
4. *Safety and health training.* Training that addresses the responsibilities of both management and employees is an essential component of all safety and health programs. The extent of training depends on the complexity of the worksite and characteristics of hazards. On-site consultation should be used to assist employers in voluntarily fulfilling their responsibilities for workplace safety and health.

A publication from the US Department of Health and Human Services (HHS) National Institute for Occupational Safety and Health (NIOSH) entitled, Health and Safety Guide for Food Processors (NIOSH, 1975), lists frequently violated regulations. These violations will be listed here without the appended general requirements because of detail. Interested parties are encouraged to acquire the publication.

The most frequently violated regulations are: walking and working surfaces, including fixed and portable ladders, fixed industrial stairs, railings and toe board, exits and exit markings; occupational health and environmental control, such as exposure to a variety of gases, dusts, vapours, noises, hazardous materials, such as flammable and combustible liquids; lack of personal protective equipment thereof, poor quality, or not functioning; general environmental controls relating to sanitation, medical, and first-aid services not available; fire protection facilities, accessibility, and function; compressed air equipment; materials handling and storage; machinery and machine guarding; hand and portable powered tools; welding, cutting, and brazing; violations of the national electrical code (NEC); and improper record keeping.

4.12 General guidance

The US FDA's Title 9 CFR 381.48 under Floors, Walls, Ceilings, etc., states that:

1. 'Floors shall be constructed of, or finished with, materials impervious to moisture, so they can be readily and thoroughly cleaned and graded for complete runoff with no standing water.'
2. 'Walls shall be smooth and constructed of materials impervious to moisture to a height of 6 feet (1.8 m) above the floor to enable thorough cleaning.'
3. 'Ceilings must be moisture resistant and finished and sealed to prevent collection of dirt or dust that might sift through from above.'

USDA-FSIS-FESD has published a list of primary facilities problem areas referred to as 'The Troublesome Dozen':

1. Proper plant layout/separation of raw and cooked product
2. Direct outside entry into production areas
3. Strip doors in production areas
4. Official Limits designation not shown on plot plan
5. Separation of official and unofficial premises
6. Common areas of two or more firms in same building
7. Inadequate dry storage area
8. Trailers or shipboard containers for dry/refrigerated storage
9. Retail/bakery areas may be part of official premises
10. Wood usage in production areas
11. Paint usage in production areas
12. Vinyl flooring and covering in production areas

The US FDA's guidelines for plant construction are more general and not as definitive as the USDA's and are covered in 21 CFR110.20 (b) Plant Construction and Design. Plant buildings and structures shall be suitable in size, construction, and design to facilitate maintenance and sanitary operations for food manufacturing purposes. The plant and facilities shall:

a. Provide sufficient space for such placement of equipment and storage of materials as is necessary for the maintenance of sanitary operations and production of safe food.

b. Permit the taking of proper precautions to reduce the potential for contamination of food, food-contact surfaces, or food-packaging materials with microorganisms, chemicals, filth, or other extraneous materials. The potential for contamination may be reduced by adequate food safety controls and operating practices or effective design, including the separation of operations in which contamination is likely to occur, by one or more of the following means: location, time, partition, airflow, enclosed systems, or other effective means.

c. Permit the taking of proper precautions to protect food in outdoor bulk fermentation vessels by any effective means, including:
   • Using protective coverings.

d. Be constructed in such a manner that floors, walls, and ceilings may be adequately cleaned and kept clean and kept in good repair, that drip or condensate from fixtures, ducts, and pipes does not contaminate food, food-contact surfaces, or food-packaging materials; and that aisles or working spaces are provided between equipment and walls and are adequately unobstructed and of adequate width to permit employees to perform their duties and to protect against contaminating food or food-contact surfaces with clothing or personal contact.

e. Provide adequate lighting in hand-washing areas, dressing and locker rooms, and toilet rooms and in all areas where food is examined, processed, or stored and where equipment or utensils are cleaned, and provide safety-type light bulbs, fixtures, skylights, or other glass suspended over exposed food in any step of preparation, or otherwise protect against food contamination in case of glass breakage.

f. Provide adequate ventilation or control equipment to minimise odors and vapours (including steam and noxious fumes) in areas where they may contaminate food, and locate and operate fans and other air-blowing equipment in a manner that minimises the potential for contaminating food, food packaging, and food-contact surfaces.

g. Provide, when necessary, adequate screening or other protection against pests.

References

Demetrakakes, P. (1994). Time to paint? Cover all the bases *Food Processing,* (July) 67–68.

Doe. (1989). *Performance standards for new commercial and multi-family high rise residential buildings*. DOE/CE-0304T. Washington, DC: U.S. Department of Energy.

Imholte, T. J. (1984). *Engineering for food safety and sanitation*. Crystal, MN: Technical Institute of Food Safety.

NIOSH. (1975). *Health and safety guide for food processors*. DHEW Publication No. (NIOSH) 75–166. Cincinnati, OH: Department of Health, Education, and Welfare.

Ortiz, L. (1992). Locating a food plant: a case study. In *Proceedings, conference on design and engineering FAP, orlando,* FL, Oct. 11–14.

OSHA. (1992). *OSHA handbook for small businesses*. OSHA 2209. Washington, DC: U.S. Department of Labour, Occupational Safety and Health Administration.

OSHA. (1993). *General industry digest*. OSHA 2201. Washington, DC: U.S. Department of Labour, Occupational Safety and Health Administration.

SEQR. (1987). *State environmental quality review-full environmental assesment form*. New York State Publication 1416(2/87)-7c.

Sanitary design and equipment requirements

5.1 Introduction

There is no standard canning plant construction or equipment; however, some equipment manufacturers may fabricate the necessary pieces of equipment to process a specific crop (e.g. corn or a root crop) or type of product. The most economical plan is to call upon the canning machinery manufacturer, and submit your problem; have them submit layout plans and the most up-to-date equipment. Placing the responsibility upon him will save you money, worry, and mistakes.

5.2 Sanitary food plant buildings

The surroundings of food plants should be neatly landscaped and well drained. Buildings should preferably be oriented to avoid excess heating by sun during the summer.

Building foundations should be brought to 60 or 99 cm (2 or 3 ft) above ground to prevent rodents from coming into the building.

Areas in the plant where different operations are performed should be separated by partitions. Areas where equipment needs to be hosed should have tiled walls. Walls in the warehouse should be smooth and light colored. All wall junctions should be round and not square, angled to facilitate cleaning.

Ceilings in processing areas and warehouses should be washable. Joints should be sealed. This helps prevent moisture condensation on ceilings, which could be a nuisance and cause product contamination.

The processing, cooling, warehousing, and packaging areas should be separated. Doors and windows should be screened with 16-mesh screen to protect against entrance of insects and rodents. All rooms should be adequately ventilated to help prevent condensation of moisture. Dressing rooms, washrooms, and clothing storage rooms should be located outside the processing area.

The quality of air within factory buildings may be set to a specified quality (temperature, humidity, and particle concentration) and quantity (fresh air volume) for the comfort and safety of employees as well as, for some products, to reduce the possibility of contamination. Sometimes work areas need to be fitted with an air intake that brings in air free of microorganisms, especially if the risk of contamination from the environment is high. This is especially true with factories producing products in which contamination with infectious pathogens (e.g. *Salmonella*, *Listeria*, *E. coli*), toxigenic pathogens (e.g. *Staphylococcus aureus* and *Clostridia spp*), and spoilage microorganisms (e.g. yeast, moulds, and lactic acid bacteria) may be a problem.

There should be no connections between the potable water system and nonpotable water.

A Complete Course in Canning and Related Processes. http://dx.doi.org/10.1016/B978-0-85709-677-7.00005-0

A basic part of adequate sanitation is large, well-ventilated, and fully equipped toilets for both men and women employees. Most areas have regulations about this matter that should be considered in designing the building.

5.3   Equipment requirements

The kind, amount, and arrangement of equipment are determined by the items, sizes of cans, and quantities of each product to be canned. Some items such as corn, peas, green and wax beans, and lima beans require special equipment, not suitable for use with other products. Such vegetables as beets, carrots, and sweet potatoes require relatively simple equipment, which is common to most canneries using pressure cookers. Spinach needs special washers and special blanchers. Asparagus requires little special equipment. The dry pack items like pork and beans, kidney beans, etc. require soaking equipment not used for canning fresh products. A 'cyclone' and finisher, together with evaporating equipment, are required for tomato pulp (puree) and for tomato paste. A 'cyclone' is also needed for pumpkin and squash, and a finisher is also usually employed. Peaches and apricots need a lye peeler and a blancher. The filling tables, syrupers, exhaust boxes, closing machines, and cookers are the same. An exception among the fruits is olives, which require special equipment throughout. Milk, meats, and soups are specialties that cannot be combined with other lines of canning to any advantage.

Although it is possible to can limited quantities of most items without much special equipment, it should be noted that adequacy of equipment is essential to efficient plant operation and the lowering of costs.

Catalogues from various manufacturers of canning machinery might lead to the conclusion that very elaborate outfits are essential. Certain pieces are necessary to do special kinds of work, others are economical as part of a unit, and still others may be desirable to reduce the number of workers.

All machines should incorporate certain qualities:

1. It should perform the function for which it is designed.
2. Its work should be done cleanly, and the machine should be easily cleaned.
3. The machine should be as simply designed as possible and of such sturdy construction that few repairs are needed.
4. It should be economical in operation.

A shortcoming is rating the equipment at large capacity so that it has to be operated at maximum speed to take care of the production. Each machine has its maximum capacity for correct performance; crowding any machine beyond this point causes trouble. It is far better to attain maximum output by avoiding breakdowns by careful attention to adjustments of machines and intelligent operation at all times.

The leading manufacturers of canning machinery have prepared factory plans or designs of the most up-to-date machinery showing their proper arrangement, which they will gladly furnish on request. The layouts given herein are examples that cover particular products.

5.3.1 Small plant

Some products require much hand work in their preparation and only the simplest mechanical equipment, and these may be packed in a small unit. These products include

Apples	Asparagus
Peaches	Kraut
Pears	Sweet potatoes
Berries	Carrots
Tomatoes	Spinach

By providing a roller grader, apricots, plums, and possibly others may be added to this list.

The basic equipment for packing 200 cases or more per day for most of these lines is

3 × preparation tables	1 × open cooker
1 × filling table	1 × cooling tank
100 × 10 L pans	1 × hoist
1 × tank and sprays for washing	18 × retort crates
1 × scalder and blancher	1 × 40-H.P. boiler
1 × kettle	1 × engine and pump
3 × 200 L tanks for brine/syrup	Scales
1 × exhaust box	Trucks
1 × closing/seaming machine	Buckets
3 × retorts	Small accessories

Tables should be 4.6 m (15 ft long and 39 inches (1 m) wide, with the top pitched from front to back and a shelf placed below to receive refuse buckets.

The closing machine should be fully automatic and of sufficient capacity.

The boiler capacity should be large, as it requires far more steam for pumping and heating water than is ever indicated by consulting the catalogue of supplies.

A small group of about 20 persons can pack a surprisingly large amount of material in a season, provided they are well directed and work steadily. By adding to the number on the preparation and filling tables, a considerable increase in volume can be affected without adding to the equipment.

5.3.2 Large plant

In larger plants, the use of tables with sinks, conveyor systems for both the product in preparation and in the can, graders, syruping machines, high-speed can-closing machines, and continuous cookers all add to the speed and volume that can be handled in a given space. Such equipment should be laid out by an expert after all facts have been ascertained.

5.4   Sanitary construction of food plant equipment

Certain general considerations must be considered with reference to materials used in equipment. Stainless steel is the most desirable material, No. 316 stainless steel being more corrosion resistant than No. 304. Parts made of cast iron should not be used in contact with food, nor is galvanized iron acceptable, as zinc wears out and eventually causes problems. Monel metal, an alloy of copper and nickel, is acceptable as it is quite resistant to corrosion, but it should not be used for products that are subject to discoloration by traces of copper such as corn, lima beans, or peas. Monel may be used for salt brines, because stainless steel is corroded by sodium chloride (Katsuyama, 1993, p.540). Aluminum is a good heat conductor but is quite readily corroded by fruit acids and by alkalis used for cleaning. Copper, brass, or bronze parts are not recommended. If used, they should be scrupulously clean, as copper corrodes, even in contact with air, forming a film of soluble copper compounds that dissolve into foods. This is undesirable, both from public health and product quality standpoints. Rubber is suitable for materials used for conveyor belts, but canvas is preferable.

Equipment should be constructed without angle corners, as these are difficult to clean; if not clean, they can be a source of food spoilage micro-organisms that may contaminate food. There should be no dead-end pipes, as these are usually difficult to clean. Pipes should be joined with sanitary-type joints, as these have no crevices and are easy to disassemble and clean.

High-pressure water jets, in combination successively with alkaline and milk acid detergents and other surface activators, should be used to clean equipment. Water temperature should not be above 60 °C (140 °F), as higher temperatures cause coagulation of food proteins, which then are deposited as a film on equipment surfaces. Sanitising agents, like iodophors, may be added to water, but this is not considered the best way to sanitise equipment. Almost all sanitising materials are ineffective in the presence of organic matter.

5.5   Food plant equipment design

The following are desirable characteristics of sanitary food plant equipment:

1. All machine parts must be designed for quick dismantling and reassembling – some merely by removing and replacing a nut or wing screw by hand. It is also best to construct these parts of lightweight material so that they can be easily handled for cleaning.
2. All food contact surfaces should be inert, smooth, nonporous, and nonabsorbing; they must withstand the application of chemical cleaners, sanitisers, and pesticides; they must be easily cleaned and readily accessible for inspection (Imholte, 1984, p. 283).
3. Open seams in cooking kettles, mixers, blenders, storage vats, and filling machines must be eliminated.
4. Surfaces of equipment in contact with food must be smooth and continuous; rough spots and crevices must be avoided.
5. All junctions, particularly pipelines and ducts, must be curved or rounded. Cooking kettles, storage tanks, holding vats, and similar units must have long curves at the juncture of the bottom and sidewalls instead of sharp corners.

6. All machine parts in contact with food must be accessible for hand-brush cleaning and/or inspection.
7. Dead-end areas in all machines must be eliminated.
8. Metals like lead, antimony, and cadmium must not be used in fabricating equipment. Copper or copper-containing alloys are not recommended.
9. Stuffing boxes or glands in which food might accumulate and decompose should not be used.
10. Pipe fittings must have a sanitary thread and threaded parts must be accessible for cleaning.
11. Sanitary-type valves, such as plug type, should be used.
12. Runoff valves should be installed as close as possible to mixers, kettles, vats, and tanks.
13. Coupling nuts on piping and valves must have sufficient clearance and must be easily taken apart.
14. Food products should be protected from lubricants and condensates as moisture condensing on ceilings may pick up dirt and peeled paint, later to drop into open cooking kettles or holding vats.
15. Mixing blades should be welded to the drive shaft or both should be in one piece. Shaft and blades should be removable from the mixer at a point above the surface of the product.
16. Machine parts in contact with food should be constructed of noncorrosive metal.
17. Equipment like kettles, certain mixers, and holding vats and storage bins should have sectional covers, which are free from seams, hinges, crevices, and heads in which dirt might collect.
18. Drive shafts must be sealed so that lubricating grease does not work its way into the food.
19. Any horizontal parts of machines or supports should be at a minimum of 15 cm (6 in) above the floor. Tubular supports are preferred, but if squared tubing is placed horizontally, it should be rotated 45° to eliminate flat surfaces (Katsuyama, 1993, p. 540). Equipment occupying large floor areas should have a clearance of 46 cm (18 in) or more to facilitate cleaning (Imholte, 1984, p. 283).

5.6 Sanitation criteria for pipes, valves, and pumps

Basic criteria for piping, valves, and pumps in food plants established by the National Food Processors Association apply to the design, materials, and construction of piping, valves, and pumps handling food products. They may also serve as guides for new developments in materials and design.

5.6.1 Materials

1. Stainless steel or similar corrosion-resistant metal
   a. Selection of alloys should be based upon the manufacturer's recommendations for the type of food and use.
   b. Surfaces must be free of cracks and crevices, and all welds must be ground smooth.
   c. Sheet or plate metal should be at least a No. 2B mill finish or equal, properly applied, depending upon usage.
   d. Sanitary tubing should be a No. 4 finish or equal internal diameter (ID).
2. Glass
   a. Heat-resistant glass piping of a borosilicate type should be used.
   b. Glass should not be used where unusual stress may occur.

3. Cast iron, cast and forged steel, and nickel alloys
   a. Selection of alloys and surface finish should be determined by exact usage. Food contact surfaces must be smooth and nonporous.
   b. Interior surface should be free of cracks, crevices, and recesses.
4. Plastics
   a. Choice of plastic material will depend upon type and temperature of the product.
   b. Abrasion-resistant and shatterproof plastics should be used.
   c. Heat resistance and softening of the material should be determined before use.
   d. It should not contain any constituents that may migrate to the food, such as free phenol, formaldehyde, etc.
   e. Fiberglass-reinforced plastics should not be used as food-contact parts where they may become broken or abraded.
5. Gaskets and packing materials
   a. The exact material should be determined by the type of product and temperature to be used.
   b. All should be nonporous, nonabsorbent, and nontoxic. No lead or zinc fillers should be used in gasket materials.
   c. Types may include neoprene synthetic rubber, polybutadiene synthetic rubber, or Teflon plastics.
6. Alternate materials
   a. Additional materials that prove equally satisfactory from the standpoint of sanitation and product protection would be acceptable.
   b. Laboratory and pilot plant use should establish that a material is cleanable, resistant to product, and does not create a food-additive problem.

5.6.2   Design and construction

1. Piping
   a. Piping, fittings, and connections should be of ample diameter to permit easy cleaning; the design should provide for total drainage.
   b. All interior surfaces should be smooth, continuous, free from pits and threads, and contain no open or rough seams.
   c. Permanent Cleaning-in-Place (CIP) systems shall be of corrosion-resistant stainless steel, heat-resistant glass, or equivalent.
   d. Any permanently installed piping system must be provided with a return cleaning line to allow complete recirculation of detergent and sanitiser solutions during cleanup.
   e. Where inspection openings are needed in CIP systems, sanitary elbows with CIP fittings should be used.
   f. Inspection openings are usually recommended at all changes of direction of pipe. However, with heliarc-welded pipe, combined with proper cleaning and sanitising, this would not be necessary.
   g. Pipes or chutes using gravity flow should be readily demountable for cleaning and inspection. When possible, easily removed covers should be used.
   h. Sharp corners should be avoided, and no dead ends, blind tees, or inside threads are permitted.
   i. All permanent joints of metal piping must be continuously welded and be free of cracks or rough surfaces.
   j. All other joints shall be ground or gasketed (with a self-positioning gasket) to form a sanitary joint.
   k. When cleaning glass pipe, materials that scratch or etch the surfaces should not be used.

2. Pumps

    a. Heads must be readily removable for cleaning and inspection with all product-zone parts conforming to proper material requirements.

    b. Sanitary-type fittings should be used on the head to connect to piping.

    c. All bearings should be located outside product zones, and if adjacent thereto, must be constructed with a readily removable seal at the entrance of the shaft.

    d. Impeller, rotor, and/or valve assemblies must be readily removable for cleaning and inspections.

    e. Pumps must have no internal bypasses in the head that are not self-draining. The pump head itself should be self-draining.

    f. Pumps and solution tanks for CIP pipeline systems should be corrosion resistant and pumps should be sized to give a velocity of not less than five feet per second (1.5 m/s) through the largest pipe or fitting.

    g. In CIP systems, product pumps should be designed to have all parts that contact the product completely self-cleaning, including the seal.

    h. Packing-gland assemblies must have a readily removable seal and should be constructed of material described under 'Gaskets and Packing Material'.

    i. Exterior surfaces should be easily cleanable.

3. Valves

    a. All valves in food-conveying systems should be of sanitary design without pockets or recesses and completely self-draining. Valves should be self-flushing during normal operation.

    b. Internal bypasses should be avoided, their usage to be determined by ability to be self-draining.

    c. All runoff or drain valves should be placed as near equipment as possible to prevent pockets or dead areas in piping.

    d. Gaskets, diaphragms, stem packing, and other nonmetallic parts should conform to the description of gasket materials.

    e. Lubricants for valves and plugs should be of sanitary quality, nonflammable, odorless, and nontoxic.

5.6.3   Placement and installation

1. Piping

    a. Should be installed a minimum of 2 ft (60 cm) away from walls and ceilings to provide access for cleaning and painting.

    b. Piping should be a minimum of 8 in (20 cm) above the floor to provide for cleaning.

    c. All sections must be provided with adequate hangers or supports to prevent sagging or buckling at joints.

    d. All hangers or supports should be easily cleanable and should not provide a collecting surface for dust or soil.

    e. All pipe should be installed to permit self-draining. A minimum slope of ⅛ in per foot (1 cm per metre) should be provided.

    f. Systems should be designed to operate with all pipelines full. Partially filled lines are a sanitation hazard.

    g. All pipelines should be identified as to use or material carried.

    h. Cold pipes should be insulated to prevent condensation. Insulating materials should be inert and impervious to water and not likely to peel, flake, or break.

    i. Consideration must be given to the possibility of electrolytic corrosion in the placement of piping, as well as in the selection of clamps and hangers and where piping attaches to electrically driven devices.

2. Pumps
   a. Should be installed away from walls and other equipment sufficiently to permit disman-
      tling and cleaning.
   b. Pumps installed on floors should be on a solid base that is sealed water tight.
   c. In mounting, minimum clearance for cleaning should be considered.

5.7   Care and use of temperature controllers and recorders

Controllers and recorders for processing should be installed so that they can be easily
read and free from heat and vibration. Bends in the thermal tube should be kept to a
minimum and sharp bends or kinks avoided. The tube should be protected so that it
will not be damaged. Only qualified persons should make repairs on thermal or control
mechanisms.

The pen arm, as it moves up and down, should follow along the arc of the
time intervals on the chart for proper interpretation of the temperature curves. If
it does not, the arm is not the correct length. This can be easily adjusted on most
controllers.

The temperature-recording pen is connected to the thermal system but as installed
does not necessarily indicate the correct temperature; it must be standardized against
the retort mercury thermometer. This adjustment is made by means of a screw that
changes the position of the pen arm in relation to the thermal pointer. When the
recorder is not in use, the pen should be left resting on an old chart. Most pens are
plated, and the plating may wear off and ruin the pen if it is allowed to rub on the
faceplate.

At the start of a day's operations, the chart should be placed in the proper position so that the pen indicates the time of day. The air filter should be drained daily to
remove oil or water.

The greatest problem connected with pneumatic controllers for processing is the
maintenance of a clean, dry-air supply at a constant pressure. Moisture, oil, corrosive
liquids, or foreign particles carried into the pneumatic system from the air supply will
eventually cause trouble. Pneumatic controllers operating with clean air require little
cleaning and maintenance.

A separate air supply system should be provided for instrument operation only.
The compressor should be large enough to provide air at a low-capacity rating; an
overloaded compressor is more likely to pump oil. The system should be designed
to supply at least 0.03 m^3 (one cubic foot) of air per minute per instrument. If a
separate compressor is not feasible, a separate distribution system should be provided for air-operated instruments and adequate filter systems should be installed
to supply clean, dry, oil-free air. The main supply line should be at least 13 mm
(0.5 in) pipe of corrosion-resistant material such as copper or brass. Black iron pipe
is not suitable. The main should be sloped to drip lines. It has been estimated that
80% of all pneumatic instrument troubles are caused by dirty air.

5.8 Equipment corrosion

Corrosion is the wearing away or alteration of a metal or alloy, either by direct chemical attack or by electro–chemical reaction. A prime contributor of corrosion is water in contact with metal. Other corrosion agents are oxygen, carbon dioxide, chlorine compounds, and acids: both organic and inorganic.

5.8.1 Galvanic corrosion

Galvanic corrosion may occur when two different metals in contact are exposed to a conductive solution such as tomato juice, pineapple juice, or even highly chlorinated water. The difference in electric potential existing between different metals serves as a driving force to pass current through the corrodent. This potential difference between metals can be measured and is arranged in what is called a galvanic series. The further apart two metals are separated in the series, the greater the likelihood that galvanic corrosion will occur. To minimize galvanic corrosion, similar metals should be used wherever possible; if dissimilar metals are used, they must be separated either by insulation, paint, or coatings. When dissimilar metals are used, a sacrificial anode such as magnesium or zinc may be employed, which will corrode faster than the other two dissimilar metals.

5.8.2 Stress corrosion

Stresses in metal, caused by internal or external pressure, will create paths within the grain of the metal that tend to corrode more readily under certain conditions. It is important to utilise preventative methods, such as stress relieving or selecting a more resistant material. The presence of nitrogen in iron or steel tends to make metals more prone to stress-corrosion cracking under certain tensile strengths and in certain environments. If steel contains aluminum, there is a better resistance to stress corrosion. It is important that periodic maintenance and inspection be utilised to detect this type of corrosion.

5.8.3 Erosion corrosion

When a corrodent moves over a metal surface, the mechanical wear and corrosion is called erosion corrosion. This may occur at high-velocity conditions, such as in heat exchangers, which may be minimised by using larger pipe sizes and streamlining bends to minimise impingement effects. Erosion or corrosion that occurs on pumps, impellers, agitators, and piping can be avoided by selection of more resistant material or by protective coatings.

Cavitation and fretting corrosion are special forms of erosion corrosion. Cavitation is caused by rapid formation and collapse of vapor bubbles on a metal surface; the high pressures thus produced can deform the underlying metal and remove protective films. Fretting corrosion occurs when metals slide over each other causing mechanical damage to one or both surfaces; this has been shown in bearings and may be minimised by using harder materials to reduce friction.

5.8.4   Crevice corrosion

Whenever scratches or crevices develop at joints, bolts, and rivets, local corrosion may occur in crevices. Environment plays an important role, because the presence of moisture, acidity, or lack of oxygen influences the degree of corrosion. It is important in equipment design to minimise crevices and take proper steps so that surfaces may be kept clean.

5.8.5   Pitting corrosion

This type of corrosion is evident in the interior walls of a pipe in which holes are formed. This pitting may continue at a very slow rate, suddenly resulting in failure that may not be visible. The selection of a metal and the cleanliness of the surface are very important factors in avoiding this problem. Another type of corrosion is exfoliation, but this differs from pitting in that the attack has a laminated appearance and is subsurface corrosion. The surface is usually flaky and sometimes blistered but can be minimised by removal of one element (zinc) in an alloy.

5.8.6   Recommendations

Once the various types of corrosion are understood, the means for detecting and controlling the condition become important. Sometimes selection and material or operating conditions are uncontrollable, but diagnosing the reason for corrosion remains important. Because it is sometimes impossible to alter the temperature and environment to which metals are subjected, preventive maintenance can do much to help minimise corrosion. Recognising the situation and using protective coatings, as well as controlling acidity and temperature within certain ranges, are deemed worthwhile. Downtime and replacement parts that fail because of corrosion can be very costly; adequate records of when to replace parts, as well as preventive maintenance are important considerations.

Because water is used in rather large quantities in most processing plants, it is important to have test analysis on the water so that the exposure of metals to electrolytes is known. Selection of certain inhibitors in the water that are approved will retard some forms of corrosion and extend the life of the material used.

By understanding corrosion and taking preventive maintenance, as well as controlling corrosion, the life of equipment may be extended and shutdown costs reduced due to corrosion failure.

5.9   Aseptic processing facilities

Aseptic processing and packaging is the filling of a commercially sterile product into sterilised containers followed by hermetical sealing, with a sterilised closure, in an environment free from microorganisms (EHEDGE Update, 2006). An aseptic processing facility is a building containing clean rooms in which air supply and equipment are regulated to control microbial contamination, and the products are processed and then packaged without any further contamination. The air in defined areas in an aseptic processing facility should be appropriately filtered to attain the required degrees of air quality. Clean and aseptic rooms should be under positive pressure, so that no contamination can take place. It must be possible to sterilise the equipment in aseptic processing facilities with both heat and chemicals. Aseptic plant design and construction is highly specialised, and experienced consultants should be used.

References

EHEDGE Update. (2006). Guidelines on air handling in the food industry. *Trends in Food Science & Technology*, *17*, 331–336.

Imholte, T. J. (1984). *Engineering for food safety and sanitation*. Crystal, MN: Technical Institute of Food Safety.

Katsuyama, A. M. (Ed.). (1993). *Principles of food processing sanitation* (2nd. ed.) Washington, DC: The Food Processors Institute.

Water

6.1 Introduction

The suitability of water for product preparation and canning is measured by consideration of two general aspects of water quality:

- The bacterial content indicates its fitness for human use and influences plant sanitation requirements.
- The composition with respect to organic and inorganic impurities affects its use in cleaning and may affect the physical characteristics of the food being canned.

Water is in intimate contact with most foods during preparation and canning operations because it is used for washing fresh raw foods at various stages during preparation and serves commonly as a carrier of raw materials between unit processes, although vibrating conveyors are replacing this function. It is the medium through which many unit processes, such as soaking, blanching, quality separation, and preheating are applied. It serves as an ingredient of formulation and is the principal portion of the brine or syrup composing the packing medium. Through its continuous contact with raw foods in preparation and canning, water exerts influence upon the efficiency of the cannery operation and upon the quality of the finished canned food.

6.2 Supply

The best source of water, from a public health view, for a food-processing plant is usually from a municipality. The municipality monitors quality and treats the water, including chlorination. An alternative source would be to develop a private water supply from surface or ground water. Surface water comes from a stream, river, or an open body of water such as a pond or lake whereas ground water comes from wells, either dug, bored, driven, or drilled, or from springs. Whatever the source, a sufficient supply of good quality must be available. Because of possible contamination, surface water would probably need more extensive treatment than ground water (see Figure 6.1). Examples of two Permutit (now Evoqua Water Technologies) treatments found in the US are described below, using an Automatic Valveless Gravity Filter (AVGF) and a Precipitator.

The AVGF system is a gravity filter that operates automatically without a valve, backwash pump, flow controller, or other instruments; nothing moves but the water. The unit operates on the loss of head principle. Raw water enters the filter chamber and flows down through the filter media into the collection chamber and up through the effluent duct to service (Figure 6.2). As the filter bed collects dirt during the filter run, head loss increases and the water level rises in both the inlet and backwash pipes.

A Complete Course in Canning and Related Processes. http://dx.doi.org/10.1016/B978-0-85709-677-7.00006-2

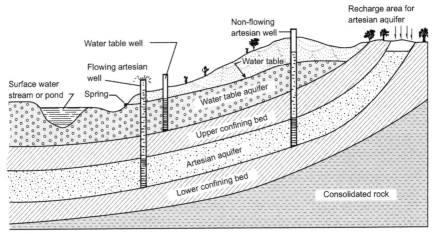

Figure 6.1 Ground and surface water source, *MWPS-14*.

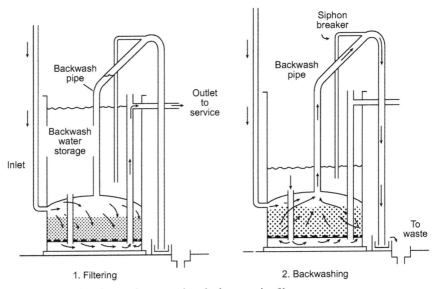

Figure 6.2 US. Filter/permutit automatic valveless gravity filter.
Source: Courtesy of United States Filter Corporation.

When the water starts flowing over and into the downward section of the backwash pipe, a siphon action occurs and backwash begins.

Flow through the backwash pipe reduces pressure immediately above the filter bed. This draws water from the backwash storage compartment down through the ducts and up through the strainers expending the filter bed and cleaning it, then discharging to waste. Backwashing continues until the backwash water level drops below the end of the siphon breaker, admitting air to the top of the backwash pipe, terminating the siphon action and backwash.

At this point, inlet water resumes its flow down through the filter chamber automatically rinsing, settling, and leveling the bed. This rinse water then flows up into the backwash storage chamber where it is held for the next backwash. When water rises to the effluent outlet level, all filtered water then flows to service.

Standard filter sand or anthracite is generally used as the filter media. However, other types of filter media, such as activated carbon, manganese zeolite, or combinations may be specified to meet particular treatment requirements.

The Precipitator offers a highly efficient means for removing impurities from a liquid by precipitation, adsorption, settling, and upflow filtration.

Chemical precipitation and hydraulic separation are the principles involved in the design and operation of the Precipitator. Most efficient and complete precipitation is achieved when a raw water–chemicals mixture is assured intimate contact with suspended sludge particles from previously treated water. This contact also promotes precipitate growth as smaller particles agglomerate into larger, heavier particles.

The principle of hydraulic separation involves an upflow design and suspended sludge blanket through which the water moves. This suspended sludge bed not only mechanically entrains or absorbs impurities and particulate matter in the rising water but also furnishes the nuclei for continuing particle agglomeration.

The top level of the sludge blanket is determined by the rising water's rate of flow and by the size and concentration of the particles. As the particles in the sludge blanket become coated with the finer particles rising into this section, or as these finer particles agglomerate with the existing particles, the larger particles are no longer suspended by the upward velocity and fall back into the mixing zone. Therefore, a continuous flow of fine suspended particles from the mixing zone into the sludge blanket zone occurs as does a counter flow of larger particles in the opposite direction. This results in a condition of equilibrium that maintains approximately the same concentration in both zones.

This process is based upon the fact that a particle is supported by an upward-moving stream of water if the velocity is high enough so that the friction of the water on the particle exceeds the pull of gravity. If the velocity is first high and then gradually decreases, a point is reached at which the particle becomes too heavy to be supported by the frictional effect of the passing water. It remains behind, becoming separated from the water that previously supported it.

In the Precipitator, this is accomplished by varying the cross-sectional area of flow (Figure 6.3). After the incoming water has been thoroughly mixed with chemicals, it passes through the sludge-blanket chamber. This chamber has sloping sides so that as the constant volume of water moves upward, its velocity is continuously decreased in inverse ratio to the increasing cross-sectional area. Because of this design principle, the precipitates separate from the water at a sharply defined level. Additionally, this suspended body of precipitates, or sludge blanket, acts as a filter to catch other rising particles.

Because sludge is in continuous motion throughout the precipitator, it may be blown off from any point within the sludge zone and may be conducted to either an external concentrating tank or internal concentrating chamber. The blowout from the internal concentrator section may be controlled by either a time clock or flow meter.

Water is pumped from a source to a holding tank to insure a constant flow and then into the top of the precipitator equipped with an inverted cone that sits off the bottom

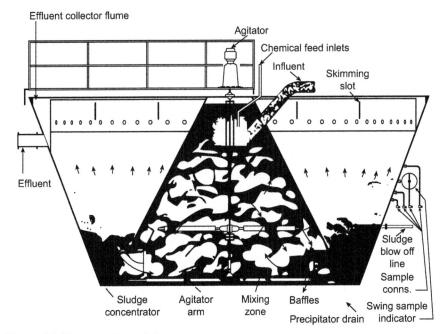

Figure 6.3 The permutit precipitator.
Source: Courtesy of United States Filter Corporation.

of the tank a few inches. Added to the cone in a continuous slurry form is aluminum sulphate or alum (for floc formation), lime or bentonite (for pH control), and a coagulating agent or polymer in a quantity dependent upon the flow rate of the water. The mixture of water and treatment chemicals causes a floc blanket to be formed, the first step in filtering the water. The treated water flows out under the bottom of the cone and overflows the top perimeter of the effluent collector flume. Organic solids form large particles and fall to the bottom of the tank that, after a time, are drained off.

After the precipitator, water goes to a series of sand and anthracite (activated carbon) filters. These filters further purify the water and remove off-flavors. After filtration, the water is chlorinated to at least 3 ppm and is pumped to a holding tank for plant use.

This water may be used for all purposes except for drinking and personal hygiene because it may not meet all public health standards, including those of recordkeeping. It is acceptable to add this water to the canning operation as brine water that will later receive a sterilisation treatment.

6.3   Microbiological content of water

Potable water should only be used in the preparation of foods intended for human consumption. 'Potable' water, by definition, contains no microorganisms capable of causing human intestinal diseases and is aesthetically satisfactory for drinking purposes, which means that it is free of undesirable odours and flavours. The fitness of water for drinking purposes, with reference to microbiological content, should be no less strict than those contained in the latest edition of the 'World Health Organization's International Standards for Drinking Water' (WHO, 2006).

Analyses determine the presence or absence of coliform organisms that indicate the probability of human or animal contamination of the water supply. Generally, public health regulations require that municipal or other public water sources meet prescribed standards for drinking water. Regulations governing cannery operations usually require similar adherence to drinking water standards for water used in cannery preparation of foods intended for human consumption. Where local standards do not exist, the latest edition of the WHO International Standards for Drinking water should be used.

Although heat-sterilised canned foods receive thermal processes more than sufficient to destroy all pathogenic or disease-causing organisms, only potable water should be used in preparation and canning. In any food-processing plant, the use of water that fails to meet public health standards also presents a possible health hazard to workers. Water-borne parasites can be a problem in some places, as they may be resistant to chlorination. Troller (1983) reported that *Giardia lamblia* is the parasite most frequently involved in water-borne outbreaks. Like other helminths, this parasite is highly resistant to the chlorine concentrations normally employed in water supplies to kill bacteria. Additionally, culture tests for coliform bacteria (the indicators of fecal contamination and its associated bacterial pathogens) will not predict the presence of this organism. *Giardia* contamination is most frequently encountered in private or semipublic water supplies, such as might be found in parks, in which water pretreatment (flocculation, settling, and filtration) is inadequate or nonexistent. Under most circumstances, pretreatment conditions are sufficient to remove cysts of *Giardia* from water supplies; however, in cases in which a threat may exist (such as a surface-water supply subject to contamination from human or animal waste), most conventional chlorine treatments are neither of sufficient concentration nor adequate duration to kill this organism. As an indication of the prevalence of giardiasis, about 27% of all reported cases of water-borne diseases in 1977 were attributed to this disease. Generally, however, this organism has not been a problem from the standpoint of direct or indirect food contamination. It constitutes a hazard principally because of direct water contamination.

Municipal water supplies, which are usually purified, are free of coliform contamination.

Deep wells usually supply water that is potable with respect to its bacterial content, but this should be checked and monitored.

Shallow wells are frequently subject to contamination from surface drainage, and both shallow well water and surface waters usually require purification to meet local drinking water standards. Pollutants are able to flow long distances through limestone and other porous soils. Well water may be either dug, drilled, or driven.

Dug wells are 1–2 m (3–6 ft) in diameter and up to 10 m (35 ft) deep, whereas a driven well is commonly 2.5–5 cm (1–2 in) in diameter and less than 15 m (50 ft) deep. A drilled well is generally 15–20 cm (6–8 in) in diameter and up to 6 m (200 ft) deep. Dug wells are most susceptible to surface-water pollution and frequently go dry during drought, making them of little value for a commercial operation.

Drilled wells should have a concrete casing or curbing around the top to prevent surface pollution, and the casing should be of new-wrought iron or steel. The pipe sections should be joined by threaded couplings or welds; sleeve joints are not satisfactory.

The total bacterial content of preparation and canning water, regardless of the type of organisms present, is an equally important consideration, because it influences plant sanitation without bearing upon the public health aspects of canned foods. One of the purposes of washing food in preparation is to reduce its load of microorganisms, to prevent loss of quality prior to canning, and to help insure effective sterilisation. Obviously, water used for this purpose should have a low bacterial content. Water used for equipment cleaning, in which contamination must be reduced or maintained at low levels, must likewise contain few microorganisms.

6.4   Disinfection of cannery water using chlorination

The most practical means presently available for insuring purity and low total bacterial content of cannery water is by chlorination. Chlorination of the entire cannery water supply, commonly called 'in-plant' chlorination, insures potable water for preparation and canning and is an aid toward better plant sanitation. It is not a cure-all to replace sound plant cleaning practice and sanitary equipment operation but is an effective means of promoting physical and chemical cleanliness.

Chlorination of the plant water supply contributes to better plant sanitation in several ways. Microbiological contamination is reduced, both in product and on equipment, wherever water containing free chlorine is used in preparation and canning. The deposition of organic materials from products being handled, and the growth of microorganisms thereon, are largely eliminated by continuous chlorination. Odours developing from organic matter on belts, washers, flumes, other equipment, and on floors are prevented by adequate chlorination. The quality of the raw food products handled in preparation and canning equipment where water chlorination is practised cannot fail to be affected favorably.

The two methods of water chlorination most commonly used in food plants involve the use of hypochlorite solutions: sodium hypochlorite ($NaOCl$) or gaseous chlorine (Cl_2), with the latter usually preferred for in-plant treatment.

Under ambient conditions, chlorine is a pale green gas, but under pressure it exists as a yellow-green liquid.

When elemental chlorine or hypochlorites are added to water, they undergo the following reactions:

$$Cl_2 \text{ (g)} + H_2O \leftrightarrow HOCl + H+ + Cl^- \tag{6.1}$$

$$NaOCl + H_2O \leftrightarrow HOCl + NaOH \tag{6.2}$$

$$(Ca)_2OCl + H_2O \leftrightarrow HOCl + 2CaOH \tag{6.3}$$

The HOCl, being a weak acid, partially dissociates relative to solution pH:

$$HOCl \leftrightarrow H+ + OCl^- \tag{6.4}$$

At pH 7.5, the concentrations of HOCl and OCl$^-$ are approximately equal. Above pH 7.5, the OCl$^-$ will predominate and vice versa. The relative concentration of HOCl is therefore dependent on solution pH. The higher the pH, the lower the HOCl concentration (relative to the OCl$^-$ concentration). The relative concentration of HOCl to OCl$^-$ will also vary with temperature. As the temperature increases, the relative concentration of HOCl decreases. The antimicrobial efficiency and sporicidal effectiveness of chlorine solutions decrease with increase in pH. HOCl is approximately 80–100% more effective than OCl$^-$ in killing bacteria. It follows that the activity of chlorine solutions are greatly affected by pH, with the greater activity at lower pH values.

Hypochlorite solutions are added through proportioning devices and are sometimes used where chlorination of preparation and canning water is employed periodically. Equipment costs are relatively low in comparison with gaseous chlorinators, but hypochlorites are considerably more expensive as a source of chlorine than chlorine gas. Hypochlorinators require constant attention for efficient operation and are somewhat less versatile than gas chlorinators.

The production of hypochlorite solution directly from weak salt brine by electrolytic generators is also possible. Such generators produce solution concentrations from a few ppm up to 6000 ppm, with small-capacity units employed for direct chlorination of equipment, floor, and wall surfaces and large-capacity units for continuous chlorination of plant water supplies of processing water. The advantages of such equipment over the use of conventional chlorination compounds involve convenience, economy, and safety. There is no need to store unstable chlorinating compounds, because hypochlorite is produced as required; solution pH is near neutrality, contributing to a more effective kill of organisms with less equipment corrosion than gas chlorine or alkaline hypochlorite compounds; and solutions do not give off irritating free chlorine gas.

Some precautions must be observed when in-plant chlorination is practised. Excessive amounts of free chlorine may result in corrosion of canning equipment, due to its oxidising activity, particularly if used in waters that themselves may display corrosive tendencies. However, water having an abnormally acid or alkaline reaction or a high concentration of corrosive salts is ordinarily not desirable for preparation and canning and needs corrective treatment. Furthermore, the advantages of effective in-plant chlorination may be realised at levels in which corrosion, due to chlorine activity, is not a danger.

It is most important to keep chlorinated water free from any contact with phenolic materials because of the disagreeable and extremely penetrating nature of chlorophenolic flavors and odours that result.

Processors have found that the use of chlorinated water for preparation and canning materially reduces cleanup time required for good plant sanitation. Furthermore, it has been reported that corrosion of canning equipment is reduced by water chlorination.

Experience and research have adequately demonstrated that in-plant chlorination has no adverse effect upon flavour and odour of canned foods washed, conveyed, blanched, or otherwise in contact with water during preparation and canning. However, water chlorinated up to 5 and 10 ppm and used for syrup make-up has produced off-flavours in several canned fruits, squash, and sweet potatoes. When chlorine has been used in all other preparation and canning operations, but eliminated from syrup make-up water used for these products, no objectionable off-flavours have resulted (see Table 6.1).

Marginal chlorination means the addition of sufficient chlorine to water to satisfy partially the total chlorine demand of the water being treated. It is attained by adding sufficient chlorine to produce temporary chlorine residuals of somewhat less than 0.5 ppm, measured at a point in the water system usually not far beyond the point of addition. This treatment serves to destroy coliform-type organisms if their numbers in the treated water are not excessive. Because the chlorine demand of the water itself is not satisfied, the chlorine residuals produced in marginal treatment are dissipated quickly. There is then no available chlorine in the water at the point of its delivery into preparation and canning lines, and the treatment has little effect upon bacterial development and therefore upon plant sanitation and product quality. Furthermore, impure waters may develop objectionable flavours when treated by marginal chlorination.

Breakpoint chlorination is the addition to water of more than sufficient chlorine to completely satisfy the total chlorine demand of the water being treated. This type of treatment produces and maintains free available chlorine residuals and largely eliminates chloramines or other combined available chlorine compounds. The addition of chlorine beyond the amount required to satisfy the water's total chlorine demand, i.e. beyond the breakpoint, merely increases the free chlorine residual in direct proportion to the amount added. Breakpoint chlorination usually eliminates undesirable odours and flavours commonly associated with marginal chlorination and supplies persisting free chlorine residuals. For these reasons, it is the preferable treatment for preparation and canning water. During canning operations, in-plant chlorinating systems are ordinarily controlled to supply water having approximately 2–5 ppm available free chlorine at the point of delivery to the canning line. During cleanup, the chlorine level is frequently boosted to 10 or 20 ppm or higher.

6.4.1 Chlorination of cooling water

The cause-and-effect relation between the microbial population levels of water used for cooling cans after processing, and the rates of spoilage that may occur in those cans, has been described many times. The canning industry has learned through experience that increased contamination of container cooling water invariably brings about a proportional increase of product spoilage.

During the period when cans are being cooled after processing, they are somewhat susceptible to the ingress of spoilage bacteria, even when can structure is normal in every respect. At the end of the process can seams may be slightly distorted because of the heat and strain of the process, and the sealing material is in a softened condition. In an effective cooling system, internal pressure developed during processing is dissipated and a partial vacuum is formed before cans leave the cooling water. At this point,

Table 6.1 **The effect of chlorine treatment on flavour of canned foods (Somers, 1951)**

Product	Lowest concentration which produced off-flavour when 2, 5, 10, and 50 ppm were added	
	Partial treatment Chlorination of all water except brines and syrups	Complete treatment Chlorination of all water including brines and syrups
	Chlorine (ppm)	Chlorine (ppm)
Applesauce, Rome Beauty[a]	10	5
Applesauce, Gravenstein[a]	(None at 50)	10
Apricots, halves and peeled	(None at 50)	50
Apricots, whole peeled	(None at 50)	50
Asparagus, all green	50	50
Beans, green cut	50	10
Beans, green limas	50	10
Beans, with pork (recanned)[a]	–	50
Beets, red sliced	50	10
Carrots, sliced	(None at 50)	10
Carrots, pureed[a]	(None at 50)	50
Cherries, Royal Anne	(None at 50)	50
Corn	–	(None with 15)
Figs, whole Kadota	50	5
Grapefruit juice (recanned)[a]	–	50
Orange juice (recanned)[a]	–	50
Peaches, Clingstone halves	(None at 50)	5
Peaches, Elberta halves	(None at 50)	10
Pears	50	2–5
Peas	–	(None with 10)
Pineapple juice (recanned)[a]	–	10
Potatoes, sweet, solid pack[a]	(None at 50)	50
Pumpkin, solid pack[a]	(None at 50)	50
Prunes, Italian	(None at 50)	10
Spinach	50	10
Strawberries, whole	(None at 50)	10
Tomato juice[a]	–	10
Vegetable juice, cocktail (recanned)[a]	–	5
Yams, syrup packed	–	5

[a]Chlorine added directly to the product.

seams of some cans may admit minute amounts of the medium in which the cans rest. The material that may be admitted is immeasurably small and has no effect upon can vacuum, and yet it may contain several viable bacteria that can initiate spoilage. The probability that one living microorganism may gain entry into a sound double-seamed can in this manner becomes greater as the contamination level of the cooling medium increases.

Laboratory experiments and plant experience have demonstrated that the degree of spoilage of processed cans increases as the numbers of bacteria in the cooling water increase. Ideally, cooling water should meet Public Health standards for total bacterial count, and in many canneries where effective chlorination is practised, cooling water does approach such standards. However, in plants where chlorination of cooling water is not practised, bacteria counts greatly in excess of these limits are often encountered. Evidence obtained in controlled plant and laboratory work indicates that cans cooled in water containing over 100,000 bacteria per milliliter exhibit rates of spoilage five to 10 times greater than when the same cans are cooled in water having no more than 100 organisms per milliliter. The importance of maintaining low levels of microbial population in cooling water is obvious.

Factors of design and operation of cooling equipment often contribute toward increases of microbial levels during operation. When cooling is carried out in still retorts and the cooling water is not reused, there is no danger of microbial buildup and total counts are low unless the original water source bears contamination. Cooling canals are susceptible to microbial buildup during operating periods unless chlorination is employed or adequate fresh water is supplied continuously, and the canal is constructed and piped to insure a water circulation through all parts to avoid dead ends where water may stagnate. Continuous rotary cooler waters have been found to be very susceptible to relatively rapid increases in microbial content during operation. Spray cooling systems or any other type of cooling equipment in which cooling water is recirculated are likewise susceptible to the bacterial buildup during operation. It is important in any type of cooler to prevent an accumulation of organic material washed from the outsides of cans into the cooling water. Such material provides nutrients for the growth of microorganisms and thereby promotes rapid bacterial buildup. It also absorbs free chlorine and interferes with the maintenance of proper chlorine levels where chlorination is practised.

Spoilage of processed canned products by entry into sound cans of microorganisms borne by cooling water is aggravated by rough handling of wet cans immediately after cooling. Denting of double seams in automatically filled can-handling equipment increases the rate of spoilage anticipated in any lot of cans. Rough handling in combination with excessive content of microorganisms in cooling water usually results in serious spoilage problems.

The microbial level of cooling waters may be controlled by chlorination as a common cannery practice. The maintenance in cooling water of free chlorine levels between 2 and 4 ppm is ordinarily adequate to keep the total bacteria count within safe limits. Chlorine is usually applied to cooling water in the form of solutions of hypochlorites, either by simple drip addition or through proportioning pumps.

6.5   Other methods of cannery water disinfection

6.5.1   Bromine

Bromine exhibits biocidal properties in water and is frequently used as a water disinfectant, especially in can-cooling waters. The bromine hydrolyses in water, analogous to chlorine, to give hypobromous acid (HOBr).

Bromine is usually added to water in the form of NaBr. This dissociates in solution to form Na^+ and Br^- ions. Whenever chlorine or hypochlorite is added to water containing Br^-, there is a rapid formation of HOBr according to the reaction:

$$Br^- + HOCl \leftrightarrow HOBr + Cl^- \tag{6.5}$$

This resulting HOBr is also an electrophilic agent but one that tends to react more rapidly than HOCl. It is therefore even more bacteriocidal than HOCl. Under alkaline conditions, hypobromous acid dissociates to form the hypobromite ion, just like hypochlorous acid forms the hypochlorite ion. However, because the dissociation to OBr^- and H^+ occurs at higher pH values, bromine is therefore still effective (as a biocide) at much higher pH values than chlorine.

6.5.2   Chlorine dioxide

Chlorine and chlorine dioxide are similar in many respects, including the fact that both are powerful oxidising agents. ClO_2 has 2.5 times the oxidising capacity of Cl_2. ClO_2 is an effective biocide in water with low organic content. However, it is less effective than chlorine at equal concentrations when the organic content of the water is high. Chlorine dioxide does not hydrolyse in aqueous solutions and is therefore not subject to dissociation in the manner of hypochlorous or hypobromous acid. The ClO_2 is the active biocidal compound.

6.5.3   Ozone

Ozone is a more powerful disinfectant than chlorine for the deactivation of a large number of microorganisms. It is considered the most potent oxidising agent available for water and wastewater treatment. However, it does not remain in the water for a very long period and can thus be considered a process rather than an additive, with no safety concerns about residual ozone. The advantage of ozone is that it leaves no residues. Excess ozone will rapidly disappear through its spontaneous decomposition in solution, and low concentrations (0.1–4 ppm) are required to effect practical disinfection, although the amounts vary according to the amount of organic matter present. The disadvantage is that, in terms of water treatment, ozone is very expensive to use on a large scale in place of chlorine or bromine. It should be noted that ozone reacts with rubber and various synthetic rubber compounds. Teflon seals should be used in water systems in which ozone is used as disinfectant. Ozone can also cause corrosion of some metals.

6.6   Canning water requirements

The importance of adequate water volume for canning purposes has been emphasised. The water demand of a cannery depends largely upon its production volume and the water requirements of the individual products being packed. However, water requirements are influenced greatly by methods of handling and types of equipment employed and so are found to vary widely for the same products in different plants. Alterations in plant layout, variations in equipment and methods, and changes in cannery technique usually alter the water volume required for the products canned. These facts account for the wide variation in rates of water usage reported for standard canned fruit and vegetable products in various plants.

6.7   Water analysis

A typical water analysis form is shown on the following page. Various water tests and their significance are outlined below.

6.7.1   Alkalinity

Alkalinity is the quantitative capacity of water to neutralise an acid. Alkalinity is a combination of the naturally occurring alkalis: CO_3^{2-}, HCO^{3-}, and OH^- salts of Ca, Mg, K, and Na. Most natural waters have an alkalinity in the range of 10–500 ppm. The combination of carbonate alkalinity and bicarbonate alkalinity indicates total alkalinity. High total alkalinity will cause rinsing problems. Samples testing over 200ppm alkalinity often require acid rinsing to prevent water spotting and a general dull appearance on equipment surfaces.

6.7.2   Hardness

Hard water is water that has high mineral content. Hard water is formed when water seeps through deposits of calcium and magnesium containing minerals such as limestone, chalk, and dolomite. See Table 6.2 for classification of hardness.

Table 6.2 Hardness classification

Class	Hardness ppm (as CaCO₃)	Grains/gallon (17 ppm = 1 grain/gallon)
Soft	0–60	0–3.50
Moderately hard	61–120	3.56–7.01
Hard	121–180	7.06–10.51
Vey hard	>180	>10.57

From US Geological Survey.

1. Temporary hardness is attributed to calcium and magnesium bicarbonates that decompose upon heating into insoluble carbonates or scale; a good example is teakettle scale. Hardness over 100 ppm will quickly form films in heating equipment.

   Boiling promotes the formation of carbonate from the bicarbonate and precipitates calcium carbonate out of solution, leaving water that is softer upon cooling.

2. Permanent hardness represents soluble calcium and magnesium salts that are not decomposed by heat but are precipitated with certain dilute alkalis; an example is the scale in the alkali-removal section of a bottle washer.

   Tests show that over 150-ppm hardness usually causes rinsing problems and excessive chain-lubrication costs.

6.7.3 Chlorides

High chloride content causes rapid corrosion, increases soap consumption, and reduces suds and foam. Tests showing over 100 ppm chlorides may indicate the need for a de-mineraliser or corrective maintenance of the water softener.

6.7.4 Iron

As little as 0.3-ppm iron can be expected to cause staining on equipment surfaces, because iron precipitates with the addition of some alkalis or oxidisers such as chlorine. Additional complex phosphates may be required to prevent this problem.

6.7.5 Sulphate

A positive sulphate test indicates sediment or turbidity problems that may require filtration.

6.7.6 Sulphide

The presence of sulphides is directly associated with a sulphur-like odour, which is objectionable in food plants.

6.7.7 Appearance

Appearance refers to sediment, colour, and clarity of the sample.

6.7.8 pH

The pH of water can generally be related to its mineral composition. Where carbonates and bicarbonates predominate, pH values are usually above 7.0, whereas sulphides and sulphates cause pH values below 7.0. The pH measurement is made with a pH metre. Low pH (below 7.0) increases the corrosion rate. High pH often indicates the need for an acidified rinse.

Many additional water tests may be considered, but food plants will find microbiological analysis necessary to maintain superior quality.

A complete water analysis is important because it can serve as an excellent guide for the plant operator to

1. Select the proper cleaning materials and procedures.
2. Get better heat transfer from heating and cooling equipment.
3. Reduce lubrication costs on conveyors.
4. Prevent unsightly films on processing equipment.
5. Control bacteria counts where water is in direct contact with foods.

6.8   Chemicals in water

Water used in canning vegetables and fruits must be soft and free from any appreciable amount of organic material, obnoxious odour or taste. The presence of sulphur or a salty taste, for instance, will affect the finished product and may cause discolouration or hardening as well as objectionable odours. If the water is too alkaline, the product may disintegrate or become mushy; if the water is too hard, it will harden vegetables, often making them difficult to process.

Hard water is objectionable for the boiler supply, the brine, syrups, and all canning purposes. Water hardness due to the presence of calcium (lime) and magnesium compounds has the effect of hardening certain vegetables, especially peas and beans.

The zeolite method is one that has been used satisfactorily for softening water for canned foods. In this method, water is passed through a tank or cylinder filled with a granular substance belonging to the class of compounds known as zeolites. This substance has the property of taking up calcium and magnesium from water passing through it and yielding up to the water an equivalent amount of sodium. When the zeolite is no longer sufficiently active, it can be regenerated by passing a strong salt-brine through it. It removes from the zeolite the calcium and magnesium taken up from the water and leaves in their place an equivalent amount of sodium. In this way, the tank or cylinder apparatus can be used for softening water during the daytime and can be regenerated with salt at night, ready for the next day's operations.

All commercial salt contains calcium and magnesium compounds in some quantities, depending on the source of the salt and the degree of its purification. Therefore, it may happen that the salt used will contribute more to the hardness of brine than the water supply. In this case, it is advantageous to use as pure a grade of salt as possible for brining products. Canners should use a salt for this purpose that contains less than 0.3% calcium (equivalent to 0.75% calcium carbonate). It is also advisable to use as weak a brine as is possible without materially sacrificing flavor to minimise the hardening effect of calcium and magnesium. Table 6.3, 'What a Water Test Report Means', is helpful for interpreting the results of chemical analysis of water.

Table 6.3 **What a water test report means**

Item	Recommended	Problems from excessive amounts
Acidity	pH 7.2–7.5	<pH 7 causes corrosion of copper and iron pipes (see copper and lead).
Sediment, odour, colour	Nothing permitted	Contributes to bad taste, odours, and appearance.
Sodium	No specified limit	Has possible health effects in persons with heart, kidney, or circulatory ailments.
Potassium	<20 ppm	Contributes to growth of small organisms in water (plankton).
Iron	For drinking: <0.3 ppm	Causes poor flavoured water, brown stains on plumbing and laundry (see acidity).
	For dairy use: <0.1 ppm	Contributes to oxidised flavours of milk.
Manganese	<0.05 ppm	Is possible health hazard. Causes poor taste in coffee and tea, 'brown laundry', brown stains on fixtures.
Copper	For drinking: <1 ppm	Is health hazard. Causes poorly flavoured water, blue or green stains on fixtures and laundry (see acidity).
Lead	<0.05 ppm	Causes serious illness or death (see acidity).
Fluoride	0.7–1.2 ppm 1.7 ppm – recommended maximum 2.2 ppm – reject	Large amount may cause mottling of teeth.
Turbidity	0–5 units	Increases in turbidity after rainfall may indicate surface or other introduced pollution.
Hardness (as CaCO$_3$)	<100 ppm	Increases detergents and soap need, causes scale on pipes and heaters. Soap curd shortens life of cloth fibres.
Calcium (as CaCO$_3$ magnesium)		The two elements calcium and magnesium compose the majority of hardness.
Sulphate	<250 ppm	May cause laxative effects.
Phosphate		May stimulate the growth of algae.
Ammonia nitrogen	<0.01 ppm	If two or more of these four items exceed the specified limits, sewage contamina-
Albuminoid nitrogen	<0.05 ppm	tion is suspected.
Nitrite nitrogen	0	
Nitrate nitrogen	<0.5 ppm	
Chloride	<10 ppm	If chloride content is normally low, an increase may indicate sewage contamination.
Residual chlorine	Palatable water will not contain enough to cause harm.	
Total dissolved solids	<500 ppm	Excessive amounts may be associated with poorly flavoured water and health hazards.
	For dairy use: <0.1 ppm	Contributes to oxidised flavours of milk.

6.9   Water quality in vegetable canning

In the initial washing of peas, or in the rinsing of peas by means of a spray after blanching, the hardness of the water has no appreciable effect on the hardness of the canned product. However, in the blanching of peas, hard water does have a pronounced hardening effect on the canned product. When blanching is done in a continuous blancher, it is found that the peas first passing through the blancher remove the greater part of the calcium and magnesium compounds from the blanch water and thus soften it. Nevertheless, if very hard water is employed and the flow of the water through the blancher is considerable, the hardening effect on the peas will be greater and proportional to the flow of water. It may be stated then that the use of softened water for blanching is advantageous when the hardness of the water supply exceeds 200 ppm (expressed as total calcium).

Because the hardening effect of calcium and magnesium compounds on dry beans is even greater than that on peas in packing dry beans, it may be advantageous to soften water that is only slightly hard. Softening of hard water used for soaking dry beans and for the preparation of brine and sauce is therefore important. If the soaking water is changed during the soaking period, such treatment is all the more important.

The hardening effect on dry beans by calcium and magnesium compounds of hard water or salt may be partially corrected by longer processing. Nevertheless, the appearance of a product so treated is sometimes less satisfactory than that of beans treated with soft water and relatively pure salt and given merely the processing necessary for sterilisation. The use of soft water and pure salt gives a superior product, resulting in time saving and therefore the amount of steam consumed in processing.

Calcium and magnesium compounds in hard water and salt do not appear to exert a hardening influence on green or wax beans. The softening of water for use on this product does not, therefore, appear to be of any great value.

Calcium and magnesium compounds in hard water and salt do not exert a hardening influence on either cream style or whole grain canned corn. The use of softened water in the brine of cream style canned corn appears to produce a somewhat darker product than does un-softened water. Its use therefore is regarded as disadvantageous.

Calcium and magnesium compounds in hard water and salt will combine with soluble oxalates naturally occurring in beets and may produce a white coating on their surface. This is especially apparent on cut beets and sometimes detracts from their commercial value. The softening of hard water and the use of relatively pure salt are therefore advantageous in avoiding this trouble.

6.10   Water use and conservation

Water is an import ingredient and tool in food processing; however, it must be used sparingly where possible. The primary consideration in the use of water is to accomplish the task for which the water is being used. In addition to product conveyance, such as in flumes, two of the more important uses are in washing raw product and cooling the finished product. These primary purposes must be accomplished without any sacrifices relative to the amount of water being used but considering conservation and reuse of water, as indicated below.

1. Research has shown that microbial counts for foods in the final container, as it relates to spoilage probability, is influenced by the sanitary condition of the plant and the quality of the water in which the product was handled and conveyed.
2. Water is a useful tool for removal of excess soil and debris that are characteristic of mechanical harvesting.
3. Water conveyance systems result in greatly increased water usage.

Whether enough water is being used, or whether more water will be required to accomplish certain tasks in the future, there is always room for water conservation in most plants through: (1) new processing procedures; (2) more efficient equipment; (3) avoiding unnecessary use or waste of water; and (4) reuse or recirculation of water.

All water released in a plant, regardless of how small the volume, must be disposed of in some manner or another, resulting in an additional load on the waste disposal system. There is a responsibility to conserve the quantity and preserve the quality of the waters.

6.10.1 Ways to reduce the total load of water on waste disposal systems

A survey conducted by National Food Processors Association (NFPA) revealed several interesting facts:

1. As operating rate increases, the use of water per unit product decreases. Any factors that increase the flow of product through a plant lead to more efficient use of water. Factors that improve case yield/per ton of product decrease the waste load per unit of raw product processed.
2. New and improved equipment has a tremendous impact on water utilisation. Continuous cookers and hydrostatic sterilisers result in significant decreases in water used per unit of product compared to still retorting. Improved and more efficient fillers that have less spillage conserve water and reduce the biological oxygen demand load of the waste.

In general, it appears that some of the more advanced and automated equipment, coupled with better methods of handling water and waste, such as better spray-washing systems, improved cooling systems, automatic monitoring and control of water flows, separation of concentrated from dilute waste, improved clean-up operations, and recirculation of water can contribute significantly to the efficient utilisation of water.

6.10.2 Water storage

If storing potable water is practised, the storage vessel should be constructed to have a smooth inner finish to facilitate cleaning. The storage system should be designed with no dead ends so that water may circulate or flow freely. Water should not be allowed to stagnate (Imholte, 1984). The tank should be so constructed as to be covered to prevent the growth of algae and access to rodents, birds, dust, and rain. If the tank has an overflow, the pipe should not be connected directly to a sewer line. The tank should be vented; the vent should be protected from insects, dust, and rain. Storage tanks should be inspected and cleaned regularly (Katsuyama, 1993).

6.10.3   Unnecessary use or waste of water

Some suggestions to prevent unnecessary use or waste of water

1. All water hoses should have automatic shut-off valves to prevent waste of water when hoses are not in use. A running hose can discharge up to one million litres of water per hour.
2. The use of low-volume, high-pressure nozzles rather than low-pressure sprays for cleanup; a high-pressure system uses less water and does a more efficient job of cleaning.
3. Avoid unnecessary water overflow from equipment, especially when not in use.
4. Avoid using water to flume a product or solid waste when the material can be moved just as effectively in a dry state by conveyors.
5. Avoid using water in excess of the amount needed to accomplish the job, such as reducing cooling-water flow to the minimum needed to accomplish the necessary temperature drop.
6. Certain water used in the plant, especially can-cooling water that is not reused and that meets the purity requirements of applicable state and federal regulations, may be discharged directly into streams without prior treatment through a waste disposal system. In some cases, this may amount to over 50% of the total water used.

6.10.4   Reuse of water

Indiscriminate reuse of water in a processing plant can result in costly spoilage losses. Reuse of water under recommended practices can greatly reduce total water consumption. In general, the reuse of water in certain operations is permissible if certain water quality factors are met and certain guidelines followed (see Table 6.4, 'Water Conservation Check Table').

6.10.4.1   Where only freshwater should be used

All prepared or partially prepared products, such as blanched peas, beans, asparagus, dried fruit, peeled tomatoes, etc. should come in contact only with freshwater. Only freshwater should be used in blanchers or in final flumes and product washers.

6.10.4.2   Where reuse of water is permissible

By using the counterflow principle of water usage, i.e. the cleanest or final water may be reused with a given operation, such as in a sequence of washers or flumes, if it is counterflowed in the direction of movement of the product. For example, where a raw, unprepared product is given two successive cold-water immersion washings, water from the second washing may be used as make-up water for the first washing if it is not heavily contaminated with organic matter and insects. The same principle would apply to flumes used for conveying and washing raw unprepared produce.

The water from can coolers, in general, is uncontaminated and may be reused as make-up water in can cooling as well as at numerous other locations within the plant.

6.10.4.3   Precautions in the reuse of water

1. Recirculation of washer and flume water (using the counterflow system) must be carefully controlled to prevent excessive accumulation of soil and organic debris.
2. Reclaimed water for washers and flumes should be effectively chlorinated if the temperature is not held at 27 °C (80 °F) or below.

Table 6.4 Water conservation check table

		Operation or equipment	May recovered water be used?	May water from this equipment be reused elsewhere?	Source of water for reuse
1		Acid dip for fruit	Yes	No	Can coolers
2		Washing of product			
	A	First wash followed by second wash	Yes	No[a]	Can coolers
	B	Second or final wash of product	No	No[a]	
3		Pump elevators (peas, etc.)			
	A	Raw product suspended in water	Yes	No[a]	Can coolers
	B	Pumping of partially prepared product suspended in water	No	No[a]	
4		Flumes			
	A	Fluming of unwashed or unprepared product (peas, pumpkin, etc.)	Yes	No[a]	Can coolers
	B	Fluming prepared or partially prepared product	No	No	
	C	Fluming of wastes	Yes	No	Any waste water
5		Lye peeling	Yes	No	Can coolers
6		Product holding vats, product covered with water in brine	No	No	
7		Blanchers – all types		No	
	A	Original filling water	No	No	
	B	Replacement or make-up water	No	No	
8		Salt brine quality graders followed by fresh water wash	No	Reused in this equipment	
9		Washing pans, trays, etc.		No	
	A	Tank washers – original water	No	No	
	B	Spray or make-up water	No	No	
10		Lubrication of product in machines such as pear peelers, fruit size graders, etc.	No	No	Can coolers

Continued

Table 6.4 Water conservation check table – cont'd

	Operation or equipment	May recovered water be used?	May water from this equipment be reused elsewhere?	Source of water for reuse
11	Vacuum concentrators	Yes	Reused in this equipment after cooling and chlorination	
12	Washing empty cans	No	No	
13	Washing cans after closing	Yes	No	Can coolers
14	Brine and syrup	No		
15	Processing jars under water	Yes	Reused for processing	Can coolers and processing or cooling water from other retorts used for processing glass.
16	Can coolers			
A	Cooling canals: original water	No	Water from these coolers may be used for cooling cans, if careful attention is paid to keeping down microbiological load. This water may be reused as indicated.	
	Make-up water	Yes		
B	Continuous coolers where cans are partially immersed in water: original water	No		
	Make-up water	Yes		
C	Spray coolers with cans not immersed in water	Yes		
D	Batch cooling in retorts	Yes		
17	Clean up purposes			
A	Preliminary wash	Yes	No	Can coolers
B	Final wash	No	No	
18	Box washers	Yes	No	Can coolers

[a] A certain amount of this water may be reused for make-up water in this equipment if the counterflow principle is used with the recommended precautions.

3. All water for reuse should first be screened. If used for cooling cans, it must be chlorinated; chlorination is recommended for all reused water. (A chlorine residue of 0.5–1 ppm increased to 4–5 ppm for a short time about every two weeks.)
4. Water must not be reused for any purpose other than cooling cans if it has been treated with dichromate or other corrosion inhibitors.
5. If cooling water is reused in equipment other than can coolers, a separate piping and pumping system is required with no cross connections to the potable water supply.

6.11   Protecting the water supply

An important sanitary precaution often overlooked in food plants is the possible contamination of a potable water supply by polluted water where it is either injected directly into the potable water line, mixed with steam, or anywhere there is a submerged potable water inlet.

Water entry into cleaning-in-place tanks, case washers, parts washers, etc. requires preventative measures to stop contaminating backflow into the main water supply. Injectors to feed sanitisers or proportioning pumps to feed lubricants also require similar preventive measures.

Most city codes require that equipment and fixtures connected to a potable water supply, in which the water supply enters below the flood rim, be protected against back siphonage of polluted water by an air gap, atmospheric vacuum breaker, or other approved method.

An air gap is the physical separation of the potable and nonpotable systems by an air space. The distance between the supply pipe and the flood lever rim should be twice the diameter of the supply pipe, but never less than 2.5 cm (1 in).

Atmospheric vacuum breakers may be used only when connected to a nonpotable system in which the vacuum breaker is never subjected to backpressure and is installed on the discharge side of the last control valve. These cannot be used under constant pressure and must be installed with the bottom of the breaker body at least 15 cm (6 in) above the flood rim of the fixture or appliance. Where a portable appliance is used, the breaker should be installed at least 15 cm (6 in) above the highest point the outlet can be raised. The most simple and effective type of vacuum breaker consists of an open valve or petcock between the cutoff valve and water outlet. When the water is on, there will be a stream of water flowing through this opening.

Pressure-type vacuum breakers that are spring loaded may be used where the vacuum breaker is not subjected to backpressure. They may also be used under continuous supply pressure. They must be installed at least 15 cm (6 in) above the usage point. Backflow preventers with atmospheric vent may be used as an alternate for pressure-type vacuum breakers and to provide protection against backpressure.

A potable water supply should never be solidly piped to a drainage ditch, sewer, or sump and should be terminated 30 cm (1 ft) above the ground or through an air gap to a drain.

The plant's water supply should be protected. Improper plumbing usually contaminates the plant's water first and then that of neighbors. Safety should be a prime consideration. The local plumbing code should be followed.

Table 6.5 **Colour codes as suggested by the USDA for meat and poultry plants**

Piping/line carrying liquid	Suggested color
Fire lines	Red
Sewer lines	Black
Edible brine lines	Green plus name
Inedible brine lines	Black
Air lines	White
Potable water lines	Green
Nonpotable water lines	Black
Edible product lines	Green plus name
INEDIBLE product lines	Black plus name
Ammonia lines	Blue
Curing pickle lines	Green plus name

6.11.1 Separate water supplies

Many food plants use both potable and nonpotable water. Nonpotable supplies include untreated surface water, reused plant water, seawater, and other water of questionable quality. Potable and nonpotable waters should be distributed in distinctly separate pipelines. Often processing plants identify pipelines by colour codes (see Table 6.5).

References

Imholte, T. J. (1984). *Engineering for food safety and sanitation.* Crystal, MN: Technical Institute of Food Safety.

Katsuyama, A. M. (Ed.). Principles of food processing sanitation (2nd ed.). Washington, DC: The Food Processor Institute.

Somers, I. I. (1951). In *Plant chlorination. Information letter to Industry national Canners association/national food processors association.*

Troller, J. A. (1983). In *Sanitation in food processing* (p. 347). New York, NY: Academic Press.

World Health Organization. (2006). *Guidelines for drinking-water quality. incorporating first addendum. Vol. 1, Recommendations* (3rd ed.). WHO Library.

Energy supply and requirements

7

7.1 Introduction

Several sources of energy are available to the food industry. The main ones used are gas (natural or liquid petroleum gas), liquid fuel oil, electricity, and coal.

The type of equipment to burn a particular fuel varies according to physical state (gas, liquid, or solid). Fuel choice is based on one or more of the following:

1. Fuel factors
   * Availability and dependability of supply
   * Ease of use and storage
   * Cost
2. Combustion equipment factors
   * Operating requirements
   * Capacity
   * Service requirements
   * Ease of control
   * Cost

It is also important to check national and local regulations related to fuel use, environmental problems associated with specific fuels, and safety requirements.

7.2 Steam supply

Since the beginning of commercial canning, steam has been the principal medium of heat transfer for the preparation and sterilisation of canned-food products. Other means of heat treatment, like radio frequency, induction and infrared heating for preparation, and cold sterilisation by ionising radiation, have been developed and have specific uses, but generally those processes cannot economically compete with steam for the purposes mentioned. There is, therefore, no doubt that steam will continue to be the most acceptable means of heat transfer for many years to come. This competition, however, increases the necessity for emphasis on the efficient production and use of steam in the canning industry. It is important that an adequate supply of steam suitable to the needs of the various operations be available at the time at which it is required. Some characteristics that make steam the energy source of choice for canning are as follows (Sing, 1986):

* Steam is efficient and economic to generate. It is made from water that is readily available.
* Steam generation is efficient and easy to control.
* Steam can easily be distributed to the point of use.
* Steam has high-heat content and heat transfer from steam is very efficient.
* Steam is clean and pure.

A Complete Course in Canning and Related Processes. http://dx.doi.org/10.1016/B978-0-85709-677-7.00007-4

7.3   Forms of steam

There are three forms of steam, depending on the amount of heat and moisture contained. As the temperature increases and water approaches boiling point, some molecules have enough energy to momentarily escape from the liquid into the space above the surface, before falling back into the liquid. When the number of molecules leaving the liquid surface is more than those re-entering, the water freely evaporates and the water has reached its boiling point or its saturation temperature, as it is saturated with heat energy (Sarco, 2007).

If the pressure remains constant, adding more heat will not cause the temperature to rise any further but causes the water to form saturated steam. The temperature of the boiling water and saturated steam within the same system is the same, but the heat energy per unit mass is much greater in the steam. At atmospheric pressure, the saturation temperature is 100 °C. However, if the pressure is increased, this will allow the addition of more heat and an increase in temperature without a change of phase.

Dry saturated steam is steam formed when water evaporates. It has a temperature equal to the boiling point at that pressure and contains no droplets of water or any superheat; i.e. if it was cooled, some free moisture would condense out, and if it were heated at all, the temperature would rise, making superheated steam at the same pressure.

One kilogram (by weight) of saturated steam at atmospheric pressure occupies 1.67 m^3, and its latent heat (enthalpy of evaporation) is 2257 kJ/kg.

Wet steam is a mixture of dry saturated steam and water. The water is at the same temperature as the steam with which it is mixed. The water might be in any form, but is usually found as drops or films of water. Saturated steam, if cooled slightly, will become wet; i.e. if steam is carried for some distance in a pipe, it loses heat, and some of the vapour condenses out to form moisture droplets. This is the kind of steam usually found in food plants.

The quality of wet steam is the weight of the dry saturated steam in this wet steam expressed as a percentage by weight of the wet steam. If the wet steam has a quality of 95%, then 95% of every pound of the mixture is composed of dry saturated vapour and 5% by weight is in the form of water droplets. In the above sample, in every kilogram of mixture there would be 0.95 kg of dry saturated vapour and 0.05 kg of water. The total heat of wet steam at a given pressure is less than that of dry saturated steam.

If 1 kg of dry saturated steam is condensed at atmospheric pressure, it will give up 2257 kJ, but, if the steam is only 95% quality, it will give up only $0.95 \times 2257 = 2144$ kJ, because 5% of the so-called steam is already in the form of water.

Superheated steam is saturated steam that has been heated at constant pressure so that its temperature is higher than that of saturated steam at the same pressure. The temperature depends upon the amount of heat that has been added. If superheated steam gives up some heat, its temperature will fall to the temperature of saturated steam at the same pressure before it begins to condense. If additional heat is removed, the steam will condense at the same pressure. Superheated steam is obtained by means of special superheating coils in a boiler or by reducing the pressure of high-pressure steam through a throttling valve.

7.4   Properties of steam supply

Steam at pressures consistent with the needs of various unit operations is necessary if the operations are to be carried out efficiently. In general, line pressures of 700–900 kPa (100–125 psi) are adequate for most operations, if the boiler is properly sized, and if steam lines to the various units are adequate to carry the amount of steam required. Attempts to operate at pressures substantially under these amounts usually result in questionable practices or inefficient operation. The retorting operation is particularly vulnerable, as it is almost impossible to accomplish adequate venting at pressures under 500 kPa (70 psi), unless exceptionally large and expensive lines are installed.

'Pure' saturated steam is steam free of air or other noncondensable gases, condensable volatile materials other than steam, excess condensate, and entrained solutes, such as boiler compounds and salts. Pure steam is required to preclude the possibility of faulty retort operation, contamination of the products involved, and corrosion or other damage to the container. Although it is extremely unlikely that any steam supply would carry enough air or other gases to affect the efficiency of a thermal process, some steam supplies may become sources of product contamination when they carry certain volatiles, excess condensate, or soluble materials carried in entrained boiler water.

Soluble corrosive materials carried from boilers in entrained condensate are all too often responsible for container corrosion, discoloration, and damage to outside enamels or lithography during retorting. The presence of these materials is usually due to faulty practices in boiler operation, such as excessive use of boiler compounds, high concentrations of salts in the boiler water, or high water levels. All of these practices are conducive to 'priming', the main cause of liquid carryover. Periodic draining and flushing of all steam lines with freshwater will usually prevent an accumulation of corrosive materials that may have been carried over into the steam lines.

7.5   Steam production

In the US, Australia, and New Zealand, boiler steam generator capacities are often listed in terms of boiler horsepower (BoHP). This value may be defined in terms of the ability of the unit to change approximately 15 kg (34 lbs) of water at 100 °C (212 °F) to steam at 100 °C.

Some manufacturers give a boiler rating in kilo Watts (kW). This is not an evaporation rate and is subject to the amount of steam in kg/h that the boiler can create from and at 100 °C, at atmospheric pressure. Each kilogram of steam would then have received 2257 kJ of heat from the boiler.

With improved firing methods, it was found that a boiler could develop considerably more than 'rated capacity'.

The most common term used today is kilograms of steam per hour that the boiler will produce under a specified set of conditions.

Steam generation equipment may be classified into two general types: firetube and watertube boilers. These, in turn, may be obtained in several different styles.

The firetube type, either in the horizontal return tubular or Scotch marine style, is the most commonly used, particularly in seasonal operations and primarily because of its lower initial cost. The firetube type is generally operated at rather low pressures, usually not over 1000 kPa (150 psi).

The watertube type is usually used in larger installations involving year-round operations in which large amounts of steam are required and in which very high boiler pressures can be used.

Either type may be operated at something over its nominal rated capacity, the degree of overload depending upon design, setting, fuel, and other factors. One advantage of firetube boilers is the large water storage capacity. Because of this feature, wide and sudden fluctuations in steam demand are met with little change in pressure. Watertube units, on the other hand, are capable of considerably higher overloads, both continuous and intermittent. The knowledge that boilers can be operated in excess of rated capacity quite often results in an installation that is undersized, as too much faith is placed in the possible overload and not enough consideration is given to loss of efficiency over extended use and expansion of operations.

Undersizing may occur where steam 'generators' or 'package' units are used because, in general, they are compactly constructed for limited space requirements and ease of installation but usually with no provision for overloads in excess of rated steam production capacity. However, some self-contained boiler units are designed to provide for considerable overload capacity.

Fuel is an important consideration in selecting a boiler. The choice of fuel usually resolves itself into a matter of economics and no general rule appears to be applicable for all localities. In some circumstances, a change in fuel and/or firing method can increase steam output. Gas is preferable to either oil or coal, if cost permits, because there is no problem of storage or waste disposal. Oil and gas firing systems require less attention and maintenance and are cleaner than most coal-fired installations.

Gas for boiler fuel is easy to burn, with very little excess air. Fuel gases are natural gases (naturally produced underground), which contain a high percentage of methane, and liquefied petroleum gases (LPG), which are mainly propane and butane.

Coal is the generic term given to solid fuels with high carbon content. There are several types of coal, each relating to the stages of coal formation and the amount of carbon content. The bituminous and anthracite types tend to be used as boiler fuel.

Oil for boiler fuel is produced from crude petroleum after it has been distilled to produce lighter oils like gasoline, paraffin, kerosene, diesel, or gas oil. Various grades are available, each being suitable for different boiler ratings; the types are as follows:

- Diesel or gas oil.
- Light fuel oil.
- Medium fuel oil.
- Heavy fuel oil.

A question arises whenever new steam plants are being considered of whether single or multiple installations should be installed. This will require careful analysis of the intended operations, particularly with respect to possible demand variations. Usually, if over 2722 kg (6000 lbs) of steam per hour are required to handle peak demands, with very low off-season demands, multiple installations should be considered. If a boiler should fail, in the case of dual installations, it is usually possible to operate with one boiler until the failure is rectified. This usually requires very careful juggling of certain operations, but it can minimise losses, which may result from a complete shutdown.

Most steam boilers, other than those classified as 'steam generators,' are capable of producing more steam than their rated capacity, provided they are in proper condition and properly operated. Horizontal Return Tubular Boilers (HRT) and Marine types may be operated consistently at a 50% overload and even up to 100% for very short periods. Watertube boilers are capable of an even greater overload. The amount of overload possible depends primarily on the type of firing and the rate of recovery after heavy demands. Automatic firing systems – whether the fuel be coal, oil, or gas – are desirable for uniform steam supply. An extra reserve of steam for periodic peak demands, such as occur in bringing up a retort, can sometimes be obtained by placing a pressure-reducing valve in the main steam line and operating the boilers at a pressure higher than is supplied to the operations. For example, boilers may be operated at 1000 kPa (150 psi) and the line pressure reduced to 700 kPa (100 psi). Careful maintenance of steam supply systems is essential as failures often result in lost production, as well as loss of perishable products, during extended shutdowns. It is usually desirable to install two small boilers rather than one large unit, because it is often possible, by careful operation, to operate for a short time on one unit while the other is under repair.

A careful analysis of cannery steam requirement and the operating requirements of individual units by competent personnel is necessary to insure the selection and installation of a steam plant that will most efficiently supply an adequate amount of steam at the time needed. A well-engineered steam plant can contribute much in reducing production costs.

7.6   Cogeneration

The term 'cogeneration' refers to the simultaneous production of several forms of energy (usually electric power and thermal energy) from fuel. Historically, it has been utilised when low-cost electric power is not available or when the supply is unreliable or nonexistent. Nevertheless, cogeneration may be a feasible alternative to food plants with constant demands of steam or heat and electricity because of the savings in electric bills (Angelino & Foster, 1994). In the food industry, cogeneration has been used mainly in beet/cane sugar processing, sugar refining, and corn wet milling and in a few cereal, gelatin, and canning-processing operations (Texeira, 1986).

Texeira (1986) defined three basic criteria needed initially to evaluate the possibility of cogeneration for a food-processing plant:

- Appropriate balance between demand profiles for steam and electricity.
- Economic justification based on electricity price and capital investment needed.
- Establishment of an adequate agreement with the local utility.

Three common cogeneration systems are the steam turbine, the gas turbine, and the reciprocating turbine. Selection of the type of system depends on the amount of electrical power and thermal power generated by the turbine. In the steam turbine system, a high-pressure boiler produces steam that drives the turbine generating electricity; the exhaust low-pressure steam from the turbine is then used for food processing. In the gas turbine system, fuel combustion gases drive the turbine to produce electricity first, and exhaust hot gases are used to generate process steam in a low-pressure boiler. The reciprocating turbine system consists of an electric generator driven by an internal combustion engine to supply electricity to the plant and of a low-pressure boiler to produce process steam, heated by the exhaust gases supercharged with compressed air (Texeira, 1986).

7.7 Consumption demand

In determining the size of boiler installations, consideration must be given principally to the peak demands of various operations, with respect to the number and frequency of occurrence of these peak demands. These peaks may differ greatly from operating demands; it is imperative that an adequate steam supply be available to preclude the possibility of affecting the efficiency of other operations that may be going on at the time of peak demands. Besides having ample boiler capacities, steam lines adequate for the peak demands of various operations must be installed with proper insulation to protect against undue amounts of condensation.

Installation of steam traps, where possible, will help in reducing the demands on boilers, because they allow for utilisation of all the potential heat in the steam. It is very important that traps be properly sized to insure adequate condensate removal.

Probably the best example of the wide variance between the peak demand and operating demand is the retorting operation. Peak demands for retorting occur during the venting period when air is being expelled from the retort. At this time, the peak may vary between 1134 kg (2500 lbs) and 2722 kg (6000 lbs) per hour for a standard three- or four-crate retort, depending upon the size of the steam inlet line. Peaks for 2.5 cm (1 in) inlets will reach a rate of approximately 1134 kg (2500 lbs); 3.2 cm (1¼ in), 1588 kg (3500 lbs); 3.8 cm (1½ in), 2041 kg (4500 lbs); and 5.1 cm (2 in), approximately 2722 kg (6000 lbs) per hour.

Roughly ¼ to ½ of the total steam required per charge is used during the venting period and will vary, depending on the venting cycle required for each installation (see Figure 7.1). Peak demand drops off rapidly to an operating demand of 45–68 kg (100–150 lbs) per hour after the vent valve is closed and the retort reaches operating temperature. Average steam consumption for a full three- or four-crate retort for processes up to 60 min will vary from 113 to 136 kg (250–300 lbs) of steam or approximately 2.3 kg (5 lbs) of steam per case of 24 × (81 × 111 mm) cans (#303 × 406 cans).

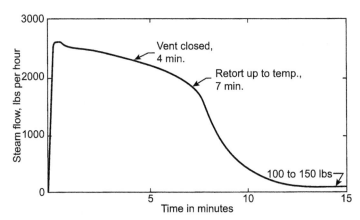

Figure 7.1 Typical steam consumption curve for retorting (still retorts).
Source: From the former Continental Can Company.

There is no significant difference in the requirements of three- and four-crate retorts, or between hot and cold retorts. Because of the high peak demands for individual retort comeup, care must be exercised in timing these operations to prevent undue demands upon the steam-generating system. Unless the system is overburdened by regular operating demands, an efficient generator will usually recover after such demands with no appreciable pressure drop at the boiler. To minimise the effect of pressure drops in boilers operating at near-peak capacity, it is sometimes desirable to install a reducing valve in the boiler header, thus allowing the boiler to be operated at something greater than line pressure. This acts as a reserve steam supply.

Other equipment, such as blanchers, may have rather high peak demands at the start when it is necessary to heat up a sizable amount of water to blanching temperature. Peaks of this nature offer no problem, as they are usually met before actual plant operations start.

7.8 Steam requirements of products

The following steam consumption values are based on actual flow metre measurements made in canneries during normal operations. All values are expressed in pounds of steam per case (24) of #303 × 406 (81 × 111 mm) or #10 (157 × 178 mm) cans or product, using still retorts. The values can be used to compare the relative use of steam per unit operation.

Steam consumption values listed here for the retorting operations are quite consistent for the products involved, with an average of 2.4 kg (5.2 lbs) of steam per case of 24 × #303 × 406 cans. This has proven to be a good value for estimating steam requirements and is also consistent with the total of 113–136 kg (250–300 lbs) of steam per still retort (Table 7.1).

Table 7.1 **Steam consumption in pounds per case during processing of several products**

| Product | Pounds steam per case | | |
	Preparation	Retorting	Total
Asparagus cuts, #303 (cannery A)	13.0	6.2	19.2
Asparagus cuts, #303 (cannery B)	10.5	4.1	14.6
Asparagus cuts, #10	17.2	6.7	23.9
Corn, whole grain in brine, #303	2.5	5.0	7.5
Corn, cream-style, #303	8.0	5.3	13.3
Peas, #303	6.6	5.2	11.8
Apple slices, #10	-	-	32.0
Applesauce, #10	-	-	20.0

7.9   Steam requirements: unit operations

Approximate steam consumption requirements based on either measured or reported values for some of the more significant operations are summarised in Table 7.2.

Once unit demands are determined, a complete analysis of the various operations must be made with respect to the occurrence and timing of these demands. With this information available, it is a relatively simple matter to sum up all peak and operating demands that occur at any given time. The highest value will then be the maximum demand for the operation; from this requirement, the steam-generating and distribution systems can then be determined. At this point, engineering assistance should be obtained to review the consumption data, determine the best type of boiler and fuel to be used, and to plan an adequate distribution system. Some consideration should also be given to possible future expansion of canning operations so that provisions can be made for additions to the steam plant. All too often, installations are made with no consideration of future needs. Too much consideration may be given to utilising all possible capacities from overload with the mistaken thought that the cost of the next size boiler would be prohibitive. This may result in inadequate steam supplies and expanded production costs. It is far better to play safe and oversize the steam plant.

7.10   Estimation of steam requirements

In estimating steam consumption requirements for a cannery, knowledge of demands, both peak and operating, for each individual piece of equipment using steam is important. If this information is not available, some means of estimating these demands is necessary. The heaviest demands usually occur during retorting operations, particularly during retort comeup; this is especially significant if two or more retorts are coming up at the same time. Continuous processing systems tend to eliminate excessively high peak demands, thereby resulting in a lower boiler-rating requirement.

Table 7.2 **Steam consumption for several unit operations in canning**

Unit operation	Peak demand (lbs/hr)	Operating demand (lbs/hr)	Average pounds used per case
Retort			
• 1 inch steam inlet	2500	100–150	5.2
• 1 ¼ inch steam inlet	3500	100–150	5.2
• 1 ½ inch steam inlet	4500	100–150	5.2
• 2 inch steam inlet	6000	100–150	5.2
Continuous pressure cooker	6000	1000–1500	3.0–3.5
Open kettle (100 °C) for 60 min	2000	100–200	1.6–4.0
Blancher, reel	3000	1000	4.0–5.0
Blancher, tubular	3000	1200	4.0–5.0
Flash steriliser (tomato juice 30 g/min)	950	750	1.6
Open kettle concentration tomato puree: 1.045	6000	5000	49 (6–10's)
Brine heating (15–93 °C)	–	–	2.5
Double batch mixer, cream-style corn	1800	750	2.5
Exhaust box, steam (1.2 × 6 m)	500	500	3 (6–10's)

All values are expressed in pounds of steam per hour, with averages based on a case (24) of #303 (303 × 406) cans or as otherwise indicated.

7.11   Suggestions for proper boiler maintenance

Food processors use boilers to generate steam for cooking, heating, pasteurisation, and sterilisation. A major portion of plant energy consumption could very well be accounted for by this operation. Therefore, proper boiler maintenance should be of great interest to the food-processing industry.

A boiler that is properly maintained is said to be 'clean' and 'tight'. 'Clean' is a condition in which the boiler watersides are free from scale and corrosion, and the firesides are free from soot and ash deposits. 'Tight' refers to a boiler free from water, steam, and air leaks. Under these conditions, a boiler will perform at higher efficiencies, absorbing more heat and reducing operation costs.

To maintain a 'clean' boiler, periodically analyse the feed water, steam condensate, and blowdown. Impurities of dissolved minerals and gases in the water supply tend to cause scale, corrosion, carryover, and caustic embrittlement in boilers. With proper water treatment, these undesirable conditions can be effectively reduced.

7.11.1   Water quality

Good boiler water is free from impurities that may cause detrimental effects within the boiler. Several tests can be made to determine water quality. They include odour, colour, turbidity, total solids, total dissolved solids, dissolved gases, pH, acidity, alkalinity, hardness, and chloride content. However, in many cases, depending on conditions and specific situations, not all tests are needed.

Normal boiler water should not have odour, discolouration, or turbidity problems. Odour in the water is a result of contamination; immediate attention should be given to tracing and solving the problem. Excessive amounts of suspended matter, which either diffuse or interfere with the passage of light, cause turbidity and give color to the water; filtration or sedimentation can usually eliminate the suspended solids in water.

Total solids include dissolved and insoluble solids that are separated by filtration. Insoluble solids are usually referred to as suspended matter. Because of the rather small amounts of suspended solids in most water supplies, it is rarely a concern for boiler maintenance. On the other hand, total dissolved solids is a very important test; it can be determined by either evaporating a water sample and drying the residue or by means of a conductivity meter. Total dissolved solids normally range from 20 to 1000 mg/L of potable water and increase with hardness.

pH represents the intensity of the acid or alkaline condition of water. A pH of 7.5–9.5 for feed water and 10.2 to 11.5 for boiler water is generally considered good. Corrosion may easily develop if the pH of feed water is outside these ranges. In this case, the pH can be properly adjusted and controlled by the addition of certain chemicals.

Boiler water below pH 8.5 is considered acidic. Acidity is due to dissolved carbon dioxide, which frequently causes corrosion in boilers; corrosion is also caused by dissolved oxygen in the boiler water. The water can be chemically treated to adjust pH; the dissolved gases, CO_2 and O_2, can be eliminated by installing a deaerator or by adding chemicals to the feed-water system.

7.11.1.1 Alkalinity

Alkalinity is a measure of water's power to neutralise acids; it is determined by titrating a water sample with a standardised solution of sulphuric acid. Boiler corrosion may be the result of low-alkaline water. As corrosion continues, metal thickness is reduced; this results in greater boiler stresses and unsafe operating conditions. Water supplied by a city main may very well have low alkalinity and should be analysed and chemically adjusted for proper water quality. On the other hand, long exposure to highly alkaline waters can cause caustic embrittlement, that is, the development of cracks below the water line and under rivets, welds, and seams. Maintaining an appropriate level of alkalinity is essential in boiler care.

7.11.1.2 Hardness

The degree of hardness is indicated by the concentration of calcium carbonate found in a water sample (a table of water hardness classification is included in Chapter 6). Hard water is undesirable, because it contains excessive scale-forming impurities; the impurities can be sulphates, chlorides, and carbonates of calcium and magnesium. These dissolved minerals are found in varying amounts in all natural waters (wells, springs, reservoirs, etc.); when they react with soap, they precipitate to form scale. As steam is continuously generated and removed from the boiler water, the concentration of these impurities increases; when the concentration reaches a certain level, the impurities settle and form scale on the heating surfaces. This scale has the same effect as insulation by retarding the flow of heat. A thin layer of scale can reduce boiler efficiency considerably and can cause dangerous overheating of the firesides. To alleviate this problem, feed water must be made free of the scale-forming impurities, either by chemical or ion exchange techniques; periodic blowdowns of the boiler water also tend to decrease impurity concentrations and reduce precipitation.

7.11.2 Reuse of condensate

'Carryover' is a continual discharge of undesirable particles (such as scale, corrosion, oil, grease, or dirt) with the steam. When condensed steam is returned and used as part of feed water, poor water quality can result, causing 'carryover'. 'Carryover' can be either 'foaming' or 'priming'.

'Foaming' is the existence of a layer of foam on the water's surface in the boiler drum; 'priming' means slags of water are discharged with the steam. Both are caused by high concentrations of insoluble particles in boiler water and require chemical treatment for prevention.

For specific purposes, tests to determine metal ions, sulphate, nitrate, grease, etc. can also be conducted to evaluate water quality; they are not normally considered routine procedures in testing boiler water.

Chloride content is used as an index in determining seawater intrusion and also as a tracer for pollution of wells. Chloride concentration generally increases as the mineral concentration increases. In boiler maintenance, it can be used to determine the need and frequency of blowdown.

7.11.3 Periodic cleaning of boiler firesides

A periodical cleaning of boiler firesides is recommended. Soot and ash corrosion on firesides can be as detrimental as scale and corrosion on watersides. Burners should be cleaned and repaired, whenever necessary. When the boiler is out of service, check the ignition system for proper flame performance; improper flame adjustment (i.e., incorrect fuel–air mixture) causes boilers to operate at lower efficiencies, increasing operating costs.

7.11.4 Maintenance procedures to insure tightness

A properly maintained boiler is also a 'tight' boiler. Three general maintenance practices are suggested to insure tightness: check steaming pressures, investigate abnormal water losses, and check for leaking tubes.

Observe steaming pressure to check for sudden drops or increases in pressure; dependable gauges are very important. Replace old, worn out, or faulty gauges with reliable ones. A common safety hazard of boilers may involve safety valve(s). Often a valve will fail to open at the set pressure; usually, this is due to a buildup of corrosive deposits around the bottom of the valve. The valve will then leak, rather than open.

Abnormal leaks should be investigated immediately. In general, if a boiler loses more than 7.6 cm (3 in) of water per month, a leak is probably in the system. Water losses can be determined by the amount of makeup water used. If possible, boiler makeup water should be held to a minimum, because it contains oxygen and carbon dioxide.

Watch for leaking tubes. An unusual 'hiss', or a sudden demand for feed water without a corresponding increase in load, may indicate a leak; one leaking tube could mean other faulty tubes. For this reason, it is important to thoroughly inspect boilers – inside and out.

A properly maintained boiler can provide maximum efficiency with no abnormal safety hazards. Boiler care means the difference between saving and wasting; do not let your boiler problems get ahead of you. A routine checkup allows an early warning so that boiler troubles can be corrected early, eliminating excessive maintenance expenditures.

7.12 Air pollution

In a given country, national and local regulations establish the maximum concentration of pollutants allowed in the air. In the US, these limits are dictated by federal and state regulations. Management must be aware of these levels and take the appropriate measurements to keep the air clean.

Air pollutants are produced during fuel combustion. Pollutants emitted may be grouped in four categories:

1. Products of incomplete fuel combustion
   a. Combustible aerosols (solid and liquid), including smoke, soot, organics, etc. but excluding ash
   b. Carbon monoxide, CO
   c. Gaseous hydrocarbons, HC
2. Oxides of nitrogen
   a. Nitric oxide, NO
   b. Nitrogen dioxide, NO_2
3. Emissions resulting from fuel contaminants
   a. Sulphur oxides, primarily Sulphur dioxide, SO_2, and small quantities of Sulphur trioxide, SO_3
   b. Ash
   c. Trace metals
4. Emissions resulting from additives
   a. Combustion-controlling additives
   b. Other additives

Levels of products of incomplete combustion can be reduced using the following methods: insuring adequate excess air, improving mixing of air and fuel, increasing residence time in the furnace, increasing combustion zone temperatures, and avoiding quenching of the flame before completion of reactions (ASHRAE, 1981).

7.13 Energy conservation

US agriculture and the processing, packaging, transportation, storage, merchandising, and preparation of food are among the most energy intensive of such operations in the world. Energy wasted in the food industry has been estimated to range from 10% to 45%, depending on the user (Potter, 1986). Potentially, the main energy savings in food processing are associated with boiler operation, the supply of steam or hot air, and reuse of waste heat (Fellows, 2009).

Line design and equipment selection should be undertaken with energy conservation aspects a factor. Many simple measures to conserve energy are often overlooked. For example, continuous retorts and sealed hydrostatic blanchers are more energy efficient than other equipment used for the same purposes; retorts, as well as steam lines, should be properly insulated.

Systems that recover waste heat may provide significant energy savings. Examples of waste-heat recovery systems that are in use in the canning industry include (Lund, 1986, chapter 17)

1. Using retort condensate to provide boiler feed water, to preheat topping water, and to defrost freezing tunnels where freezing operations are run parallel to canning operations;
2. Using waste heat from refrigeration compressors to preheat water used to defrost freezing tunnels;
3. Using boiler blowdown to preheat feed water to the boiler;
4. Using can cooling water to provide boiler feed water and to preheat can-topping water.

The canning industry can apply its experience in conservation and reuse of water to a stepped-up energy-saving effort. Methods canners may use are given in Bulletin 36L, issued by the National Food Processors Association (1974). The US canning industry uses millions of dollars worth of various fuels for processing fruits and vegetables, specialty products, and seafood; using the steps outlined in the publication, some canners have reported savings of 10–15% in energy requirements.

The energy conservation methods were developed by a team of production and engineering managers from the food-canning industry and its equipment suppliers. NFPA Bulletin 36-L (Appendix 1) points out possible savings in housekeeping operations, boiler and power plant operations, agricultural practices, raw-product handling and cleaning, processing equipment and operations, transportation, and equipment maintenance.

Areas of demonstrated energy savings were

1. Securing optimal performance of existing equipment.
2. Energy savings through redesign/retrofit of existing equipment.
3. Optimisation of process criteria and evaluation of new technology.

Appendix 1: NFPA Bulletin 36-L on energy conservation

The following text is presented in NFPA Bulletin 36-L with some modifications and additions introduced in this edition of this book. The section on Agricultural Practices was considered to be beyond the scope of this book and thus was not included.

Introduction

The canning industry recognises that maximum efficiency in the utilisation of all energy-consuming utilities is a moral obligation of the food-producing community to merit the high priority of energy resources assigned to food manufacturing and distribution. Important also is the fact that there is a practical business necessity for individual companies to remain competitive in the face of spiraling energy and other costs.

In canning-plant operations, energy conservation and water conservation are closely associated. In many canning plants, water conservation programs have been an important part of management considerations for several years. The necessity to focus managerial efforts on intensive campaigns to conserve energy resources is relatively new to most companies. Until recently, energy costs in the US were not sufficiently high to repay intensive conservation efforts on the part of food plant operators. The single responsibility of the canning industry that overrides all others in importance, however, is the safety and wholesomeness of canned-food products. All utility conservation efforts must be reviewed in the light of this responsibility.

Organisation for utility management

No utility conservation program can be any stronger than the commitment of the corporate management to see that the job gets done. Establish conservation goals. One canner reports close to a 15% reduction in energy (power and fuel equivalent) expenditures per unit of production already achieved as a result of a corporate group program. A 30–40% reduction is achievable.

Organise a management team to originate, coordinate, and evaluate utility conservation ideas.

Some multiplant operators have organised their utility conservation teams from representatives of the various divisions, plants, or profit centers within the corporation. Energy conservation ideas are exchanged in a committee approach and then each division within the company is given the responsibility for an energy conservation program within that division, setting up specific goals that each can hope to achieve. Utility managers and energy and water conservation coordinators can be appointed at appropriate plant or division levels to supervise conservation programs in all operating divisions.

Quality assurance or quality control personnel must have the final word in evaluating any utility conservation ideas affecting canned product safety or wholesomeness.

Install an effective system of metring utility input at the plant and devise an appropriate feedback system for reporting and evaluating utility conservation results.

Utility management priorities utilised to deal with a fuel emergency in several corporations may be summarised as follows:

1. First Priority
   - Emergency fuel-allocation considerations – make whatever fuel-allocation decisions may be necessary to stay in production.
2. Second Priority
   - Enforce good housekeeping utility conservation measures – do immediately whatever can be achieved by altering plant practices to conserve utilities and motivate personnel to participate actively in such steps.

3. Third Priority
   - Engineering considerations for the expenditure of capital money to recover utilities – utilisation of waste heat and reuse of water.
4. Fourth Priority
   - Evaluate opportunities for alternate fuel systems in the event that this should become necessary.

Utility audits

To measure effectiveness in utility conservation programs, the following information elements are necessary:

1. An accurate metring of utility (electric power, fuel, and water) input. All forms of energy input must be converted to appropriate common units: kilowatt hours, BTU's, fuel oil equivalents, etc.
2. Evaluation of canned-food production in terms of utility usage per standard or actual case.

To assess the effectiveness of utility conservation measures, data on water and energy usage must be expressed in terms of suitable production units. Utility costs are likely to fluctuate rapidly, and for this reason, cost figures are not a convenient yardstick for evaluating the success of conservation measures. Utility costs per case are, unfortunately, likely to show increases, even in the face of substantial achievements in increasing the efficiency of utility utilisation.

If total energy input is to be evaluated, the audit should cover all energy sources utilised by the company, preferably before the start of the conservation program and at relatively frequent intervals thereafter. This will, of course, include electric power from metre records, gas utility purchases from metre records, purchases of bottled gas for any in-plant uses, power plant (boiler) fuels of all types, vehicle fuel purchases, including truck, automobile, and forklift truck operations required in production, and metring of water use for unit operations. Local utility companies may be contacted for accurate metre and energy conversion factors. Average energy equivalent factors are available from handbooks and other sources.

Housekeeping operations: a good place to start

Lighting

1. All lighting, except for security and safety lighting, should be turned off when not in use. Time switches or motion detectors for interior lighting and photocell switches for exterior lighting should be considered.
2. Although adequate lighting for work areas is essential, peripheral areas should be surveyed with the objective of removing bulbs or fluorescent tubes or replacing with bulbs of lower wattage. Phase out unnecessary lighting in office, corridor, cafeteria, plant, ingredient receiving and storage, and warehouse areas.
3. Use fluorescent, mercury, or sodium fixtures where feasible, rather than incandescent, because they deliver more light per kilowatt hour, respectively. The initial cost is higher in the same order.
4. Install separate, independent lighting circuits and switches where practical to allow localised lighting of work areas.

5. Replace age-yellowed prismatic panels and louvers. Up to 15% improvement in lighting efficiency may be realised.
6. Consider the policy of group, instead of singular, bulb replacement. Lamp output drops up to 15% after two years.
7. Keep lamps, fixtures, and reflecting surfaces clean. Post instructions for operating, cleaning, and maintenance of light fixtures and audit for custodial compliance.
8. Whenever possible use natural lighting.

Heating

1. Office temperatures should be held to no more than 20 °C (68 °F). The use of personal electric heaters is wasteful and should be permitted only under extreme conditions. Keep windows free of obstruction for maximum sunlight and keep windows and doors closed. Consider time clock controls to reduce heating after working hours.
2. Plant and warehouse temperatures should be reduced to the lowest temperature at which employees wearing jackets or sweaters can work efficiently.
3. Maintain clean filters on heating and ventilating equipment.
4. Control the makeup air temperatures and quantities in ventilation systems to the minimum.
5. Warehouses that are only intermittently used should be heated to no more than 4 °C (40 °F), whereas 10 °C (50 °F) may be sufficient for finished product warehousing.
6. Obvious air leakages should be eliminated by weatherstripping or other means. Expenditures for extensive insulation and weatherstripping are not always justified in existing buildings; they should be considered in severe climates.
7. Particular attention should be paid to keeping warehouse doors closed to the maximum extent possible, consistent with necessary lift-truck traffic. Consider split canvas curtains, air curtains, and self-closing doors.
8. Construct entrance ports for large doors leading into plant or office. Install seals at truck loading doors.
9. Consider use of carpeting to reduce floor heat loss.
10. With the implementation of heating savings in buildings heated by steam, it may be practicable to operate a boiler only intermittently to maintain the minimum necessary temperature. This should include consideration for nighttime and weekends and shutting down of all boilers not required for other uses during mild weather.

Water

1. The supply of water requires energy for production, transportation, purification, and waste treatment. Consequently, saving water will save energy and dollars as well.
2. Stop leaks and use automatic-off faucets, or shut off water lines left running for no reason.
3. Consider reducing the temperature of hot water for personnel use and turning water heating down or off on weekends.
4. Make a thorough study of the use of processing water with the objective of accomplishing necessary washing and cooling without waste and with the maximum reuse of water. Consider the installation or extension of the counterflow use of water and the reuse of empty can wash water for washing of filled and sealed cans.
5. Check the frequency of cleanup needed for maximum efficiency and use high pressure–low volume for quick cycle cleanups.
6. Consider clean-in-place systems for specific applications.
7. Minimize water use for clean-up operations.

Electric power

1. A systematic review of the entire electric power system should be made. Adequate instruments, e.g. ammeters, voltmeters, and RPM indicators should be used.
2. Loose drive belts waste power.
3. Overloaded motors waste power in the form of heat and are obviously undesirable because the unnecessary stress will shorten the service life of the motor. All driven equipment should be well lubricated and checked for free operation to minimise frictional loading and reduce risk of overheating.
4. Underloaded motors waste power. They also lower the power factor, which can increase billing charges. Improving the power will definitely save energy.
5. In most areas, improvement of the power factor by installation of capacitors cannot be justified on a savings basis. Capacitors will free up system capacity for expansion. The possibility should not, however, be summarily written off.
6. Reducing the maximum demand load by possible shifting of heavy loads to offload hours will be reflected in lower power bills but will not in itself save energy, except in the unlikely instance in which the facility is a major customer of a small utility. It may be of value in an overall area electrical shortage.
7. Do not operate standby equipment when the primary equipment can carry the load and turn off electric motors during nonproduction periods.

Steam

1. Check steam distribution systems from boiler to end-of-line for losses and remove unused or unnecessary steam piping.
2. Repair valve seats to prevent steam leakage, such as into empty retorts and steam kettles.
3. Repair valve packing in controllers and hand valves to prevent steam leakage.
4. Install and maintain steam traps at the end of steam manifolds at retort installations. It may be worthwhile to pipe the steam exit of the traps to a return line to be used to warm makeup water for boiler operations. Similar arrangements may be practicable in some agitating cooker installations.
5. Keep heat transfer surfaces clean.
6. Keep insulation in good condition.
7. Check condensate return system for malfunctioning traps and leaks in the line.
8. Check for excessive steam vented to atmosphere.
9. Watch usage of heated makeup water; excessive usage will indicate losses in the system, which should be corrected.
10. Periodically, check steam-using equipment, e.g. blanchers, heat exchangers, exhaust boxes, cookers, retorts, and kettles for operation at the proper temperature and the absence of leaks.
11. Reduce operating steam pressures during production periods to a minimum suitable to the demand.

Boiler and power plant operations

1. Continued training of boiler operators is essential, particularly in full-load operation with alternate fuels.
2. Each boiler installation should include an onstream gas analyser to measure directly the volume by percent of oxygen and combustibles in flue gases. If there is excessive air in the flue gases, the boiler fuel settings should be corrected immediately.

3. Each installation should have and use a stack gas temperature-monitoring device.
4. Instrumentation and auxiliary equipment must be kept in first-class condition.
5. Boilers should be frequently checked for cleanliness and condition, including:
   a. Dirty burners – both air and fuel passages; replace inefficient burners.
   b. Cracked or loose refractory, especially around drumheads.
   c. Loose linkages on stack dampers and secondary air controls.
   d. Lag in boiler control leading to improper oxygen content in flue gas.
   e. Clean internal and external surfaces of drums and tubes.
6. Watch for unequal loading when boilers are operated in parallel on a common pressure-controlled fuel line.
7. Reduce boilers to low pressure on weekends or during low or nonproduction shifts, using the minimum pressure and number of boilers possible.
8. Use waste steam or hot water from production operations for preheating boiler makeup water, or use clean, wasted hot water for boiler makeup. Some examples:
   a. If a vent blowdown manifold is used at retort installations, a water line installed in the manifold could utilise heat from the steam for preheating boiler makeup water. The vent manifold would have to be sized large enough to meet requirements in the Good Manufacturing Practices regulation 21 CFR 113, subtracting the area occupied by the water line.
   b. If cans are cooled in the retort, suitable piping arrangements can be made to utilise the initial hot water for boiler makeup water.
   c. In water processing of glass containers, the steam–water mixture from the pressure regulating valve overflow could be used for boiler makeup water or for heating boiler makeup water.
9. Investigate alternate sources of boiler fuels, such as filtered compressor lubrication oil or recycled motor oil.
10. Check steam distribution system for losses.
    a. Keep insulation in good condition.
    b. Check condensation return system for malfunctioning traps and leaks.
    c. Check for excessive steam vented to the atmosphere.
    d. Monitor use of treated makeup water; excessive use indicates losses needing correction.

Raw product handling and cleaning

1. Review product receiving, cold storage, handling, and cleaning methods to determine whether engineering changes are feasible for energy or water conservation.
2. Utilise gravity flow wherever possible.
3. Determine the need for unit operations that may be obsolete because of production line changes (e.g. multiple product inspections).
4. Schedule full and continuous production loads whenever possible.
5. Minimise water use consistent with proper cleaning and investigate dry-cleaning possibilities. Reuse water by counterflow where possible.
6. Determine whether reduction of washwater temperature is practicable.
7. Avoid preheating blanching equipment before necessary.
8. Investigate alternative blanching procedures and avoid unnecessary cooling of blanched product.
9. Use insulation to minimise heat loss.
10. Institute an incentive system to encourage suggestions by operating personnel.

Processing equipment and operations

1. Adequate venting of air from thermal processing (retorting) equipment and continuous free steam flow from the bleeders is essential for safe processing. Do not attempt steam conservation by decreasing retort venting or closing retort bleeders.
2. Avoid preheating and venting continuous retorts before the time indicated as necessary by the production schedule.
3. Consider insulation of retorts to prevent loss of radiant heat and minimise employee heat exposure.
4. Check air equipment for leaks to reduce compressor time. Compressed air is a costly utility. Inspect and review all plant operations to locate and eliminate unnecessary or wasteful uses.
5. If economically possible, use heated water or venting steam for regeneration, plant heating, etc. (e.g. in continuous cooker-coolers, the atmospheric cooler water might be cycled to the pressure cooler and then cleaned for boiler makeup use).
6. For vacuum production, mechanical pumps have a higher initial cost but are more economical to operate than steam-jet ejectors.
7. Product-to-product heat exchanger regeneration may save energy in aseptic canning operations.

Transportation

Detailed energy conservation considerations involved in the transportation and distribution of the finished canned product are beyond the scope of this book. All companies own some fleet vehicles, however, and many may have substantial private trucking operations. These must not be overlooked in any conservation efforts.

1. Set up a record system to measure 'miles per gallon' for each vehicle. Establish a procedure to take action (tune up, etc.) whenever a vehicle's mileage performance falls below a predetermined point.
2. Conduct a 'purchasing specification' analysis of both tractors and trailers to avoid:
   a. Under or over powering of tractors (horsepower, axle, and transmission).
   b. Unrealistic trailer capacity requirements.
3. Consider advantages of diesel versus gasoline.
4. Make use of route engineering to reduce empty miles (improvement in productivity). This may involve: Sales Policy, with respect to order minimums, increments, delivery frequency, back hauls, 'trade-offs' with reduced level of customer service.
5. Extend use of rail and piggyback where possible.
6. Emphasise 'driver education'-speed limitations and shifting speeds. Consider programs for driver evaluation. The AAA says 40% can be lost through inefficient practices.
7. Review and update maintenance programs on a regular basis.
8. Maintain close relationship with truck and engine manufacturers for industry-related fuel-savings ideas and methods.

Equipment maintenance

Preventive maintenance of all cannery equipment is essential in achieving peak efficiency. Although this is undoubtedly already a major feature of cannery operation, it must be doubly emphasised at this time.

The preceding sections provide a suggested checklist relating to utility conservation, and proper implementation of the procedures suggested requires that all equipment be maintained through a continuous, preventive program. This will not only ensure efficient operation but will also preclude costly and energy-consuming production losses caused by equipment breakdown and motor failures.

Acknowledgements

The author wishes to thank Olga Padilla Zakour for her contribution on this chapter in 13th Edition.

References

Angelino, R., & Foster, S. (1994). Sugar processor benefits from co-generation. *Food Processing*, *55*(5), 53, 56–58.

ASHRAE. (1981). *ASHRAE handbook. Fundamentals. American society of heating*. Atlanta, GA: Refrigerating and Air-Conditioning Engineers, Inc.

Fellows, P. J. (2009). *Food processing technology* (3rd ed.). United Kingdom: Woodhead Publishing Limited.

Lund, D. B. (1986). Low-temperature was &-heat recovery in the food industry (Chapter 17). In *Energy in food*.

Potter, N. (1986). *Food Science* (4th ed.). Westport, CT: AVI.

Sarco, S. (2007). *The steam and condensate Loop*. United Kingdom: Published by Spirax Sarco Limited.

Sing, R. P. (1986). Energetics of an industrialized food system. In R. P. Sing (Ed.), *Energy in food processing*. Amsterdam: Elsevier.

Texeira, A. A. (1986). Co-generation in food processing plants. In R. P. Sing (Ed.), *Energy in food processing* (p. 283). Amsterdam: Elsevier.

Cleaning and sanitising

8

8.1 Introduction

Sanitary conditions are important in food plants, because sanitary practices contribute to better quality, better keeping characteristics, and lower incidence of spoilage of processed foods. Present-day science and technology afford the food processor effective and economical means of processing food under sanitary conditions.

The preservation of food products depends upon preventing contamination by spoilage-causing microorganisms before, during, and following processing operations. In addition, certain microorganisms are the causative agents of food-borne infections and intoxications and may render a food product contaminated with them unfit for consumption.

In canning, it is important to consider raw materials as potential carriers of processed-food spoilage microorganisms. Heat-resistant bacteria present in soils can cause 'flat sour' and other spoilage of canned vegetables if soil is not thoroughly removed by washing. Fresh fruit may be contaminated with mould. If the infected portions are not removed during the early stages of processing, both the final product and the air inside the plant may become heavily contaminated and infect subsequent lots of sound fruit. Mould growth, in addition to indicating that the product is not sound, can impart off-flavours and odours to the raw products which can be carried into the processed food. Shipping containers for raw fruits and vegetables should be thoroughly washed and treated with a fungicide to prevent spread of mould infections.

8.2 Sanitation program

The main components of a sanitation program are

1. A commitment by top management to have a sound sanitation program.
2. Employees who follow good personal hygiene practices and who are trained and retrained in the use of cleaning products and procedures, pest control, and housekeeping.
3. Routine audits/monitoring programs and corrective action programs.

Two of the critical components of a sanitation program are cleaning and sanitising. The aim of cleaning and sanitising is to remove unwanted soils, microorganisms, and other agents of disease and product spoilage. Although these practical components of sanitation will be discussed here, it should not be forgotten that sanitation is more than just removing soils. Food processors have a responsibility to do all that is necessary to assure that products are safe and wholesome and do what they are supposed to do.

A Complete Course in Canning and Related Processes. http://dx.doi.org/10.1016/B978-0-85709-677-7.00008-6

Failure to implement and maintain an appropriate sanitation program can result in:

- Poor product quality or even spoilage.
- Reduced product stability/shelf life.
- Injury/death to consumers.
- Lack of consumer confidence or damaged company image.
- Government recalls, shutdowns, closures.
- Legal actions, imprisonments, fines.
- Bankruptcy.

Cleaning and sanitising usually involve the use of different chemical agents and different application procedures, and they have distinct roles. Neither cleaning, nor sanitising alone, is sufficient to assure that the desired level of sanitation will be achieved.

8.3   Cleaning

Cleaning can be defined as the removal of soil particles from surfaces by mechanical, manual or chemical methods.

The purpose of cleaning is to

1. Prevent the transfer of ingredients in one batch of product entering another. Cross contamination of raw material can obviously cause serious quality, and hence, economic problems for manufacturers.
2. Remove soils which can harbour microorganisms.
3. Remove soils which can provide nutrients for microorganisms which can cause quality and/ or public health problems.
4. Eliminate soils which can build up and ultimately reduce the efficiency of processing equipment such as heat exchangers, pumps, etc.
5. Prevent the entry of foreign matter into the product which might directly reduce product quality and/or cause harm to consumers.
6. Comply with good manufacturing practices/regulatory demands.
7. Protect product quality and thereby prevent economic losses.

8.3.1   The cleaning process

Soil removal is a complex process and is dependent on many variables. No single cleaning compound can satisfy all of the requirements throughout the environment. The choice of the appropriate cleaner will depend on several factors:

1. Type of surface(s) to be cleaned
2. Type of soil(s) to be removed
3. The method by which it will be applied
4. Characteristics of the water being used to dissolve the detergent.

Ideally, a good cleaner should be: economical, nontoxic, noncorrosive, noncaking and non-dusting, easy to measure, stable during storage, and easily and thoroughly dissolved.

8.3.2   Types of surfaces

The characteristics of the surface being cleaned must be considered when selecting cleaning compounds. Stainless steel is the material of choice for food-manufacturing equipment and has greater corrosion resistance than softer metals. However, stainless steel is available in a wide array of grades and finishes and, therefore, familiarity with the specific limitations of the stainless steel surface being cleaned is necessary.

For soft metals such as brass, aluminum, and zinc, specific corrosion inhibitors are incorporated into the cleaner.

Plastics are resistant to most cleaning compounds. However, it is important to be familiar with the types of plastics being used in the environment which may be susceptible to attack by certain cleaning compounds. A sample of any plastic material suspected of being damaged by chemical cleaners should be provided to a chemical supplier for an evaluation.

8.3.3   Types of soil

Soil types can be classified in different ways. A classification based on their chemical nature and the recommended detergent for each type is shown in Table 8.1. When a mixture of soils is present, it may be necessary to use two separate cleaning steps. For example, a combination of protein and hard-water film is satisfactorily removed with an alkali, followed by a water rinse, and finally an acid cleaner.

Table 8.1 Types of soil and their characteristics

Soil	Solubility	Ease of removal	Change on heating	Recommended detergent
Protein	Water insoluble, alkali insoluble	Very difficult	Denaturation Chlorinated Much more stubborn	Alkaline
Fat/oil/grease	Water insoluble	Difficult	Polymerisation More difficult to clean	Alkaline
Sugars/ carbohydrates	Water soluble	Easy	Caramelisation More difficult to clean	Alkaline
Mineral salts (e.g. Ca^{++}, Mg^{++})	Variable water insolubility	Easy to difficult	Generally easy to clean unless interacting with other components	Acid

8.3.4 Method of application

The method by which detergents are applied will also have a strong bearing on the selection of the cleaner.

1. *Manual Cleaners*: Because these products contact the skin of users, the choice of products is necessarily limited to materials which have mild alkalinity, or acidity, or are neutral. Furthermore, if the soil load is high, manual cleaners must be applied with strong mechanical action such as scrubbing.
2. *Circulation Cleaners, i.e. Clean-in-Place (CIP)*: Cleaning solutions applied by circulation equipment are subjected to high turbulence. Because certain products can, under such agitation, produce undesirable foam, circulation cleaners must produce little or no foam to be acceptable. These types of products are normally strongly alkaline or acidic.
3. *Clean-out-of-Place (COP)*: For this application, small equipment parts/dismantled equipment and utensils are placed in a tank and the detergent is circulated within the tank. Because this is normally a closed system, strong alkaline or acid detergents are used.
4. *Soak Cleaners*: The cleaner is simply placed in the equipment to soak without agitation, or the components of equipment, e.g. gaskets, valves, connectors, and utensils, are soaked in tanks containing the cleaner. This is the least efficient cleaning method, but the application of some agitation using small stirrers or passing through air markedly improves the efficiency of cleaning.
5. *Spray Cleaners for High-Pressure Systems*: As with cleaners for CIP systems, cleaners for high-pressure systems should have low-foaming characteristics. They should be relatively mild cleaners, however, to minimise the dangers of overspray or bounce-back of solution.
6. *Foam Cleaners*: Compatible foaming agents have been added to these cleaners. The product is vigorously aerated by injecting compressed air. Such products are becoming increasingly popular for external environmental cleaning, because foams by nature tend to adhere to surfaces and thereby provide prolonged contact time.

8.3.5 Characteristics of water

Water is the solvent most commonly used for cleaning operations. It is used before detergent cleaning as a prerinse to remove gross surface soils and after cleaning to rinse away loosened soils and residual cleaning solution. Because of its important role in the cleaning process, water quality is a critical factor which must be considered when selecting a detergent. In particular, the amounts of calcium and magnesium carbonates (water hardness) are the most significant components of water which affect the cleaning process. When hard water is heated, carbonates will precipitate, causing scale formation. In addition, without proper chelation of the calcium and magnesium salts, the use of soap-based cleaners can also result in precipitation. Thus, for cleaners to function properly when dissolved in hard water, it is necessary to incorporate sequestrants or chelants which control calcium and magnesium, preventing scale formation and precipitation. If scale formation occurs, it is desirable to use acid cleaners specially formulated to remove these mineral deposits.

8.3.6 Sequence of events during cleaning

The cleaning of a surface actually proceeds in a series of actions rather than in one simple step. First, the detergent is dissolved in water. In so doing, a cleaner containing hard water-controlling sequestrants will first react to prevent calcium and magnesium from reacting with the soil to form 'soap scum'. Next, the solution is applied to the surface and soil separation begins. Water-soluble soils, such as sugar, are readily dissolved in the aqueous cleaning solution. In addition, the cleaning compound aids in the separation of difficult to remove soils by reducing the surface tension of water to the point at which soils are in intimate contact with the water. Surfactants are used for reducing surface tension. The cleaner then loosens the soil by reducing the energy involved in binding the soil to the surface.

Soil separation is enhanced through the physical methods by which cleaners are applied. These include scrubbing, agitation, foaming, and high pressure spraying. Furthermore, increasing the temperature of the cleaning solution will aid in loosening soil.

Once separated from the surface, soil dispersion begins. Soluble soils remain emulsified in the cleaning solution as long as the limits of solubility of the soil have not been exceeded. Clumps of insoluble soils are broken down into smaller particles or droplets which are held in suspension. Oil and fat droplets are, for example, broken down into smaller droplets which are dispersed evenly throughout the cleaning solution in a process called emulsification.

Finally, the cleaning solution and the dispersed soil are removed, preventing redeposition of soil. The surface is then rinsed with potable water to remove residual cleaning solution and soils.

8.3.7 Factors which affect cleaning

Four key factors affect the cleaning operation: (1) time, i.e. the amount of time the detergent is in contact with the surface; (2) action, the mechanical action used to apply the detergent; (3) temperature of the cleaning solution; and (4) concentration of the detergent. Other factors considered are amount and nature of soil and water hardness.

Cleaning Time: Clearly, the effectiveness of the detergent greatly depends on the amount of time the product is in contact with the surfaces being cleaned. Detergents must have enough time to wet/coat the soil, penetrate it, lift it off the surface, and suspend it in solution.

Amount of Mechanical Action: The use of mechanical action provides mechanical energy, which greatly improves soil removal. This is largely why high pressure, CIP, and spray cleaning are much more efficient than soak cleaning.

Concentration of the Cleaner: In general, as the concentration of the cleaner is raised, cleaning efficiency is improved. Increased concentration provides chemical energy which can be used to break the bonds holding soils onto surfaces. However, too much detergent can actually reduce cleaning efficiency. The concentration recommended by the chemical supplier should be used.

Temperature of the Cleaner: In general, as the temperature of the detergent is elevated, the rate of action increases. It is unusual, however, to use detergents above 60 °C (140 °F), because doing so risks baking soils onto surfaces, which then become extremely difficult to clean. Soils suspended in the detergent may precipitate onto surfaces, whereas ingredients in the detergent itself, e.g. chlorine, become less stable and sometimes more corrosive at high temperatures.

Amount of Soil: The performance of the detergent diminishes as the soil load increases; soil ties up the detergent to a point at which less alkalinity or acidity is available to clean. As the soil load builds, precipitates can form which eventually are redeposited on the surface, thereby requiring additional cleaning.

Nature of the Soil: The type of soil dictates the kind of cleaner required. In general, alkaline detergents are used for organic soils, and acidic detergents are used for mineral/scale deposits. The decision as to what specific formulated product is needed is largely based on a knowledge of the properties of the specific soil attached to the surface.

Water Hardness: Mineral salts can react with the soap or alkali in a detergent to produce a precipitate. Not only does this reduce the alkalinity and thus detergency, but the precipitate represents additional soil that must be removed from the system. Mineral scale can also form from the precipitation of salts from warmed rinse water.

8.3.8   Cleaning compounds and their composition

The two main types of detergents are alkaline and acid detergents. Alkaline detergents are used for removal of organic soils such as fats/oils, proteins, and carbohydrates. They remove soils by emulsification, saponification, and/or peptonisation of organic materials. These types of detergents must contain various types of ingredients, such as detergency boosters, wetting and penetration agents, water conditioners, emulsifiers, and dispersing and rinsing aids. They may also need to contain corrosion inhibitors, foam boosters, and foam suppressors.

Chlorine is often used in conjunction with alkaline detergents to aid in protein removal. However, it is not present in detergent as a sanitising agent.

Acid detergents are used for removing inorganic soils such as hard water (e.g. calcium and magnesium) and other mineral films such as iron, sulphates, etc. These types of detergents also must often be formulated with various ingredients to enhance cleaning power and reduce risks of damage to equipment.

Water is the most abundant component of most detergent-use solutions. It is also used alone as a prerinse to remove gross surface soils and as a postdetergent rinse to remove loosened soils and cleaning solution. Water is not, however, very effective by itself for removing proteinaceous and lipid-type soils.

Water quality is a critical factor, which must be considered when selecting a detergent. Only potable or drinking water should be used for cleaning; nonpotable water is not a suitable substitute. Some plants find it necessary to chlorinate their water supplies or to supplement the chlorination provided by municipalities. Supplemental chlorination is usually done when microbiological problems exist or are anticipated. Examples are can-cooling canals in food-canning operations or areas of continuous water application in which a concentration of 5 to 8 ppm of chlorine is recommended.

Water hardness may be the most important factor influencing the effectiveness of a detergent. Cations (magnesium, calcium, and manganese) naturally present in water can deposit residues and films (carbonates) on equipment surfaces, especially under warm conditions. The addition of soap-based cleaners and alkalis can also result in precipitation of insoluble mineral/soap films (i.e. soap scum). In addition to being unsightly, mineral films reduce the effectiveness of many surfactants used in detergents. This problem can be overcome, however, with the use of 'built' detergents, which contain sequestrants or chelators in addition to the surfactant. Usually these either phosphates or organic chelators function by binding the cations and keeping them in a soluble form. Heavily built detergents are required in areas in which water is very hard; the presence of these chemicals in soft-water areas may not be necessary. Hardness may also be reduced by mechanical softeners which usually contain an ion exchange resin for this purpose. Such devices should periodically be checked by submitting samples of water effluents to a reputable laboratory for total bacteria counts.

The chemicals used in cleaning compounds fall into one of several categories: alkaline compounds, acids, chelators, wetting agents (surfactants), solvents, and chlorine. A description of these chemicals and their major functions in detergents follows.

1. *Alkaline compounds*: Used for removing organic soils, such as oils/grease/proteins, and carbohydrates. Their strength varies greatly depending on use application.
   a. *Caustic soda or sodium hydroxide (NaOH)*: A highly alkaline material, used for removing heavy soils, especially where high-temperature processing occurs. It has high germicidal activity, protein dissolution, and deflocculation/emulsifyng properties. Because of its nature, it is not used as a manual cleaner. Caustic-based product use is limited to stainless steel and black-iron equipment; corrosive to aluminum, brass, and other soft metals.
   b. *Soda ash or sodium carbonate*: Low in cost. It is used as buffering agent to maintain proper alkalinity/pH control of use solution, widely used in heavy duty and manual cleaners.
   c. *Trisodium phosphate*: Highly soluble, good soil emulsification, good detergency, good rinsing, available as chlorinated TSP, poor chelator.
   d. *Sodium metasilicate*: High alkalinity, good buffering agent, primarily used as corrosion inhibitor, enabling the use of caustic-based and chlorinated products on soft metals.
2. *Acids*: Used for removal of inorganic soils, such as hard water and other mineral films.
   a. *Phosphoric acid*: Most widely used acid in US. Relatively low-corrosive properties, compatible with many surfactants and used in manual and heavy-duty formulations.
   b. *Nitric/sulphuric acid*: Not used in manual cleaners due to corrosive properties. Useful in high-temperature processing operations for heavy-scale removal. Fuming may be a problem in some formulations.
   c. *Hydrofluoric/hydrochloric acid*: Generally not used because of their corrosiveness to stainless steel, the exception being for the removal of extremely heavy scale in steam-producing equipment, boilers, or certain processing equipment.
   d. *Organic acids*: Most commonly either sulphamic, citric, or hydroxyacetic. Much lower corrosive properties than inorganic acids, enabling use on soft metals and in manual cleaning products. Used for water softening and mineral control. Non-irritating but much more costly than inorganic acids.
3. *Chelators/Water conditioners*: Used to sequester calcium and magnesium salts on the alkaline side; prevents 'soap scum' from forming.
   a. *EDTA (ethylenediaminetetraacetic acid)*: Used in formulations as water conditioner to tie up water hardness and minimise scale buildup. Used in many conveyor lubricant formulations.
   b. *Gluconates*: Includes sodium gluconate, glucoheptonates, and gluconic acid. Formulated into highly alkaline products as water conditioners.

   c. *Complex phosphates*: Are highly effective water conditioners. Group includes sodium
      and potassium salts of pyrophosphates, tripolyphosphates, and hexametaphosphates.
      Choice in formulations depends on needs, including calcium activity and temperature
      stability.
   d. *Polyelectrolytes*: Long-chain polymer of acrylic acid. A large variety of molecular weights
      from 1500 to 10 million atomic mass units. Component in nonphosphated detergents.
4. *Wetting agents*: Also known as surfactants, these are organic compounds useful in both alkaline and acidic formulations to provide detergency, increase soil wetting/penetration, solu-
   bilisation, emulsification, and dispersion. They are also used to improve rinsing and control
   foaming properties. Foam control may be 'defoaming' or 'foam producing', depending on
   the wetting agent used and the concentration and temperature of the solution. Surfactants
   come in three types, namely, cationic, anionic, and nonionic. Detergents used in food-processing applications are usually formulated with anionic or cationic surfactants, rather than
   nonionics, which are incompatible with chlorine.
   a. *Anionic surfactants* are any surface-active compounds which hold a negative charge in
      solution. They are essentially soaps or neutralised organic acids (e.g. alkylbenzylsulpho-
      nates). They have good wetting and rinsing properties, and some have biocidal activity
      within a specific pH range. They produce variable amounts of foam upon agitation. Not
      generally compatible with cationic surfactants.
   b. *Cationic surfactants* are quaternary compounds, tertiary and heterocyclic amines, which
      produce positive ions in solution. They produce low to moderate amounts of foam, are
      difficult to rinse, and have good biocidal properties within a particular pH range. Not
      generally compatible with anionic surfactants.
   c. *Nonionic surfactants* are condensation products of ethylene oxide and fatty acids/alco-
      hols, commonly used in laundry detergents. They produce neither positive nor negative
      ions in solution. They have excellent wetting and penetration properties, and the amount
      of foam produced upon agitation varies with the surfactant. They have no biocidal activ-
      ity. Compatible with both anionic and cationic surfactants.
5. *Solvents*: Typically a glycol ether compound, not the familiar organic solvents like acetone,
   methylene chloride, or chloroform. Depending on compound, the solvent may be approved for
   use on food contact surfaces, floors/walls, and various environmental cleaning operations.
6. *Chlorine*: Used as an additive in alkaline cleaners (never added to acids) to remove soil,
   particularly to improve protein film removal. Not in detergent for sanitising.

8.4   Sanitising

The objective of sanitising is to destroy vegetative cells of microorganisms of public
health significance and to significantly reduce the number of other undesirable microorganisms. Proper sanitation will, therefore, help to assure that the product will be safe
and wholesome and will have an acceptable shelf life.

A requirement to achieve satisfactory sanitation is to first prepare a clean surface.
Sanitising a dirty surface is not possible. Soil left on a surface provides a medium
on which microorganisms can attach and become difficult to remove. Such residues
also provide nutrients on which microorganisms can survive and grow. Soils become
harbourage sites, which shield microorganisms from direct contact with the sanitiser.
Furthermore, soils can react chemically with the sanitiser, reducing its efficiency.

Several methods of sanitising can be used. Among them are

1. High-temperature sanitising, the use of hot water or steam.
2. Radiation – ultraviolet light is frequently used for water treatment.
3. Chemical sanitising – this is the most common method of sanitising.

8.4.1   Basic requirements of a sanitiser

The number of different types of sanitisers acceptable for use in industrial manufacturing plants is somewhat limited due to the potential toxicity of many antimicrobial agents. A sanitiser should meet the following criteria:

1. Be nontoxic to humans.
2. Have antimicrobial activity against a wide spectrum of microorganisms.
3. Destroy microorganisms, rather than merely inhibit them.
4. In the US, EPA registration as a sanitiser.

8.4.2   Factors affecting the efficacy of a sanitiser

The efficacy of chemical sanitisers is affected by physical and chemical factors, including

Contact Time: The time required to kill microorganisms varies with the types of microorganisms, their age, and physical and chemical properties of the cells. To be approved for use on previously cleaned food-contact surfaces, sanitisers must be capable of killing 99.999% of certain bacteria in 30 s or have killing power equivalent to 100-ppm chlorine. Longer contact times help to destroy additional survivors, although a point is reached at which no further biocidal action occurs.

Temperature: The rate of kill increases as the temperature is raised. However, the stability of some sanitisers, e.g. halogens, declines rapidly at high temperatures. Chlorine and iodine rapidly decompose at about 140 °F (60 °C) and 120 °F (48 °C), respectively. Some sanitisers, e.g. chlorine, become more corrosive to surfaces and/or dangerous to handle at high temperatures.

Concentration: Increasing the concentration of a sanitiser enhances the rate of biocidal activity. Again a point is reached wherein 'more is not better', and other problems, e.g. corrosion and excessive foaming, can be exacerbated at high concentrations.

pH: The activity of most sanitising agents is highly dependent on the pH of the use solution. Acid sanitisers are much more effective at or below pH 3; iodine sanitisers are usually more effective at acid pH levels versus alkaline pH levels.

Water Hardness: Certain sanitisers are less effective at elevated levels of water hardness. The efficacy of some quaternary ammonium salts (quats), for example, is significantly reduced in hard water. Chlorine sanitisers and iodophors are, however, less affected by water hardness.

Soil: The efficacy of sanitisers is reduced by the presence of organics to varying degrees, depending on the type of sanitiser. Iodine and chlorine sanitisers, in particular, are adversely affected by organic soils. Quats and acid sanitisers are less affected.

Chemical Incompatibility: The presence of residual cleaning chemicals can also greatly reduce sanitiser efficacy. Anionic surfactants from a detergent, for example, can neutralise the cationic surfactants in a sanitiser. Residual alkalis can similarly neutralise or raise the pH of an acid sanitiser to a suboptimally active level.

8.5   Sanitisers

Sanitisers can be grouped into two categories: (1) halogen or oxidising, in which the active component, e.g. chlorine or iodine, changes oxidation state, and (2) surfactant or non-oxidising, e.g. quaternary ammonium compound.

8.5.1   Chlorine sanitisers

General properties:

1. Strong oxidising agents.
2. Disinfecting power is pH dependent.
3. Best spectrum of kill among common biocides.
4. Relatively cheap.

8.5.2   Types of chlorine sanitisers

Two basic types of chlorine sanitisers are available: inorganic or hypochlorite types and organic chlorine-releasing sanitisers. The hypochlorite sanitisers are most commonly formed by injecting chlorine gas into an alkaline solution such as sodium hydroxide and water. The net reaction forms sodium hypochlorite (NaOCl), sodium chloride (NaCl), and water:

$$2NaOH + Cl_2 \rightarrow NaOCl + NaCl + H_2O \tag{8.1}$$

Concentrated, liquid hypochlorite has a shelf life of about 1 year, but diluted solutions will degrade rapidly, within minutes or hours, especially at elevated temperatures. NaOCl will break down into the more germicidal hypochlorous acid (HOCl) as the pH is lowered:

$$NaOCl + H_2O \rightarrow NaOH + HOCl \tag{8.2}$$

Two precautions regarding hypochlorite sanitisers are that they are quite corrosive, particularly as the pH becomes more acidic, and that they should never be used under acidic conditions, because deadly chlorine gas will form.

Granular chlorine sanitisers are based upon the salts of an organic carrier which contains releasable chlorine. Chlorinated isocyanurate is highly stable, dissolves rapidly in solution, and releases one of its two chlorine ions to form sodium hypochlorite in water. Isocyanurate is a ring structure which contains a CN bond but does not contain or release cyanide.

Organic chlorine-releasing sanitisers are formulated with buffering agents which control the pH, corrosion properties, and the rate at which chlorine is released.

As with inorganic chlorine sanitisers, HOCl will form from organic chlorine-releasing sanitisers, the amount depending on the pH of the use solution. Because organic chlorine-releasing sanitisers contain these special agents when diluted with water, the use solution typically has a pH of around 6.5 to 8.5, at which point a significant amount of HOCl forms. Thus, it is only necessary to use these types of sanitisers at 100 ppm of available chlorine. On the other hand, hypochlorite sanitisers, upon dilution, have a typical pH of about 10 at which virtually no HOCl or high levels of the less effective hypochlorite ions remain. Therefore, it is necessary to use 200-ppm available chlorine when using an inorganic, hypochlorite type of sanitiser.

8.5.3 Advantages of chlorine sanitisers

1. Very fast acting and effective against both Gram-positive and Gram-negative bacteria, yeasts, bacterial and fungal spores, and many viruses.
2. Oxidising power enables them to help brighten surfaces, leaving them with a clean and shiny appearance.
3. Very cost effective.
4. Available in liquid or granular form.
5. Relatively stable in hard water.

8.5.4 Disadvantages of chlorine sanitisers

1. Corrosive, especially under acidic conditions. Hypochlorites are more corrosive than organic chlorines.
2. Organic soils severely reduce disinfecting power.
3. Heavy metals, elevated temperatures, and ultraviolet light accelerate the degradation of chlorine compounds.
4. Irritating to skin.
5. Unstable use solutions, especially at elevated temperatures.

8.5.5 Iodine sanitisers

Iodine is not very water soluble and tends to stain materials. To overcome these problems, elemental iodine is normally complexed with surface-active agents, which serve as carriers and solubilisers. The complex is usually prepared in an acid medium. The general reaction of an iodophor in solution is:

$$\text{Surfactant-}I_2 + H_2O \rightarrow H^+ = I_2 + HOI \qquad (8.3)$$

As with chlorine, the potency of iodine sanitisers is pH dependent. The efficacy of iodophors is greatest around 2.5 to 3.5 and declines sharply above pH 5.

8.5.5.1 General properties

1. Iodophors are very effective at low concentrations against a broad spectrum of microorganisms, including Gram-negative and Gram-positive bacteria, yeasts, moulds, bacterial and fungal spores, and many viruses.

2. The activity of iodophors is mainly due to the amount of free iodine that exists in equilibrium with the bulk of the inactive complex. As free iodine is consumed, additional iodine is rapidly released from the complex. Thus, although a relatively small amount of iodine is available at any one time, a large reservoir of iodine is readily available.
3. Because the formation of free iodine is favoured under acidic conditions, iodophors are most active as acid sanitisers. As the pH is raised, less free iodine is available, whereas the formation of hypoiodous acid, a less potent species, is favoured.
4. Typically, the in-use concentration for surface sanitising is 12.5 ppm. When applied to surfaces by recirculation, 12.5–25 ppm may be used; 25 ppm is used for disinfection or spray sanitising.

8.5.5.2 Advantages

1. Broad spectrum of kill at low concentrations.
2. Very fast acting and easily penetrates cell walls of microorganisms.
3. Maintains activity at cold temperatures better than most other disinfectants.
4. Fairly stable, although vaporisation accelerates as the temperature is raised above 50 °C (120 °F).
5. Excellent for manual sanitisers and hand disinfectants because of high efficacy and nonirritation to skin.
6. Not affected by hard water.

8.5.5.3 Disadvantages

1. Efficacy is significantly reduced by organic soils, although to a lesser degree than chlorine sanitisers.
2. Because they are readily absorbed by various materials, staining of plastic hoses and gaskets can occur.
3. Upon agitation, iodophors tend to produce foam due to the presence of surfactants.
4. Less effective than chlorine versus bacterial spores and phage.

8.5.5.4 Mode of antimicrobial action

Iodine, like chlorine, is very reactive with many materials and readily undergoes oxidation–reduction reactions. It is, therefore, fast acting and able to penetrate the cell walls of a wide variety of microorganisms. Iodine exerts its antimicrobial activity via several mechanisms, including

1. Disruption of protein bonds, primarily attacking sulphhydral groups;
2. Inhibition of protein synthetic reactions;
3. Reaction with components of cell membranes, altering their chemical and physical properties, allowing leakage of cell constituents and damage to the structural integrity of the cell.

8.5.6 Quaternary ammonium compounds (quats)

8.5.6.1 General properties

1. Ammonium compounds in which four organic groups are linked to a nitrogen atom producing a positively charged ion (cation). The negatively charged ion (anion) is usually chloride.
2. Quats are generally used under neutral or alkaline conditions. Efficacy may be reduced under acidic conditions.

3. Quats are generally more effective against vegetative bacteria, particularly Gram-positive species; they are less effective against Gram-negative bacteria. Their activity against yeasts, moulds, and viruses varies with the particular quat, but they are, in general, less effective against bacterial and fungal spores.
4. Typically quats are used at 200 ppm of active material.
5. Water hardness can depress their activity.

8.5.6.2   Advantages

1. They are surface-active agents with good wetting properties, and when combined with certain detergents, can serve as combined cleaner/sanitiser.
2. Because they are noncorrosive and have good wetting, foam producing, and penetration properties, they are very effective for sanitising floors, drains, walls, external surfaces of equipment, and disinfection of can and bottle warmers or washers.
3. Organic soils have a smaller negative effect on efficacy than on chlorine and iodine sanitisers.
4. They are nonirritating to the skin.
5. They do not contribute odours or tastes.
6. Quats leave residual films which provide longer lasting sanitising action.
7. Low toxicity at normal use concentrations.

8.5.6.3   Disadvantages

1. Residues of quats could possibly kill or inhibit starter cultures used in fermentations or other biological reactions.
2. They have limited activity against Gram-negative bacteria.
3. They can be inactivated by anionic detergents.

8.5.6.4   Mode of antimicrobial action of quats

The most widely accepted theory is that the surface-active nature of the Quat enables it to coat cells disrupting cell walls and membranes. With the structural integrity of the cell disrupted, leakage of internal cellular components and metabolites occurs.

8.5.7   Acid anionic sanitisers

This type of sanitiser has proven to be highly effective in a wide variety of industrial applications. Acid-anionic sanitisers are negatively charged surface-active sanitisers. Formulations include inorganic and organic acids plus a surfactant. They are unaffected by hard water or organic soils. The dual function of acid is that it can be used for rinsing and sanitising in one step. Acidity, detergency, stability, and noncorrosiveness make them highly effective.

8.5.7.1   General properties

1. This type of sanitiser consists of an anionic (negatively charged) surfactant in an acid medium, commonly phosphoric acid. Alkyl aryl sulphonates and sulphonated oleic acid in phosphoric acid are examples of acid anionic sanitisers. Carboxylic acids (fatty acids) are sometimes incorporated as well.

2. They are fast acting against both Gram-negative and Gram-positive bacteria but not very effective against yeasts and moulds.

3. Antimicrobial activity is usually greatest at pH below 3; activity above pH 3.5–4.0 is minimal. Activity is reduced significantly at neutral or alkaline pH levels.

4. Antimicrobial activity is enhanced as the temperature is raised.

5. They are noncorrosive and nonstaining.

6. Acid anionics tend to produce foam, although the level will vary from very low to very high foaming, depending on the nature of the surfactant and other ingredients in the sanitiser.

8.5.7.2   Advantages

1. They provide rapid antimicrobial activity.

2. They serve as an acid rinse, which helps prevent scale formation, leaving surfaces bright and shiny.

3. They have good wetting and detergency properties.

4. They are noncorrosive and nonstaining, therefore safe to leave in equipment overnight.

5. They retain their antimicrobial activity in the presence of hard water and organic soils better than most other types of sanitisers.

6. Being acidic, they are well suited for use in the presence of carbon dioxide.

7. They can be applied by either recirculation or spraying, depending on the foam characteristics of the sanitiser.

8.5.7.3   Disadvantages

1. Selective biocidal activity against yeasts and moulds.

2. Activity can be neutralised by cationic surfactants and alkaline residues.

3. Activity is limited to low pH conditions.

8.5.8   Peracetic acid

8.5.8.1   General properties

1. Disinfectants based on peracetic acid (PAA) are strong oxidising agents formed from equilibrium mixtures of acetic acid, hydrogen peroxide, PAA, and water.

2. Like chlorine, PAA has broad-spectrum biocidal activity against vegetative Gram-positive and Gram-negative bacteria, yeasts, moulds, bacterial and fungal spores, and many viruses.

3. PAA has widespread application in such diverse operations as soft-drink plants, dairies, and processed foods.

Its widespread use is due to not only its powerful germicidal action but also its suitability on most types of surfaces. Properly diluted, PAA may be used to sanitise stainless steel pipelines, tanks, fillers, crushers, filtration systems, utensils, and road tank trucks. It can also be applied to mild steel, soft metals like aluminium and copper, and certain plastics. However, concentrated PAA should not be allowed to be exposed to these types of materials (copper or other soft metals) as it is highly reactive and will cause corrosion and the rapid evolution of gas and heat. PAA, like any acid, becomes highly corrosive in the presence of chloride ions (about 50 ppm).

PAA most often is applied by CIP/circulation procedures because it is effective and does not foam upon agitation. Equipment parts and utensils may also be immersed in

PAA baths. Spraying is sometimes used, although this can be hazardous to handlers and/or workers who are at risk of being burned by spray bouncing off equipment or walls, ceilings, etc. Like other acid sanitisers, PAA is unaffected by carbon dioxide, hence its suitability for use in areas in a production plant where the gas is present, e.g. fermentation lines/tanks and beverage carbonators.

8.5.8.2 Advantages

1. Broad spectrum kill.
2. Noncorrosive to stainless steel.
3. Nonfoaming upon circulation.
4. Safe degradation products: acetic acid, hydrogen peroxide, and oxygen.
5. Contains no phosphates.
6. Active under carbon dioxide atmosphere.
7. Effective in hard water.

8.5.8.3 Disadvantages

1. Unpleasant, pungent odour.
2. Irritating to eyes, skin, and throat.
3. Rapidly decomposes at high temperature (above 60 °C), generating gas and heat.
4. Highly reactive with rust, other metal contaminants, organic soils, alkaline residues causing a loss of activity.

8.6 Successful application of CIP chemicals

In addition to understanding the chemistry of cleaning and sanitising chemicals, the successful application of cleaning compounds via Clean-in-Place (CIP) procedures also depends on an understanding of the types of soils to be removed and equipment design.

8.6.1 System design

Successful CIP cleaning of tanks, vats, lines, and processing equipment depends on the location and design of spray devices. The permanently installed fixed-ball spray is preferable over rotating and oscillating spray devices. Its advantages include: no moving parts, stainless steel construction, performance is not affected greatly by minor variations in supply pressure, and it sprays the entire surface all of the time. Fixed-ball sprays are available with a variety of characteristics in terms of flow rate, discharge pressure, and pattern of coverage. Experience has indicated that cylindrical and rectangular tanks can be adequately cleaned if sprayed at 0.035–0.106 L/min/m^2 (0.1–0.3 gal/min/ft^2) of internal surface, with patterns designed to spray the upper third of the tank. If considerable appurtenances, such as heating or cooling coils and complex agitators, exist in the tank, some special patterns may be required to cover these surfaces with resultant increases in the total flow rate required.

Vertical silo-type tanks may be cleaned satisfactorily at flow rates of 2.8 to 3.4 L/min/m (2.5–3.0 gal/min/linear ft) of tank circumference. Nonclogging disk sprays are used in vessels of this type because of the relative difficulty in reaching the spray devices for occasional inspection and cleaning.

8.6.2   Example of CIP programs

CIP programs are written for specific applications, i.e. equipment and soils; no one procedure will suit all applications. When a cleaning protocol is chosen, various factors need to be considered, including volume of product needed, cleaning time, soil type, and condition. Examples of typical programs which may be used for different soil or environmental conditions follow.

8.6.2.1   Lightly soiled

1. Prerinse with water 5–10 min.
2. Wash with a chlorinated alkaline detergent; 0.5%, 20–30 min at 60–66 °C (140–150 °F).
3. Rinse.
4. Acid wash or sanitise according to directions.
5. Rinse, if required.

8.6.2.2   Heavy burnt soil

1. Prerinse with water, 5–10 min.
2. Wash with built caustic 1–4% at 70–80 °C (160–180 °F) for 20–80 min.
3. Rinse.
4. Acid wash; 0.5–1.0% at 66–77 °C (150–170 °F) for 20–40 min.
5. Rinse.

8.6.2.3   Acid/override program for heavy soils

1. Prerinse with water; 5–20 min.
2. Circulate 0.5% phosphoric acid product for 20 min at 60–70 °C (140–160 °F).
3. Override with 2–4% caustic, circulate 30–90 min at 60–70 °C (140–170 °F).
4. Rinse.

8.6.2.4   Acid cleaning (i.e. for vessels under CO_2, counterpressure)

1. Prerinse with water; 5–10 min.
2. Circulate 1% acid 20–30 min at ambient temperature.
3. Circulate a peracetic acid product 0.25% (optional).
4. Rinse.

Note – Concentration should always be determined by titration, not pH.

8.6.3   Safety and operation of CIP systems

The use of a CIP system and the application of hot sanitation have inherent potential hazards resulting from the presence of voluminous flow rates of high-temperature water and cleaning solutions. Adherence to certain important operating techniques, procedures, and precautions will substantially eliminate these hazards. The best way to protect personnel from the dangers of hot sanitation, such as hot equipment and splashing water, is to train and educate workers and systems operators. In addition, signs should be posted to warn people about these dangers. Equipment should be safe-guarded from potentially harmful effects of hot sanitation, which occur due to igno-rance and neglect. For example, it is tempting to use hoses for a fast and easy means of connecting process equipment and piping into the desired CIP circuit. Hand-piped cup jumpers should be used instead. If hoses absolutely must be used in the circuit, hose material must be checked for high-temperature compatibility and hose fittings should be vulcanised into hose ends.

It is recommended that a temperature–time recorder/indicator, which has the tem-perature sensor in the return pipeline, is included with the CIP system. This will give a history and serve as permanent record of what the CIP circuit has experienced.

8.6.4 Automated dispensing of cleaners and sanitisers

Cleaning and sanitation chemical allocation and dispensing systems are available in the market to simplify procedures, to control chemicals and formulations, and to improve safety of the cleaning/sanitation operations. These systems are built with a programmable controller and a modular pump station. Every cleaning and sani-tation application used in a plant receives a code which is preset in the controller. The required code is entered for the specific application and the controller activates the pump station, feeding from concentrated formulas and a water and air deliv-ery station, to deliver the use solution. Daily, weekly, and monthly printed sum-maries of chemical usage can be obtained by connecting the controller to a printer (Anonymous, 1994).

8.7 Equipment cleaning procedures

The following suggestions for equipment cleaning procedures should result in reduc-tion of bacterial growth on equipment and minimised contamination of food being processed.

1. Before cleaning, dismantle or open the equipment as far as possible.
2. Disconnect lines where possible, or open cut-outs to avoid washing debris from one piece of equipment to the next.
3. Remove as much waste as possible with brush, shovel, broom, or other appropriate tool.
4. Rinse surfaces to be cleaned with water to remove food residues.
5. Clean surfaces with hot water with an added detergent specifically formulated for the removal of a particular type of soil. Use cleaning aids, such as high pressure or brushes, to remove tenacious deposits.
6. Complete the cleaning by thoroughly flushing with hot water to remove detergent residues and finally rinse with cold water to cool equipment below 27 °C (80 °F).
7. Allow equipment to drain and air dry.

8. Do not depend on high-pressure steam to sterilise equipment; in many instances, the steam spreads contamination by blowing it from crevices and cracks onto equipment which has been cleaned.

9. Before resuming operations, sanitise equipment by rinsing, or spraying with a 100–200 ppm chlorine solution.

10. In the same manner, sanitise water pipes used for recirculating wash water and for pumping peas, corn, etc. as well as brines and syrup.

11. Avoid contamination of equipment by spatter from floors or from contaminated equipment.

12. Keep hoses used for rinsing equipment off the floor.

13. Make sure that water used for brine, syrups, and for cleaning is free from contamination. Water storage tanks must be frequently drained, cleaned, and sanitised to eliminate bacterial buildup.

14. Thoroughly backwash and regularly sanitise water filters and water softeners. Microbiological buildup accumulation and actual growth is very common in this equipment.

15. Eliminate dead ends in water pipes, brine, and syrup pipes and pipes used for transferring foods from one piece of equipment to another.

16. Eliminate dead ends in flumes, sharp curves, bad solder, and welded joints.

17. Provide in-plant chlorination and maintain a chlorine residual of 1 ppm in the plant water supply. Provide controls so the chlorine content can be increased to 25 ppm or higher during cleaning operations.

18. Eliminate scale from the surfaces of pipeline blanchers, water pipes, and equipment surface. Such deposits may harbour thermophiles and other types of microorganisms.

19. Keep viners clean to reduce the contamination of peas and lima beans.

20. Boxes, bins, etc. must be in good repair and washed after each trip to the plant. Rinsing boxes and bins with a chlorinated final rinse is recommended.

21. Clean and sanitise corn huskers and cutters daily.

22. Replace wooden husker and cutter bins with metal ones, and clean and sanitise daily.

23. Keep cooling tanks clean and chlorinate cooling tanks or canals. Satisfactory chlorine residual is 2–5 ppm.

24. During canning and freezing operations, periodically rinse equipment, conveyor belts, and picking tables with water to prevent accumulation of debris, thereby physically removing large numbers of microorganisms.

25. During a breakdown, rinse off equipment and cool below 32 °C (90 °F) to arrest bacterial growth.

26. During short-period shutdowns, keep washers, dewatering screens, blanchers, and similar equipment running and cool down to below 32 °C (90 °F).

27. Use only sugar, starch, salt, spices, and ingredients which have been tested and approved by a reliable laboratory for canning or freezing purposes.

Pressure cleaning systems, either centralised or portable units, are recommended because they reduce labour, improve safety, lower operating expenses, and improve quality control.

8.8 Sanitation and plant personnel

Good health of the plant personnel is important. Testing of the personnel for communicable diseases would be desirable but is not feasible. Persons with wounds or with respiratory diseases should not be allowed to work in food-processing plants. Persons known to have had communicable diseases should be tested before they are allowed to work to have assurance that they are not carriers.

Food-processing plant personnel that handle food with their hands must, whenever possible, wear appropriate gloves. Human hands contain microorganisms that can be classified in two groups: transient flora and resident flora. The resident flora is found in hair follicles and skin crevices, whereas the transient flora is mainly on the skin surface. Transient flora is removable with detergents, whereas resident flora is more difficult to remove and requires the use of disinfectants aided by brushes with stiff bristles.

The disinfectants that are used are mainly phenols, iodophors, and quaternary ammonium compounds. Among the phenols, hexachlorophene is used and has some residual microbiological effect for up to 24 h. Phenols are most effective against Gram-positive bacteria but not effective against Gram-negative bacteria. *Salmonella* are Gram-negative. Because the effectiveness of phenolic disinfectants is reduced by the presence of food residues on the hands, wash the hands first with a detergent to eliminate food particles. Iodophors are active against most Gram-positive and Gram-negative bacteria but should be used at pH 5.0 or below. Quaternary ammonium compounds also act against Gram-positive and Gram-negative bacteria, but their action is neutralised by soaps, food materials, and hard water. Because Quats are more active in the presence of organic matter than any other class of bactericides, soft or softened water must be used, and hands should be washed first. If a detergent is used for the preliminary wash, it must be removed by thorough rinsing of the hands with abundant water.

It is recommended that the hand-washing area for plant personnel be located where the supervisor is able to observe whether the hands are properly sanitised. Plenty of hot water of at least 65 °C (150 °F) should be provided to mix with cold water. As a disinfectant, a solution of 200 ppm (0.02%) of a quaternary ammonium compound is recommended. This solution should be prepared fresh three times during each 8-h work period.

After washing hands with soap and warm water and rinsing, hands should be dipped in a sanitiser.

Drying should be done with hot air. It may be difficult to get plant personnel into the habit of following this hand washing and sanitising procedure, but it is an effective way of preventing infection of food with microorganisms that are present on the hands, and it helps prevent the infection of wounds, which could cause infection of food with pathogenic microorganisms.

8.9   Machinery mould

Machinery mould (*Geotrichum candidum*) is a ubiquitous microscopic organism. It is impossible to economically preclude it from most commercial food-processing establishments. Once introduced into food-processing areas, it will begin to grow. Optimum conditions for growth include constant moisture, temperatures between 20 °C and 30 °C (68 °F and 86 °F), a pH range of 2.0 to 8.5, and a source of nutrients, especially fruit acids and sugars, and vegetable starches. All of these conditions are commonly found in fruit and vegetable processing plants.

Geotrichum grows readily on all surfaces contacted by product residues and juices, such as conveyor belts, equipment walls and supports, floors, gutters, and building walls; hence, its name, machinery mould. Initially, it adheres to surfaces; as growth progresses, the mould develops into small greyish-white clumps. Profuse growths are visible as slimy films. It has been reported that machinery mould becomes visible when counts approach $2300/cm^2$ (15,000 per square in). In these stages, particles of mould can readily be wiped off by, and mixed with, raw product. If growth is allowed to continue unchecked, the slime will produce a rotten odour. The rate of growth will depend on environmental conditions.

Machinery mould by itself has no public health significance. However, because the organism may be visible in clumps of growth, felt as slime on equipment surfaces, or, after prolonged growth, evidenced by a rotten odour, machinery mould has been used by regulatory inspectors as a rough index of plant sanitation. Its detection on food-processing equipment has been used as confirmatory evidence to support inspectional reports of products having been packed under insanitary conditions.

Industry should view machinery mould as an indicator organism, i.e. an index of plant sanitation. Emphasis should be placed on maintaining sanitary conditions through the implementation and conduct of a well-planned program. Special attention must be given to known problem areas. Frequent and thorough sensory evaluations should be made of the entire facility. Whenever machinery mould, or any sign of insanitation, is seen, felt, or smelled in any area, reevaluation of the sanitation procedure for that area is indicated. With proper equipment, supplies, and training, all diligently applied to keeping the plant clean, machinery mould should not be a matter for undue concern.

8.10   Control of insects

Early detection and control procedures are necessary in plants, warehouses, and in transit if products are to meet the standards required by the public, the industry, and regulatory agencies. Most food packages are resistant to insect invasion. Before the shipper is accused of problems encountered, the manufacturer must assume the responsibility of preventing insect contamination in the manufacturing process. Because insects on the outside of a package are also unappealing to consumers, it is important to prevent insect infestations in warehouses, as well as in the manufacturing plant itself.

The following are suggestions for assisting the pest control program in dealing with insect problems. In most cases, these suggestions are just good housekeeping practices which will improve the overall efficiency of the operation. The entire plant should be considered as a unit when dealing with pest problems. This includes the outside storage area, lot, trees, and sometimes neighbours. Even if the pest control program cannot control all insect-producing factors surrounding a food manufacturing plant, usually any improvements will be rewarding.

8.10.1   Outside the plant

Garbage and Trash Facilities: Do these meet the needs of the plant? Is material picked up or burned at regular intervals? Is the storage area for garbage and trash adequately drained and regularly cleaned? Are the containers covered and stored on an impervious base? Does the design and construction prevent pests from developing or hiding in the area?

Immediate Surroundings: No insects should be permitted inside a food-handling plant. Because ants, certain beetles, bugs, millipedes, and other insects thrive on vegetation and live in the soil, infestation can be discouraged by maintaining a vegetation-free barrier approximately 60 cm (two ft) wide around the foundation of the entire plant. Treat this border periodically with a residual insecticide. This vegetation-free barrier can be easily incorporated into the landscaping plan.

Neighbours: Livestock operations, riding stables, and kennels may produce flies and other insects that may migrate into the plant.

Building Exterior: Although it may not be possible to prevent all insects from entering a plant, proper construction and maintenance will ease the insect control problems inside. Doors and windows should fit tightly, have screens, and be checked frequently to see that they close tightly. The fresh-air intake should have screens to keep out unwanted pests. Make sure that frequently used doorways are self-closing and that areas screened and/or protected by air curtains are in proper working order. Double doors may be of value in some situations.

Birds around any food-handling establishment are undesirable. They should be screened out and prevented from nesting or roosting on the structure. A roof that tends to accumulate food debris from exhaust fans and other equipment may attract various pests. These should be cleaned regularly. Other structures, buildings, storage sheds, etc. should be inspected frequently to determine whether they harbour insect pests.

Lights attract many flying insects. If outdoor lighting is required, place lights away from the plant so they are not directly above windows or doorways. Inside the plant, do not hang lights directly above processing areas, for insects flying to these lights could fall into the product. Place light traps so they will draw flying insects away from processing areas. Clean out the light traps regularly to avoid the buildup of dead insects which serve as food for more serious pests like dermestid beetles.

8.10.2   Inside the plant

Sanitation problems generally arise with crowded conditions or space limitations, because material is allowed to be stored 'temporarily' in corners, walk spaces, and under steps. These areas are then not cleaned regularly. In warm rooms and during the summer, this can mean the beginning of an insect problem.

Even though pest control programs usually have to deal with buildings and premises as they are, minor changes, which would simplify sanitation and insect control can often be made, during routine maintenance at little or no extra cost.

Walls and Floors: Be sure porous walls such as cement blocks, bricks, and wood are painted, preferably with enamel, concrete, or epoxy paint. Repair cracks, especially those at the floor–wall junction. Expansion joints, baseboards, drains, light switches, and other fixtures should all be tight fitting.

Ceilings: Ceilings, when improperly installed or maintained, can be a source of infestation. Voids above ceilings should be tightly sealed. If not, be sure they are cleaned regularly. Dust should not be allowed to accumulate on the topside of exposed beams. If dust is from a food product, this can attract insects.

Pits, Elevators, Motor, and Pump: These areas should be regularly cleaned. Delegate the responsibility of keeping these areas sanitary to maintenance workers.

Plumbing: In addition to following municipal and state building codes which regulate plumbing, plant pest control programs need to pay special attention to drains and broken traps. These can be sources of insect problems. Condensation from pipes can create rot and mildew on food packages and may encourage insect infestation.

Ventilation: Openings should be screened. Food residues, such as grease or dust accumulating around exhaust fans, should be cleaned routinely.

Processing and Storage Areas: Food-processing and storage areas require the strictest insect control. Nevertheless, consider the plant as a unit, because infections can originate from incoming products or by migration from other areas in the plant. By following good manufacturing practices, insect contamination of food products can be kept quite low. Even one undetected infestation can be extremely costly in terms of products condemned or diverted to animal feed.

The frequency and type of cleaning required, such as steam or vacuum, varies with the type of manufacturing. All employees should help maintain an insect-free environment in a food plant; however, designating one individual for pest control in each area might be a good idea. This individual may require additional training and equipment to perform his duties adequately. Screens, filters, and dust collectors should be serviced regularly, or daily, according to a published schedule. Food-handling machinery should be cleaned regularly, according to manufacturing procedures outlined by regulatory agencies. Dispose of trash, garbage, and refuse every day.

Receiving Areas: Handle empty bottles, packages, bags, drums, and liners as carefully as food. Because these are usually purchased from specialised suppliers, the employee receiving them should carefully inspect each shipment. Reject shipments with insect infestation or damage before the shipment infests the plant. Pay particular attention to multiwall bags and boxes, for insects can hide between them undetected.

Trucks and cars infested with insects should not be used for shipping. Even though the trucker or railroad might be liable for damages, the possible adverse publicity resulting from insect-contaminated products could be more costly in the long run.

Warehouse Insect Control: Allow sufficient space in storage areas so that all stacks of packaged food can be inspected. Leave at least 46 cm (18 in) of space between the walls and the stacks.

All floors, shelves, racks, cabinets, and pallets require regular cleaning, especially if food residues exist. Keep empty containers as clean as those storing food. Raw products, damaged goods that are returned for repackaging, and returned goods should be handled with the same care as the finished product. Don't overlook food material designated for animal feed, for this can be a potential source of infestation.

Non-Food-Handling Areas: Locker rooms, toilets, offices, cafeterias or lunch-rooms, and other non-food-handling areas should be routinely inspected for evidence of insects.

8.10.3 Find them early

The key to a successful pest control operation in a food plant is early detection. Most pests leave evidence of their presence, which a trained observer can detect before they multiply and do serious damage. To assist in early detection, look for the following signs:

1. Evidence of rodents: tracks, droppings, broken containers, and spilled products.
2. Evidence of birds entering the plant: feathers, droppings, nesting debris, and open or broken windows.
3. Evidence of insects: insect webbing, spider webs, live or dead insects, parts of 'skins' of insects, holes in packages.

Some insects are nocturnal (cockroaches and silverfish). It may be necessary to check the plant at night or some other time when the rooms are dark. Storage areas should especially be checked for evidence of damage. Containers, packaging material, and burlap or cotton bags can be sources of infestation.

Acknowledgement

The author appreciates the contribution of Mark J. Banner, Ph.D., Diversey Corporation, to this chapter in the 13th Ed. Diversey Corporation, Livonia, MI.

Reference

Anonymous. (July 1994). Benefits flow from detergent-dispensing system. (Sanitizing operations at food processing plants). *Food Processing* 7, 69–71. Putman Media, Inc.

Food processing residuals treatment and disposal

9

9.1 Introduction

Satisfactory treatment and disposal of wastewaters from the processing of fruits and vegetables is a complex problem. Unacceptable water conditions arising from the discharge or organic wastes have led to the passage of stronger pollution control laws and greater activity by regulatory agencies in many countries. Water is commonly considered community property; use of water from streams and underground sources is a privilege not to be abused, and water 'borrowed' from the community supply must be returned in a condition and by methods which do not cause irreparable deterioration in water quality, danger to the environment, and/or inconvenience to the community.

Food-processing facilities must provide treatment and disposal facilities adequate to protect water quality, to be good citizens of our neighbourhoods and in the environment in which we live. Failure of a food processor to plan adequately will result in excessive costs for treatment or disposal of food-processing residuals. Assistance may be obtained from academic institutions, regulatory agencies, and competent environmental engineers.

Wastewater generated from agricultural and food operations is biodegradable and nontoxic but may have high concentrations of biochemical oxygen demand (BOD) and suspended solids; these characteristics set it apart from common municipal wastewater managed by public sewage treatment. Processing of food from raw materials requires large volumes of high-grade water. Vegetable washing generates waters with high loads of particulate matter and some dissolved organic matter. Animal slaughter and processing produce very strong organic waste from body fluids, such as blood and gut contents. Processing food for sale produces wastes generated from cooking which are often rich in plant organic material and may contain salt, flavourings, colouring material, and acids or alkali. Very significant quantities of oil or fats may also be present.

Food processors must meet the standards established under the law regardless of plant location and regardless of the capacity of the stream to absorb wastes without unacceptable deterioration (Clesceri et al., 2005).

9.2 Factors considered in waste disposal problems

The high organic strength typical of food-processing wastewaters is the principal reason for the difficulty encountered in their treatment/disposal. Raw untreated cannery wastes consist of small particles and sometimes discarded whole pieces of raw

A Complete Course in Canning and Related Processes. http://dx.doi.org/10.1016/B978-0-85709-677-7.00009-8

product, skins, and seeds suspended in water, which carries in solution the juices of the product being processed. As compared to domestic sewage, food-processing wastewaters are unusually high in sugars and starches, and the contaminant strength (BOD) is typically 5 to 25 times greater.

The pollution effect of food-processing wastewaters discharged into a stream will depend on the strength of the waste, the amount of dilution afforded by the stream, and the amount of oxygen present in the stream water. As long as oxygen is present, decomposition of the organic waste will proceed without harm to life in the stream. However, if the strength and volume of the wastewater is such that dissolved oxygen falls below levels prescribed as minimum by regulatory agencies, fish and other forms of aquatic life will disappear, foul odours will arise from sludge deposits, and floating scum will be present.

The goal of the food processor in treating this wastewater is to render it suitable for discharge into a stream without causing pollution, for discharge onto a land application system for disposal without causing a public nuisance, or for acceptance into a public sewage treatment plant with the capability and capacity to handle this type of wastewater.

9.2.1 Facility needs survey

Before plans and specifications for installation of wastewater treatment/disposal facilities are prepared, the following basic considerations must be investigated and evaluated to provide the most cost-effective system.

1. *Character of residuals.* It should be determined if the residuals contain materials hazardous, or potentially hazardous, to public health. It is also essential to know the concentration of organic solids in the wastewater and the relative concentration of suspended and soluble solids. An estimate should be made of the amount of suspended solids which could be removed by screening.

2. *Waste flow measurements.* It is necessary to know the volume and pattern of wastewater discharge which will require treatment. Wastewater flow must be measured (this can be accomplished with metres or weirs or by the bucket-and-stopwatch method) and maximum flows determined. Results should be calculated in terms of daily, maximum 30 day, and annual volumes of wastewaters.

3. *Segregation of concentrated wastewaters.* The possibility should be considered of separating individual wastewater flows at the point of origin into contaminated waters requiring treatment and wastewaters with little or no contamination, which may be discharged without treatment. This latter group may include can-cooling waters, condenser waters, etc. This is an important point in the initial survey of the problem because the volume of wastewaters requiring treatment determines the size and type of the treatment plant. Reduction in the volume of the wastewaters results ultimately in capital cost savings.

4. *Study of possible reuse of water as means of reducing waste flow.* The reuse of water is discussed in Chapter 6. A study of water usage in a food-processing facility always reveals that more water is used than is necessary. Water reuse in certain operations can substantially reduce the amount of water used and thus reduce the volume of wastewaters required. This reduction in volume of the wastewater may not be accompanied by a reduction in the total organic loading but will make treatment of wastewaters more economical and effective.

Water reuse, however, should not be indiscriminate. Full regard must be given to the fact that improper reuse of water may lead to quality control problems.

5. *Study of means to reduce amount of gross solids added to wastewaters.* Unnecessary dumping of solid residuals into plant gutters must not be allowed, as this causes an unnecessary burden on initial screening, disintegration of larger solids into particles passing the screen, and leaching of soluble solids from gross solids, resulting in increased organic strength of liquid wastes which must be treated.

6. *Investigation of existing or proposed regulations governing disposal of industrial wastes.* The treatment given must render wastewaters suitable for acceptance into water pollution control plants (if the effluent is to be discharged into a public treatment facility) or suitable for discharge into an area receiving stream. If land disposal is the method of choice, then the effluent must be suitable for rapid oxidation and stabilisation without causing odour problems.

7. *Selection of treatment and disposal procedures which will accomplish the desired results.* Selection of the method of treatment and disposal should be based on a consideration of the space available for the treatment plant site, the effect of climate on the method of disposal, and whether a continuous or batch-type treatment system is desirable.

8. *Consideration of possible food-plant expansion.* Finally, the question should be answered of whether the method of treatment selected and the specifications for equipment to carry it out are commensurate with possible long-range plans for expansion of food-processor operations.

9.3 Methods of treatment and disposal of wastes

9.3.1 Primary, secondary, and tertiary methods of waste treatment

At present, there are two basic ways of treating wastes prior to discharge into streams: primary and secondary treatment. In primary treatment, solids are allowed to settle, or are screened out, or both methods may be used to separate solids from the liquid waste. Secondary treatment, a further step in purifying wastes, uses biological processes to reduce dispersed solids and soluble organic content of liquid waste. Secondary treatment removes up to 90% of the organic matter in wastes by making use of bacteria. Secondary treatment may also include modifying the pH of waste by addition of alkaline or acidic substances to bring the waste to a pH value between 6.0 and 9.0.

Other more advanced methods of treating wastes take up where primary and secondary treatments leave off. These more advanced methods are generally called tertiary treatment methods. One of these processes aims at getting rid of dissolved refractory organic substances. As the word indicates, this stubborn organic matter persists in water and resists conventional biological treatment. By passing the effluent through a bed of activated carbon granules, more than 98% of the remaining dissolved organic matter is removed by adsorption. Another tertiary treatment is that of electrodialysis, by which salts from an effluent are removed from water by the action of an electric field. These tertiary methods are, however, quite complex and costly. If cost is disregarded, it would be possible to supply any quality of water for any reuse from canning or other industrial plant wastes.

9.3.2   Removal of gross solids by screening – primary treatment

Food-processing wastewater usually contains two types of solids: one which is particulate and in suspension and one which is in true solution. Regardless of the ultimate disposal method, it is desirable to remove by screening as much of the suspended solids as is economically feasible. Failure to screen wastewaters causes unsightly conditions in streams due to floating solids from which foul odours may be produced, overloads of biological treatment systems, and problems in pipelines conveying the wastewaters.

Screening is required as a minimum by most states as a pollution control measure if the wastewater is to be discharged to public sewers, and no intentional maceration, grinding, or comminution of gross solids in order to pass them through a regulation-size screen is allowed.

9.3.3   Screen types in general use

The size, shape, and nature of the solid particles to be removed from the water have a definite effect on the efficiency of the screening operation. Wastewaters containing fibrous materials are difficult to pass through finer screens.

Rotary drum screens: These screens are essentially revolving drums covered with a screen cloth varying in fineness, depending on the product being processed. Liquids and solids enter the drum through an open end; liquids pass through the screen and solids are retained on the inside of the cloth. The solids are then scraped from the cloth as they are elevated from the water level and pass over a refuse trough; from the trough, solids are usually removed to a hopper by screw conveyor.

Vibrating table-type screens: These screens have been widely used. Finer screen cloths can be used than would be possible with drum-type screens, and the amplitude of the vibrations can be adjusted for differences in the material being screened. Vibrating screens may be purchased in different sizes; the size and number of screen units required will depend on the volume of waste to be screened.

Rotating circular type screens: These screens are drum-type screens covered with wedge wire in which wastewater passes from outside to inside with solids removed and retained on the outside surface; retained solids are scraped off the surface onto a conveyor.

9.3.4   Disposal of gross solids removed by screening

Generally, solid residuals from processing are a liability to the processor. Often, however, ingenuity can reduce the costs of their disposal. Sometimes a processor is able to arrange for the solid to be hauled away for animal feed; its value as feed is usually considered payment for the hauling. Disposal may be by spreading on agricultural land or hauling to landfills or compost systems. In some urban areas, canners have to pay on a tonnage basis to have solids removed. In these instances, it is desirable to have the waste as dry as possible, so after screening the solids should be pressed to further reduce the water content. Vibrating-type screens give drier solids than revolving-drum screens. Elevation of the solids to a hopper facilitates loading into trucks and allows considerable dewatering during storage in the hopper.

9.3.5   Treatment/disposal of screened wastewaters

The final volume of screened liquid waste which reaches the point of discharge should be as small as possible, consistent with high-quality food-processing quality control needs.

Screened wastewaters must be provided additional treatment. The most common methods are as follows:

1. Discharge to municipal sewage treatment systems
2. Biological treatment
3. Chemical precipitation of solids
4. Aeration flotation
5. Land irrigation.

9.3.6   Need for equalisation tank

Whatever the method of final treatment provided, it is desirable to collect screened wastewaters in an equalisation tank before treatment. Passage of the wastes through such a facility will furnish emergency storage in case of equipment failure, make possible a uniform flow to the treatment/disposal facilities, and allow blending of the fluctuating organic or chemical compositions of the wastes. In the equalisation facility, provision should be made for sludge removal and aeration by some means to prevent bad odours.

9.3.7   Discharge to municipal sewage treatment plant

Discharge of screened raw food-processing wastewaters into a municipal sewage treatment plant is one desirable disposal method, if satisfactory arrangements can be made. Typically, however, the seasonal nature of food-processing operations and the high pollution strength of the wastes may cause serious problems. Generally, public officials are not anxious to receive food-processing wastewaters.

The average food-processing wastewaters are 5–30 times stronger than ordinary domestic sewage. The nature of pollutants in these wastewaters may also upset conventional sewage treatment processes. Sugars and sugar-like compounds in liquid wastes from fruit and vegetable processing require a biological process for stabilisation which may not be desirable for the complete stabilisation of domestic sewage. It has been observed that in cases in which the volume of domestic sewage is not large enough to give high dilution of the food-processing wastewaters, and as municipal treatment systems are not specifically designed to treat the high strength food-processing wastewaters, attempts to treat the combined wastes are not effective. However, many food processors are discharging wastes to large municipal systems in which treatment is completely satisfactory.

In contemplating the discharge of food-processing wastewaters to a municipal system, the type of treatment process used by the municipal treatment plant by both processor and municipality should be considered. Short-term biological treatment processes and physical/chemical treatment processes are reported to be particularly adversely affected by food-processing wastewaters.

9.3.7.1   The volume of food-processing wastewaters

The relationship between the volume of food-processing wastewaters and the volume of domestic sewage should be considered; that is, whether or not dilution of the food-processing wastewater will be high enough to prevent harm to the treatment process.

9.3.7.2   The capacity of the sewage treatment plant

Seasonal production of food-processing wastewater means that consideration must be given to the cost of increasing plant capacity for short-term treatment, in relation to the costs of other methods of treatment and disposal.

9.3.7.3   The type and cost of pretreatment of the cannery waste

In nearly all cases, screening of the food-processing wastewaters is necessary and, in many cases, pH adjustment of the processing waste is required before its discharge into the sewer.

9.3.7.4   The charges which the food processor must pay

The charges to the food processor for the privilege of discharging his wastewaters to the municipal treatment system are usually based on one or more of the following considerations:

1. *Initial payment.* An outright purchase of the privilege of waste-disposal services may be based on the cost of new equipment required to treat food-processing wastewaters with further service charges to cover operating and maintenance thereafter.
2. *Monthly charge for wastewater disposal services.* The charge for service may be monthly payments of a fixed sum, which is usually an estimate of the additional expense incurred by the Municipal Sewage Plant in treating food-processing wastewaters.
3. *Charges based on wastewater volume.* The most commonly used basis for charge is volume. Wastewater flow metres monitor the food-processing wastewater discharge with monthly billing based on the actual discharges. In the event wastewater flow metres are unavailable, it is assumed that the amount of water metred into the plant is the amount of wastewater discharged. Because a considerable volume of water is lost through evaporation or addition to the finished product, a compensating reduction should be allowed. Commonly, extra charges are levied for the extra strength of the food-processing wastewaters.

In some instances, clean water discharges to the sewer, such as cooling and condenser waters, are measured and the volume excluded from disposal charges. If charges are levied for clean waters, it may prove cost-effective to remove them from the wastewater flow and discharge them to storm sewers or drainage ditches.

9.4   Biological methods for treatment of food-processing wastewaters

Screened liquid food-processing wastewater is very amenable to treatment by biological methods, and high degrees of reduction in the strength of the wastewater may be achieved. However, treatment depending on biological oxidation should be attempted only after investigation of the variability and specific characteristics of food-processing wastewaters and careful consideration of the type of treatment process selected. All types of vegetable and fruit canning wastes will support the growth of microorganisms, which utilise the sugars and other carbohydrates present. Both anaerobic and aerobic treatment of food-processing wastewaters have been accomplished and successful large-scale operations have been installed utilising both types of treatment processes.

Anaerobic treatment (treatment in the absence of free oxygen) requires equipment that is more elaborate and more careful attention; for these reasons, it may be feasible only in selected cases. It has proven cost effective for very high-strength wastewaters (typically with BOD5, 5-day biological oxygen demand, concentrations greater than 5000 mg/L).

Aerobic treatment of organic wastes is more easily carried out. Commonly, a two-stage biological process is utilised. In the first stage, fresh screened wastewater is discharged into a tank in which air is continuously diffused into it. This supplies the free oxygen required by the aerobic bacteria in their biochemical action on the organic compounds. The wastes may be detained and aerated in a batch type or continuous flow operation. The efficiency of the process depends on building up in the aeration-digestion tank unit a vigorous culture of suitable microorganisms. Conditions in the treatment unit must be controlled to prevent inhibition of their growth. Good diffusion of air throughout the tank and adequate mixing must be maintained continuously. The pH of wastewaters may require the addition of small amounts of lime (or other neutralising agent); at times, some form of ammonia and/or phosphoric acid is added to the wastewater flow to achieve a suitable nutrient balance (generally believed to be BOD5:N:P of 100:5:1) in the aeration unit.

Effluent from the aeration unit flows into a settling facility where quiescent conditions allow settling of the microorganisms to form sludge that is removed and pumped into the aeration unit in which microorganisms are returned to 'work'. The clear supernatant flows on to the next treatment unit or to discharge into the stream.

9.5   Chemical treatment to remove suspended solids – secondary treatment

Situations have occurred in which removal of suspended solids beyond that accomplished by screening would render the waste acceptable for disposal to a municipal treatment plant. In some cases, chemical treatment of food-processing wastewaters has been the most convenient method to accomplish this.

Table 9.1 **Typical dosages for chemical treatment of vegetable waste**

| Type of waste treated | Chemical dosage (kg per 1000 kL of waste) | |
	1st Chemical	2nd Chemical
Pea waste	Lime, 0.8	Alum or ferrous sulphate, 0.4
Beet waste	Lime, 1.2	Ferrous sulphate, 0.5
Corn waste	Lime, 1.0	Ferrous sulphate, 0.9
	Lime, 0.7	Zinc chloride, 0.3–0.7
Tomato waste	Lime, 0.5	Alum, 0.1

Ordinarily, chemical treatment for food-processing wastewaters should be undertaken only after careful consideration of costs, inconvenience of handling large volumes of sludge produced, and failure of the method to give a reduction in strength of wastes of more than about 50%.

Controlled chemical treatment will remove suspended and colloidal solids but will not affect solids in solution. For this reason, the degree of treatment obtained will depend on the relative concentrations of suspended and soluble solids. Wastewaters high in sugar content cannot be greatly reduced in strength by chemical treatment.

Two types of chemical treatment used are the continuous flow method and the fill-and-draw or batch-type treatment. Each method requires use of the same chemicals. Although the continuous flow type of treatment will handle larger volumes of water in a given period, certain disadvantages are inherent in the process. It is difficult to maintain the optimum chemical dosage in a system in which fresh untreated wastewaters enter at one end and treated effluent is discharging at the other. It is also difficult to remove the large volumes of sludge produced.

The fill-and-draw method largely overcomes these disadvantages. With this method, screened waters are pumped into one of two or three tanks; when this is full, the flow is turned into a second tank. Agitation is started in the first tank and the proper amount of the first chemical is added, then half the dosage of the second chemical is added while agitation is continued. After several minutes, the remainder of the second chemical is added slowly until a large, heavy flow begins to settle out. When the supernatant liquid is clear, it is discharged and the sludge at the bottom is pumped onto sludge drying beds. Examples of typical dosages for chemical treatment are shown in Table 9.1. The approximate costs of treating waste may be calculated, if the cost of the chemicals is known.

9.6   Aeration–flotation process for removal of suspended solids

A method of wastewater treatment using the principles of aeration–flotation is used by some industries, particularly meatpacking and rendering plants. For this type of wastewater, the process is reported an efficient and economical method for byproduct recovery, pollution control, and water conservation.

The theory of aeration–flotation for removal of suspended solids is because the amount of gas or air which will dissolve in a liquid is directly proportional to the absolute pressure under which the liquid is maintained. At sea level and a temperature of 21 °C, water will dissolve approximately 2% of air by volume. As the pressure increases, the amount of dissolved air will increase.

If wastewater supersaturated with air and under pressure is suddenly released into an open tank at atmospheric pressure, the volume of air contained above saturation will be released from solution in the form of extremely fine bubbles. These bubbles will attach to particles of suspended matter in the liquid and carry them to the top, forming a layer of 'float', which can be removed by skimming. Chemicals such as lime and alum may be added to the wastewater before its passage through the aeration–flotation system. Formation of chemical flow in the liquid aids in removal of certain suspended organic solids.

Experiments with the aeration–flotation process have indicated a possible removal of 80 to 85% of suspended solids from peach-canning wastewater and 60 to 65% of the suspended solid from pumpkin-processing wastewater. The use of flocculating chemicals did not increase the effectiveness of the process in the treatment of these two wastewaters.

Although the treated effluent was comparatively clear and free of visible suspended solids, removal of the suspended solids did not greatly reduce the strength of the waste. Determination of the BOD_5 Test on treated and untreated samples showed an average reduction of 17% in the case of peach waste and seven percent for pumpkin waste. These results were expected, because soluble solids were responsible for the greater portion of the strength of these wastes. However, in situations in which removal of suspended solids is indicated, to improve the acceptability of food-processing wastewater for a particular final treatment, the process of aeration–flotation may be considered.

9.7 Odour control in cannery wastewater tanks

To aid in the prevention of offensive odours from tank wastewaters, the tank itself should be as large and shallow as the situation will permit. Septic conditions quickly develop in deep tanks. A waste depth of not more than three ft is preferred, and five ft is the maximum. The growth of weeds and grass should not be permitted in tanks, because they are of organic composition; their presence in the tank contributes to the odour problem.

In the presence of free oxygen, starches and sugars of food-processing wastewaters are changed by bacterial action into stable inoffensive compounds such as water, carbon dioxide, nitrates, sulphates, and inert materials. When oxygen is exhausted from the wastes, anaerobic bacteria continue the decomposition with the production of foul-smelling gases such as hydrogen sulphide, ammonia, and mercaptans. This anaerobic condition in treatment units will occur quickly, unless fresh waste is continuously added or available oxygen is supplied by other means.

One satisfactory method for odour control in treatment units is the addition of nitrate of soda, which supplies the available oxygen required for inoffensive decomposition of the wastes. Sodium nitrate of fertiliser grade is added daily to the fresh waste going into a treatment unit in an amount necessary to satisfy 20% of the five-day BOD. Other successful odour control chemicals include lime, hydrogen peroxide, chlorine, and similar oxidising agents.

9.8 Disposal of waste by spray irrigation

In recent years, spray irrigation has been used increasingly as a means of food-processing wastewater disposal; in many respects, it is an improvement over other methods. It consists of spreading liquid waste over the surface of the ground by means of a high-pressure sprinkler system. The rate of application used should produce no damage to vegetative growth and avoid surface erosion and runoff. The system usually requires the following:

- A mechanically operated screening unit
- A collecting facility for accumulation of screened waste
- Stationary screens to prevent clogging of the outlet from the tank
- A pump which develops the required nozzle pressure
- A force line for transporting waste to the irrigation site
- Lateral lines for distribution from the main line
- Self-activated revolving sprinklers
- The land on which to spray
- An actively growing cover crop to aid in absorption, prevent soil erosion, and enhance evapotranspiration.

With the proper equipment and controlled application of the wastewater, spray irrigation will completely prevent stream pollution, will not create odour problems, and is usually less expensive than other methods of waste disposal.

Selection of the irrigation site: The selection of land for the irrigation site is of major importance in this method of wastewater disposal. The location must be within practical pumping distance of the food-processing operation. Consideration should be given to the economics of pumping long distances, as compared to the costs of disposal by other means. The topography of the land and characteristics of the soil are important factors to consider. Spray irrigation of land which is not fairly level may not be successful because of run-off and erosion. The possibility of stream pollution from runoff should be considered. Depressions in the surface of the land may cause ponding of the wastewater, with odour production resulting from its decomposition.

The amount of land required to dispose of a given volume of waste is determined by the absorption characteristics of the soil, which are, therefore, an important factor in the success of spray irrigation. Typically, the standard soil percolation rate test is used to obtain preliminary information. Persons having knowledge of local soil conditions should be consulted.

Importance of the cover crop: When screened wastewaters are sprayed over land without a cover crop, soil erosion and runoff occur almost immediately; a cover crop is absolutely essential. The type of vegetative cover may be determined by how the land is to be used, either during or after the canning season. At some installations, cattle are grazed on the vegetative growth during spray operations; the cutting of hay from the land is also a common practice.

Other considerations: Failure to properly screen processing wastewaters is frequently the cause of difficulty in the operation of a spray irrigation system. The presence of gross solids results in plugging of the spray nozzles.

If odours are to be avoided in spray irrigating, the food-processing wastewaters must be applied to land while still fresh; this is a factor to consider in constructing the collecting facility. A facility which has the capacity to provide a long detention time will require supplemental aeration to keep wastewaters fresh; this can be provided with mechanical surface aerators or by diffusing air through the stored liquid.

The costs of spray irrigation: Because many variables affect the final costs of installing a spray irrigation system for cannery waste, it is not possible to make definite cost estimates for particular areas. The amount of land required will depend on the volume of wastewater to be sprayed and the capacity of the soil to absorb it.

9.9   Valorisation

Industrial processing of foods generates large amounts of solid waste/by-products and high volumes of effluents which still may contain valuable natural components. The costs for getting rid of them are high. Sustainable byproduct valorisation, i.e. the creative application of product and process development to achieve commercially successful products of added value, makes sense both from an economic and environmental point of view. Byproduct valorisation is a derivative of waste management (Pleissner and Lin, 2013).

Recovery of nutrients from food waste can be performed by chemical and biological/enzymatic methods. Food waste is a mixture of different kinds of food and it must be turned into a liquid form to be further treated. This can be done using enzymes or chemicals. After this, it is further treated to recover the desired components. It can also be used as a nutrient source in microalgae cultivation, which in turn can be used for feed or further processed into many different products. e.g. biodiesel.

Byproduct valorisation may involve various types of treatment and processing and products produced from what could be called 'waste'. Examples include purees, pastes, mixes, and juices; cut chilled or frozen vegetable and fruit pieces; and biochemicals and bioactives (e.g. antioxidants, biophenols, amino acids, vitamins, natural flavours, functional proteins, and pigments). Nonfood applications of waste include natural macromolecules (e.g. cellulose, starch, lignin, and enzymes), biofuels, and fertilisers.

Some waste recovery is feasible from most food-processing plants. To identify the most cost-effective products to recover, an audit of the entire process should be carried out by trained personnel and/or consultants.

Acknowledgements

The 13th Ed author acknowledges with thanks the contribution of Paul H. Russell Jr., P.E., DEE, JR Engineering P.C., Newark, NY.

References

Clesceri, L. S., Eaton, A. D., & Rice, E. W. (2005). *Standard methods for examination of water & waste water method 5210B*. Washington, DC: American Public Health Association, American Water Works Association, and the Water Environment Association.
Pleissner, D., & Lin, C. S. K. (2013). Valorisation of food waste in biotechnological processes. *Sustainable Chemical Processes, 1*, 21.

Part Three

Canning operations

Canning operations

10

10.1 Introduction

All food processing can be separated into specific processes that require specialised equipment and handling. When broken down to each operation, the process flow is relatively simple and can be easily controlled (see Figure 10.1).

10.2 Receiving raw products and packaging materials

All incoming raw materials, ingredients, and packaging components should be inspected upon receipt to ensure that they are suitable for processing.

Raw-material quality and end-product quality are directly related, and this holds true for fish, meat, fruit, and vegetables. Immediately after catching, slaughtering, or picking, the quality of the raw materials starts to deteriorate. This deterioration can be slowed down by various handling factors like storage temperature (lower temperature reduces the deterioration), careful handling (reducing any physical damage), air in the environment (reducing the amount of oxygen generally slows down deterioration), and the amount of moisture (including humidity) around the product.

All of these factors should be considered when purchasing, transporting, and receiving raw materials.

Raw materials should be received in an area separate from the processing areas. Prior to being placed in inventory, ingredients susceptible to microbiological contamination which would render them unsuitable for processing should either be examined for microbiological condition or received under a supplier's guarantee that they are of microbiological condition suitable for use in processing. Products should be held prior to processing in such a manner such that growth of microorganisms is minimised.

10.3 Separation of the edible portion

Many products, as received at the factory, require some special treatment to separate the edible portion, e.g. peas must be removed from the shells (vined); fruit must be peeled and cored; corn must be husked; fish must be cleaned. For each such operation, special machinery has been developed so that large volumes can be handled quickly. The canning season for each product is usually short and the canner must install sufficient machinery to meet processing requirements.

A Complete Course in Canning and Related Processes. http://dx.doi.org/10.1016/B978-0-85709-677-7.00010-4

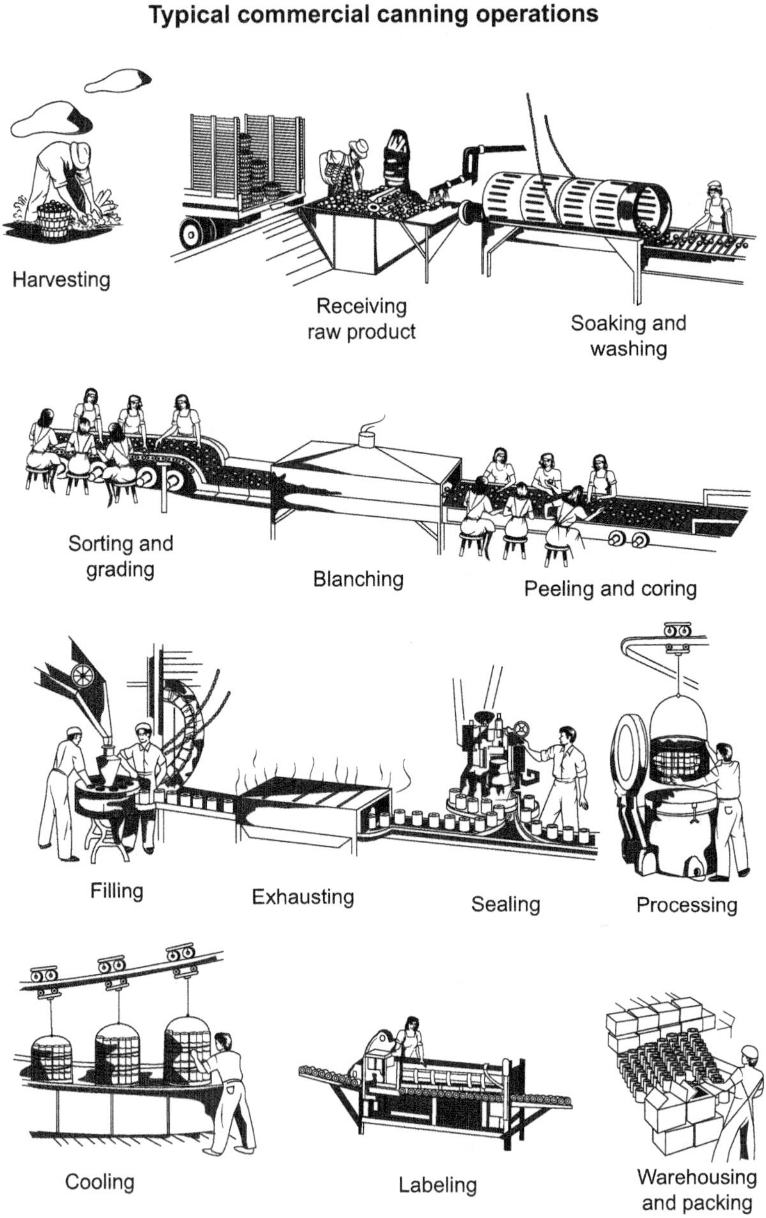

Typical commercial canning operations

Figure 10.1 Typical Canning Operations (courtesy former American Can Company) from the 13th edition of this volume.

10.4   Washing

When received at the cannery, most raw food materials are immediately and thoroughly washed. Washing obviously has the objectives of separating soil and foreign materials, but it also reduces considerably the load of spoilage bacteria naturally present in foods, which increases the effectiveness of the sterilisation process.

Washing is done by equipment in which products are subject to high-pressure water sprays or strong flowing streams of water while passing along a moving belt or while being tumbled on agitating or revolving screens. Sometimes a flotation type of washer is also used to remove chaff or other extraneous material. Sometimes washing is preceded by mechanical removal of soil and other fine materials that adhere to the food; stones and other heavy objects are separated.

Water conservation is important due to the shortage of water in some areas, the consequent increase in cost of water, and the cost of liquid wastewater disposal. The total expenditure for water may be a relatively large and significantly increasing cost item. In canning fruit, the volume of water required varies with the method of preparation of the fruit for canning and with the kind of fruit. Lye-peeled peaches require much more water than preparing plums or cherries, because very large quantities of water are needed for washing.

The rotary washer used in the lye peeling of peaches is very effective. It consists of a rotary drum, or series of several drums, each of which is equipped with an inner helical conveyor. These drums rotate in tanks of water in which the water is continuously or frequently changed. The spiral carries the fruit progressively through the different washing tanks, the first of which is contaminated with a small amount of lye from the lye-peeling tank; the last two tanks are filled with hot and cold water, respectively. The effectiveness of the rotary washer depends on the speed with which the product passes through the washer, volume of water used, the temperature of the water, distance of the sprays from the product, and the depth of the product in the washer. Many washers are overloaded when processing such products as tomatoes, with the result that much of the material does not receive the full force of the sprays. In many tomato product plants, a combination of roller, conveyor, and sprays is used. The conveyor is about 0.76 m wide and is made of bronze or stainless steel tubes about 76 mm in diameter, placed crosswise and moved by an endless link chain. As the conveyor travels, the tubes revolve, turning the tomatoes repeatedly and exposing them on all sides to the sprays. Two sets of sprays are used: one under very heavy pressure and the other under medium pressure. Beyond the sprays, the roller conveyor serves as a very effective sorting belt.

If the fruits or vegetables are agitated in water, the efficiency of the soaking process is greatly enhanced. A simple form of agitating device, often found in apple juice plants for washing apples, consists of a flume in which the apples are conveyed through a current of rapidly running water. Compressed air may also be used to agitate the water in tanks in which the fruits or vegetables are washed. In one method of washing spinach, the water in the tank is agitated by means of a pump. Some soaking vats are equipped with a propeller which may be in contact with the product; in this case, the propeller should move slowly to avoid bruising.

Washing of fruits and vegetables by means of sprays of water is the most satisfactory method; however, a product that is heavily contaminated with soil should be thoroughly soaked to loosen the soil before passing it under sprays. The efficiency of a spray of water for washing depends on the pressure of the water, on its volume, and on the distance of the spray nozzle from the product to be washed. The spray in which a small volume of water under heavy pressure is used is much more effective than the one in which a large volume of water under low pressure is employed.

Most spray washers consist of pipes that are fitted with hacksawed openings, but for pressures of water in excess of 140 kPa (20 psi), adjustable nozzles should be used to prevent unevenness and to direct sprays in the desired channels. Sprays are effective only if the water touches all parts of the surface of the product. One means of attaining this objective is to place sprays above and below the travelling woven wire-cloth conveyor. The same effect can also be attained by causing the product to roll over the spraying process. The most effective means of agitating the product under the spray is a revolving spray washing machine used on tomatoes and roots. This consists of a slightly inclined perforated drum, fitted on the inside with spirals or with corrugations. This type of washer is also effectively used in the washing of spinach.

In some canning procedures, operations not primarily intended to clean the raw material may exert a cleaning effect. Thus, blanching or scalding serves the additional purpose of cleaning the food, in addition to its primary objectives; the same applies to the water spray to cool foods after blanching.

10.5   Size grading

Many products are put over a series of screens having holes of different diameters. For products like green beans, a series of parallel bars with varying distances between them are used; some machines revolve and others vibrate. Each product has certain size-grading procedures that have been proven to give the most satisfactory results. Larger round units may be put in a long narrow trough, the bottom of which contains moving rollers, ropes, or cables which are close together at one end and further apart at the other end; the small units pass through first and larger ones toward the open end.

10.6   Inspecting

All products must be inspected before they go into the cans. Machine inspection of raw product checking for colour, texture, size, shape, and density can work extremely fast as the product flows over the process line. After foreign or undesirable objects are identified, they are ejected from a moving belt by pneumatically actuated flaps.

Although modern machines eliminate foreign or undesirable material, reduce human inspection, and improve plant productivity, personal inspection is still needed for proper assurance of freedom from imperfections. This is usually accomplished by passing the washed product over an inspection belt that is designed so that inspectors can sit comfortably close to and facing the belt. Both hands should be free for use. Special attention should be given to proper lighting; it should be ample, but diffused rather than glaring, and should be placed and shaded so that light does not shine in the inspector's eyes.

10.7   Blanching

Blanching is an operation in which a raw food material is immersed in water at 88–99° C (190–210° F) or exposed to live steam for a specified period.

The objectives sought in blanching are not always the same but vary according to maturity and product for one or more of the following reasons:

1. *Inhibition of enzymatic action.* Natural product enzymes are inactivated by blanching, and thus undesirable changes in colour and flavour are avoided, as well as reduction in the content of certain vitamins.
2. *Expelling of respiratory gases.* Raw fruits and vegetables contain intracellular gases, of composition similar to that of air, but somewhat higher than air in oxygen and carbon dioxide content. The release of gases prevents strain on can seams during heat processing and assists in the development of high vacuum in the finished product. Another desirable effect is a reduction in internal can corrosion by reducing oxygen content of can-headspace gases. Headspace oxygen acts as a depolariser in electrochemical corrosion reactions, thus increasing rate of corrosion.
3. *Softening of food.* Product becomes easier to fill in the container, and higher drained weights are obtained.
4. *Facilitating preliminary operations.* Peeling, dicing, cutting, and other preparatory steps are accomplished more easily and efficiently.
5. *Setting the natural colour* of certain products and preventing oxidative browning in others.
6. *Removing undesirable raw flavours* from food.
7. *Blanching also aids in cleaning the product.*

Blanching is usually accomplished in equipment especially designed for individual products. The equipment must be designed so that it is possible to subject raw materials to a particular temperature range for a proper period. The shortest blanching time that accomplishes the desired objectives usually gives the best product. Many vegetables and some fruits are blanched.

Continuous hot water blanchers are usually of the following two types:

1. *Immersion or conveyor type*, employing either screw or chain conveyors by which the product is moved through a tank of hot water.
2. *Hydraulic type*, called a pipe or tubular blancher, in which the product is conveyed in hot water through numerous lengths of pipe by a circulating pump and aided by steam jets.

Continuous steam type blanchers are mechanically more complex than hot-water types. They continuously move the product through a tank containing live steam using a chain or belt conveyor.

The main disadvantage of hot-water blanching is the large volume of water needed and its direct contact with the product, which may result in some leaching of water-soluble food constituents such as vitamins, minerals, sugars, and starch. Results of research on several specific nutrients indicate a wide variability in nutrient losses caused by either hot water or steam blanching.

The leaching of water-soluble substances also results in increased biological oxygen demand (BOD) of liquid effluents discharged from processing plants, increasing costs of disposal. There is obvious need for modifying processing operations to reduce and, whenever possible, eliminate the amount of pollution generated. On average, over 40% of total plant effluent BOD in vegetable processing is generated by blanching.

Hot-gas blanching has been demonstrated to reduce the volume of wastewater effluent from a blancher to less than 1% of that produced with steam or hot-water blanching.

In pilot plant studies, hot-gas blanching has been found applicable to several commercially important vegetables, with the exclusion of cob corn or beets. In hot-gas blanching, vegetables are heated by means of combustion or flue gases from a gas burner. Hot flue gases are recirculated upwards through the vegetables which are conveyed through the blancher between two wire-mesh belt conveyors. Steam is added to the recirculating flue gases to increase relative humidity, minimise partial drying, and facilitate heat transfer, which would otherwise result in weight losses. The data obtained indicate that hot gas-blanched vegetables have quality well within the range of commercial acceptability. Overall results from studies on nutrient retention show no significant difference due to type of blanching received by the vegetable samples investigated. It is reported that hot gas-blanched spinach and peas show higher retention of ascorbic acid (vitamin C) than either hot-water or steam blanching. However, product weight loss with hot-gas blanching has been found to be higher for many vegetables. Hot-gas blanching is still at the developmental stage for commercial application. (When a blanched food product is washed prior to filling, potable water should be used.)

10.8 Peeling

10.8.1 Steam peelers

In steam peeling, products are exposed to steam under pressure in a fully insulated retort which revolves slowly, mixing the vegetables while rotating. Steam requirements are held to a minimum. Once the vegetables are steamed, they are discharged into a hopper below the retort in which they are conveyed over rubber-coated rollers running at different speeds. This causes the vegetables to gyrate as they are pushed along the rollers by means of mechanical fingers. Water from high-pressure jets immediately above the rollers knocks the skins off as the product passes along the rollers and drops them into a hopper below the rollers, while the peeled product is discharged onto a conveyor for inspection and further processing.

10.8.2 Abrasive peelers

A tumbling action is utilised in abrasive peelers so that all surfaces of the vegetables undergoing peeling are exposed to a rubbing action against an abrasive surface, thus loosening the peel which is removed by water sprays. Some machines can be fitted with cylinder brushes which are interchangeable with the abrasive rollers. These brushes can be used for washing and, in some instances, will provide the only peeling necessary for thin-skinned vegetables, such as new potatoes. A combination of abrasive rolls and brushes is also an effective peeling medium for many products. In a number of operations, it is used following a steam or lye treatment.

10.8.3 Mechanical peelers

Mechanical peelers are used mainly with fruits. Equipment is specially designed for each application. Apples and pears are typical examples of mechanically peeled products.

10.8.4   Flame peelers

The flame peeler is a sloping, revolving cylinder approximately positioned on a 10–15% slope. The flame is produced at the lower end. Rotation is at 12–14 r.p.m. The product is gradually introduced at the higher end and moved by gravity turning at random, thus exposing all surfaces to direct contact with the flame. Generally, the product is exposed between 60 and 90 s in the flame peeler. The amount of fuel burned per unit of time and the draft in the chimney used to eliminate the gases resulting from combustion also affect the rate of efficiency of the peeling operation.

The charred and shrivelled skins are partially removed by passing the product, such as pimientos, through a rotary washer having strong water sprays which partially remove the peels and cool the product. Flame peeling is used almost exclusively for pimientos.

10.8.5   Lye peelers

The active agent in lye peeling is a water solution of caustic soda. Caustic soda, chemically sodium hydroxide, is often referred to as lye; it is one of the strongest commonly available alkali materials. Temperature, concentration, agitation, and the product holding time in the caustic solution should be controlled at all times.

For peeling root crops, it is recommended that caustic soda alone be used. Some small benefit may be gained by use a wetting agent to obtain a smoother peel on root crops and fruits (e.g. peaches, pears, apricots).

A lye-peeling operation requires a generous water supply, caustic soda, and a source of heat. Apparatus for continuous lye peeling is now sold by food-machinery manufacturers as a standard item. The use of copper, bronze, brass, aluminium, lead, zinc, and tin for fittings, cocks, valves, and other accessories must be avoided because they are attacked by caustic soda.

A preliminary wash of raw product is very desirable; this greatly reduces contamination of the lye bath with dirt, weeds, and other foreign material.

To condition the skin of exceptionally tough vegetables, such as storage beets, the product is preheated prior to the lye treatment. The requirements for certain types of processing make a slight precook necessary. A blancher will reduce heat consumption of the bath by raising the temperature of the product charged. This step is strictly optional for most operations and is undesirable for certain types of potato peeling. A lye peeler is essentially a heated tank for holding a solution of caustic soda with means provided for passing the produce through this tank at a controlled rate.

Two main types of peelers are used: one is a rotary or Ferris wheel type, and the other is known as a roto-screw peeler.

The rotary or Ferris wheel type peeler consists of a perforated drum equipped with angular vanes around the periphery. The lower portions of the drum assembly are surrounded by a shell-like container or tank used to hold the lye solution. Steam coils or other means are provided to control the temperature of the operation. As the drum rotates, raw produce is trapped by the advancing vanes and carried through the hot treating solution. This apparatus has the distinct advantage of requiring comparatively little floor space.

A *roto-screw-type peeler* consists of a long, narrow, shallow vat-like tank. It is equipped with a continuous link belt conveyor, to which are fixed cross members similar to that of a typical drag conveyor. The movement of these cross members or flights propels the produce through the lye bath.

After the lye treatment, vegetables must be water washed with pressure sprays to remove the lye-disintegrated peel. This wash is second only to the lye bath as a critical part of the lye-peeling process. In general, the washer is a revolving drum with a corrugated inner surface equipped with water sprays on its axis; pressure in the sprays ranges from 275 to 690 kPa (40 to 100 psi) or higher, preferably of the order of 550 to 690 kPa (80 to 100 psi) for rotary washers. The design of nozzles is critical because the water spray must loosen soft gelatinous tissue and yet not cut the product. Corrugations are built into the rotary washer to tumble the product and effect a rubbing action to help remove skin from the vegetables.

The washing operation for most root crops should be designed such that a large mass of product is retained in the washer at all times. This provides a good tumbling or rubbing action which aids in skin removal and leads to greater water economy. Properly designed and operated washers will yield product that is clean, smooth, and entirely free of caustic soda and/or other processing aids. Sometimes an acid dip is used following washing; some manual trimming may be necessary to remove major defects in the peeled and washed vegetables.

10.9   Size reduction

After peeling and washing, fruits and vegetables are often cut into a desired size and shape, such as slicing, crinkle slicing, dicing, shredding, granulating, comminuting, or strip cutting.

10.10   In-plant handling of cans and can ends

In small-volume canning plants, containers may still be handled by hand from the cases they arrive in from the container manufacturer. However, in plants producing several million cases of product a year, the containers are handled mechanically. Containers are purchased on pallets and automatically depalletised into the production line where they are washed, filled, and sealed. The closures are fed into hoppers above the filling machines and then to the closing and sealing operation.

Although magnetic handling continues to offer many advantages in the conveying of tinplate containers, no such force exists to lock nonmagnetic plastic, aluminium, and composite containers to belts or chains. Handling lines for these containers consist of horizontal tabletop chain conveyors for slow lines, with wide plastic mesh belts being used in high-speed lines. Wide belts allow low line speeds to minimise damage to containers. Single fillers branch off the wide belt lines to feed individual fillers. Vacuum technology used on conveyer belts allows for rapid transport of nonmagnetic containers.

Even the familiar cable conveyor system has been greatly improved by the use of continuous cable guides to minimise can damage and noise. Similar benefits are achieved by nonmetallic contact material inserts now used to produce useful conveying concepts. When high-volume air is drawn through holes in a belt or chain, a 'vacuum' is produced which can elevate some empty containers vertically. Blowing high-volume air through a specially constructed deck plate causes a force which both lifts and propels. This type of air conveyor is excellent in situations in which minimum backpressure is desired when containers accumulate.

10.11   Cleaning the cans

Some hygiene regulations require that cans be washed before being filled, because a bright stacked can does not necessarily mean a thoroughly clean one. The can-cleaning operation should be accomplished in three steps:

1. Cans should travel a short distance in an inverted position.
2. They should be flushed with a relatively large volume of clean good-quality water under good pressure.
3. They should again travel a short distance in an inverted position for draining excess water.

Satisfactory can washers are available from manufacturers of canning machinery. Effective can washing eliminates microorganisms which would otherwise contribute to the total microbiological product load.

If washing is not used, cans should at least be inverted and the interior blown out with high-pressure clean air.

10.12   Filling

Accurate and uniform filling of food is necessary to maintain a uniform headspace and maintain a constant product weight. The term 'filling' includes the addition of liquids, such as brine, syrup, gravy stock, etc., in addition to the product.

The amount of headspace in a can is very important.

- If it is too small, can ends will bulge, thus rendering the can unsuitable for sale. Too small a headspace can also result in understerilisation of the product, as well as allowing insufficient space for accumulation of hydrogen gas (which is produced by certain canned foods), causing can ends to 'dome'.
- Conversely, too large a headspace results in underweight net contents of can, low vacuum, and too much air, which may accelerate product deterioration and cause can corrosion during storage.

Filling may be done either by hand or machine. Delicate products, such as asparagus, broccoli, and soft fruits, are filled by hand; the fill-in weight is controlled by use of a small counterpoised weighing scale.

The simplest products to fill mechanically are liquids and semisolids, such as syrup, brine, fruit juices, jam, soup, etc.

The ideal filling machine should fulfil the following functions:

1. The quantity filled must be uniform and accurately measured.
2. There should be no spilling or drip, even when running at high speed.
3. A 'no can, no fill' device should be incorporated.
4. Changing can size or quantity of fill should be a simple operation.
5. The filler should be capable of handling a wide range of products.
6. There should be no dead spaces in the filler in which dirt and debris can accumulate and provide an opportunity for microorganisms to multiply.
7. All surfaces in contact with food should be made of noncorrosive materials such as stainless steel.

The headspace, and consequently the fill-in weight, for liquid products can be adjusted automatically:

1. By filling the cans to the top and by inclining them on the conveyor at a predetermined angle, thus allowing the surplus liquid to escape.
2. By displacing a given amount of liquid with a plunger.
3. By filling the can from a chamber of a predetermined volume.

The fill-in-weight for semiliquid or viscous materials can be automatically controlled in many cases by forcing the product into the can from a chamber of predetermined volume by use of a plunger. With a few products, however, a plunger is not needed.

Automatically controlling the fill-in weight for solid products is often done by filling from hoppers of predetermined volume or by filling from hoppers into revolving pockets which then deposit the product into the cans. Most of these fillers also have devices by which brine or syrup is added to the cans along with solid product.

An important consideration to take into account at all stages of the filling operation is that the filling of containers, either mechanically or by hand, should be controlled to ensure that the filling requirements specified in the scheduled sterilisation process are met.

A shake-while-progressively-filling method settles or orients the product within the container to achieve a dense fill, thereby minimising spaces within the 'fill' and resulting in accurate filling. Because maximum density of fill is achieved at the time of filling, no additional settling takes place in the container. The fill angle of the container also determines the final headspace within the filled container.

Mechanical fillers, based on various principles, are available for filling foods such as fruits, vegetables, seafoods, meats, snack foods, macaroni products, and others in particulate form. Mechanical fillers have also been developed for prefilling or topping broth, sauces, brine, and gravies and are used for 'portion control' dispensing, in which containers are filled either partially or completely; examples include filling meat chunks into stew or ravioli and meatballs into a can which is later sauced.

10.13   Vacuum in canned foods

The term 'vacuum', as used in the canned-foods industry, is an indication of the amount of air left in the headspace of food cans. Not all the air is exhausted from canned foods.

Table 10.1 Typical vacuums measured in canned food

	Kilo Pascal's (kPa)	Inches mercury (inch Hg)	Millimetres mercury (mm Hg)
Zero vacuum	0	0	0
Typical low vacuum product	7–20	2–6	50–150
Typical medium vacuum product	33–67	10–20	250–500
Typical high vacuum product	88	26	585
Complete vacuum in can	101	30	760

The food industry measures vacuum in terms of kilo Pascals (kPa); in of mercury (in Hg) or mm of mercury (mm Hg) indicates total vacuum. Zero shows no vacuum at all (see Table 10.1).

A vacuum of 33 kPa (10 in Hg) means that one-third of the air has been removed from the headspace of a can.

There are several reasons for obtaining vacuum in canned foods. These include: the maintenance of can ends in a concave position during normal storage; the reduction of oxygen; and the prevention of permanent distortion of can ends during thermal processing.

Techniques employed from the very beginning of canning have been such that a vacuum has resulted. Because bacterial spoilage usually results in gas formation which causes bulging of the can ends, consequently any distortion of the end from the normal concave shape is taken as an indication of spoilage by the industry and by the consumer.

Low oxygen content in canned foods is desirable to minimise adverse chemical changes in the product, such as oxidation of fats or vitamins, to prevent discolouration in some products, to reduce internal corrosion of the can, and to create anaerobic conditions in the canned food's container.

During thermal processing, there is considerable expansion of can contents. This may result in permanent distortion of the ends, particularly in larger can sizes, unless provision is made for this expansion without the development of undue pressure within the can. This is accomplished by providing adequate headspace under vacuum, whereas in some instances, counterpressure on the exterior of the can during cooling is necessary. This latter procedure is known as 'pressure cooling'.

The mechanical gauge in common use for measuring vacuum is the Bourdon tube type (See Figure 10.2). This gauge is the same for pressure as for vacuum. The Bourdon tube gauge has been adapted for measuring can vacuum by equipping it with a hollow puncture tip surrounded by a rubber-sealing ferrule.

Atmospheric pressure varies from day-to-day and decreases with an increase in altitude. The decrease in atmospheric pressure measured will be slightly more than

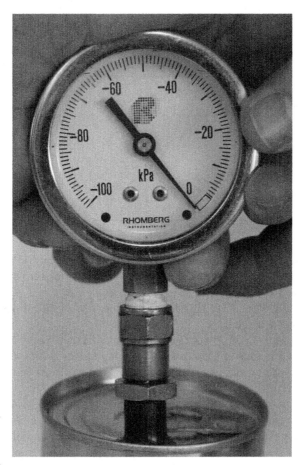

Figure 10.2 Vacuum gauge.

3.3 kPa (1 in) for each 305 m (1000 ft). It is interesting to note that a can having 33 kPa (10 in) true vacuum at sea level would only have approximately 17 kPa (5 in) in Denver, Colorado, US ('Mile High City', altitude 1609 m) and a light pressure on top of Pikes Peak, Rocky Mountains, US (altitude 4302 m (14,115 ft)). Conversely, the vacuum would increase in cans closed at high elevation and moved to lower altitudes.

10.14   Exhausting and vacuum closing

The exhausting of containers for removal of air should be controlled so that the conditions for which the process was designed are met. Heat is used to expand the product, to expand and drive out the occluded and dissolved gases in the product, and to minimise air in the headspace before closure. Vacuum in canned foods may be obtained by preheating foods prior to closing. In producing vacuum by this means, the product may be heated prior to filling, after filling, or it may be heated both before and after filling. The time of heating and final temperature attained before closure have a very important relationship to the ultimate vacuum in the can.

Heating may be accomplished by passing the filled can through a steam or hot water exhaust box. It is common to refer to exhaust box treatment as 'thermal exhaust' and to preheating before filling as 'hot fill'. Exhaust boxes are generally best adapted for canned foods that can readily be heated, such as brine- and syrup-packed fruits and vegetables. The major disadvantages of exhaust boxes are their bulkiness and large steam requirements.

In mechanical vacuum closure by high-speed vacuum-closing machines, the filled cans at a cold or rather low temperature are passed into a clincher, which loosely clinches the covers without forming an airtight seal. The cans are then transferred through a suitable valve into a vacuum chamber, subjected to vacuum for an instant while in the vacuum chamber, sealed, and then ejected through another valve. Vacuums drawn on the machine while the cans are in the vacuum chamber may vary over a wide range, depending mainly on the desired final vacuum in the can and on the temperature of the liquid contents. This method of exhausting air from canned foods subjects the contents to a vacuum for a rather short interval of time before closure. Therefore, the air is withdrawn mainly from the headspace, and only partially from the product itself, and proper adjustment of the headspace is necessary for proper performance.

A difficulty with high-speed vacuum closures is that, because the vacuum is applied over a very short interval of time, 'flashing' of the liquid contents from the can may occur when there is an appreciable amount of dissolved or occluded air in the product and in the void spaces in the product. This happens because air escaping rapidly from the contents of the can will often carry liquid with it. This is especially true with syrup-packed fruits, such as peach and pear halves, sliced and diced fruits, and pitted cherries, and it results in some syrup loss. This difficulty has been overcome by employing a prevacuuming step prior to vacuum closure. For canned fruits, particularly, prevacuuming syrupers have been developed which operate as follows: Freshly prepared fruit is filled into cans and passed into a vacuum chamber which draws a vacuum in the range of 67–91 kPa (20–27 in) on the filled can, exhausting most of the air surrounding the product and some from within the product itself. While still under vacuum, syrup is added to the can to a predetermined height and the filled can is then transferred to the open atmosphere. With this preliminary air-exhausting step, the filled cans can be sealed in a vacuum-closing machine with little variation in the final vacuum from can to can and with little or no loss of liquid content.

Another method for producing vacuum in canned foods is based on the injection of live steam into the can or glass-container headspace at the time of closure. If live steam is properly injected into the headspace of a container prior to capping, it will replace essentially all the air, and when it condenses as the can cools after sterilisation, a vacuum is formed. This, therefore, provides a simple, inexpensive method of exhausting the headspace volume of its air. This operation is known as 'steam-flow' or 'steam-vac' closing. During this operation, the product is not heated, except slightly at the top surface, and little or no occluded air within the product or in the void spaces beneath the top surface is removed by the action of the steam. Consequently, to obtain good vacuum, the product must not contain much occluded air at the time of closure, and an adequate and controlled headspace must be provided.

Preheating of the product prior to filling, or even after filling, followed by 'steam-vac' closure, has been practised for some products which contain large amounts of occluded air and other gases. Steam flow is faster and allows a hotter fill than does a mechanical vacuum.

Undercover gassing just before capping is sometimes used to reduce oxygen content in the headspace. Nitrogen gas is used. Liquid nitrogen is injected on the surface of food in cans just before capping when can pressure instead of vacuum is desired to protect the integrity of cans and can seams when using light tin plate or aluminium containers.

Low vacuum in canned foods may be the result of poor blanching, exhausting, faulty closing, or leakage through the can seam after closure. With some products, like vacuum-packed corn, container vacuum after closing has an effect on the rate of heat penetration during product thermal sterilisation.

10.15   Can closing

To ensure the prevention microbiological spoilage of the product in a sanitary can, it is essential that a hermetic seal be provided in forming the seam structure in attaching the end to the can body. Unless a good seal in this area is assured, the effort in production preparation, packing, and processing is of no value.

The objective of the double-seaming operation is to mechanically attach a compound-lined metal end onto a can body by properly interlocking their hooks to form a hermetic seal.

Food products require a hermetic seal because of susceptibility to bacterial spoilage. The double seam must provide and maintain an efficient hermetic seal under the widely varying conditions that the can must withstand between packing and final consumption of product. The quality of a can depends largely upon the quality of the double seam. The quality of the double seam depends upon

- The condition and adjustment of the double seamer
- The quality of the unseamed can body and can end
- The dimensions (thickness) of the body or end.

As the quality of double seams is subject to wide variation, it is essential that the double seams of cans being discharged from a double seamer be inspected at frequent intervals.

Good quality tinplate, adequate sealing compound, and an efficient can-closing machine combine to produce a strong and hermetic double seam that will adequately protect the canned food during sterilisation, cooling, and storage.

Double-seaming machines can operate at speeds >1000 cans per minute for particulate foods in lightweight tin plate, deep-drawn aluminium, and composite and plastic containers of up to 108 mm in diameter. Fluid and semifluid canned foods can be closed at speeds even faster.

Considerable progress has also been made in automatic closing of glass jars for canned foods. Glass-container closing machines are somewhat slower for comparable size containers than metal-can closing machines because of the fragility of glass. Small glass containers may, however, be filled and closed at speeds close to that of cans.

10.16 Dud detector

A very useful tool or piece of equipment to employ after the container closing operation is a Dud Detector, or more precisely a 'pressure/vacuum monitor'. An example of a unit is a Tap Tone (manufactured by Teledyne Taptone): a noncontact inspection system based on an electromagnetic impulse (tap) on each container that indicates the resulting frequency response (tone). This equipment automatically rejects containers according to user-determined criteria. Using digital signal processing techniques, the system analyses the frequency response and assigns a merit value which can be correlated to a specific vacuum or pressure level. This equipment is applicable to most container closures, works on full closed cases, as well as single containers, and is available in a portable unit. The portable unit could be applicable for warehouse spot-checking of goods, rework/salvage operations, and statistical process control.

10.17 Container coding

Can and glass container coding and the frequency of code changes are extremely important to food processors.

Container codes should give the following information:

1. Plant where packed – if more than one plant
2. Product and style of pack
3. Day of pack
4. Hour of pack
5. Line on which product was packed.

Important benefits come from a coding system that is properly correlated with production and shipping records.

If a problem should occur, such as underprocessing with a potential public health hazard, spoilage in the warehouse or in distribution, discovery of extraneous contamination, consumer complaints and alleged illnesses, or seizure by a regulatory agency with subsequent forced or voluntary recall, the investigation and/or recall can be confined to one day's pack, one hour of production, or to one line in the plant. Otherwise, the problem may result in restraint from sale of all, or a sizable portion, of a plant's production.

A batch code may consist of a retort load or the production during a period, preferably not more than two and one-half hours. Some canners have found it convenient to change die codes during the 'break' period.

In low-acid foods, each container should be marked with an identifying code which should be permanently visible to the naked eye. Where the container does not permit the code to be embossed or inked, the label may be legibly perforated or otherwise marked, if the label is securely affixed to the product container. The identification should identify, in code, the establishment where packed, the product contained in the package, the year packed, the day packed, and the period during which packed. The packing period code should be changed with sufficient frequency to enable ready identification of lots during their sale and distribution. Codes may be changed on the basis of the following: intervals of every four–five hours, personnel shift changes, or batches, provided the containers comprising such batches do not extend over a period of more than one personnel shift.

Acknowledgements

13th ed. Acknowledgments Anchor Hocking packaging Co., 312 Elm St., Cincinnati, OH 45202–52739, Tel-5 13-333-3400, Fax-5 13-333-340 1.

Angelus Sanitary Can Machine Co., 4900 Pacific Blvd., Los Angeles, CA 90058, Tel-2 13-583-2 17 1, Fax-2 13-587-5607.

Elbicon, Inc., 2744 W. 4th Ave., Denver, CO 80219, Tel-303-936-1457. Fax-303-936-16 12.

Magnuson Corporation, P.O. Box 590, Pueblo, CO 81002–0590, Tel-7 19-940-9500, Fax-7 19-948-9540.

Vanmark Corporation, Industrial Parkway, Creston, IA 50801, Tel-5 15-782-6575, Fax-5 15-782-9209.

FMC Corporation. Food processing systems Division, 2300 Industrial Ave., Madera, CA 93639, Tel-209-661-3200. Fax-209-661-3222.

Key Technology, Inc., 150 Avery St., Walla Walla. WA 99362, Tel-509-529-2 161, Fax-509-522-336 1.

Lyco Manufacturing. Inc., P.O. Box 31, Columbus, WI 53925, Te14 14–6234 152, Fax4 14-623-3780.

Odenberg Engineering. Inc., 6890 Luther Drive, Sacramento. CA 95823.Tel-9 16422–18396, Fax-9 16422-840 1.

Raque Food Systems, Inc., P.O. Box 99594, Louisville. KY 40269–40594, Tel-502-267-9641, Fax-502-267-2352.

Robins, 4100 Pistorio Rd., Baltimore, MD 21229, Tel-4 10–2474000, Fax4 10-247-9 165.

Solbern Division of Howden Food Equipment, Inc., 8 Kulick Rd., Fairfield, NJ 070043385, Tel-20 1-227-3030, Fax-20 1-227-3069.

Tap Tone Container Inspection Systems, Division of Benthos, Inc., 49 Edgerton drive, North Falmouth, MA 02556–02826, Tel-8004234044, Fax-500-564-9945.

Urschel Laboratories, Inc., P.O. Box 2200, Valparaiso, IN 463842200, Tel-2 19464–48 11, Fax-219462-3879.

Process room operations

11

11.1 Introduction

Good manufacturing practices (GMP) are the practices required to conform to guidelines recommended by regulatory agencies which control the manufacture and sale of food products. These guidelines provide minimum requirements that a food-product manufacturer must meet during manufacture to assure that the products are of high quality and do not pose any consumer risks. GMP guidelines are overseen by food regulatory agencies in most countries.

Regulatory bodies and food-control legislation insist on Good Manufacturing Practices (GMPs) for Low-Acid Canned Foods. Codex Alimentarius (CAC/RCP 23–1979) (Codex Alimentarius, 2011) and the US Food and Drug Administration have published 'Thermally Processed Low-Acid Foods Packaged in Hermetically Sealed Containers' (21 CFR Part 113) and, Acidified Foods Good Manufacturing Practice Regulations' (21 CFR Part 114), an excellent set of GMPs for food canning. These are available online.

Regulations pertaining to canned foods may be modified from time to time. Therefore, canners must always consult and comply with the latest regulations that affect their operations. It is outside the scope of this book to explain all the details which are included in the GMP regulations. It is very important, however, that canners of acidified and low-acid foods become thoroughly acquainted with the GMP regulations and that they make sure that their plant equipment and processing procedures conform in detail with those regulations. Further, the low-acid canned food regulations require that the operators of processing systems and container-closure inspectors be under the supervision of a person that has attended and completed an approved course.

The details of this most important operation in the successful canning of low-acid foods in metal cans (tinplate and aluminium) and in glass has been the subject of much research by laboratories connected with the canning industry. The National Food Processors Association has prepared a carefully worded document which every canner should study. Because of its importance, some of the important paragraphs of NFPA Bulletin 26-L, 5th edition and Bulletin 30-L, 5th edition have been included in this chapter.

11.2 Commercial sterility in canned foods

In acidified and low-acid canned-food sterilisation, the main concern of the canning industry is to prevent the growth of *Clostridium botulinum*, the food-poisoning bacterium capable of producing a highly lethal toxin. *C. botulinum* is a heat-resistant bacterium. A sterilisation process that assures the destruction of *C. botulinum* does not

A Complete Course in Canning and Related Processes. http://dx.doi.org/10.1016/B978-0-85709-677-7.00011-6

necessarily kill all other microorganisms capable of producing canned-food spoilage under normal conditions of canned-food handling and storage. These microorganisms also need to be destroyed.

A thermal process that produces commercial sterility in low-acid canned foods may be defined as 'that process by which all *C. botulinum* spores and all other pathogenic bacteria have been destroyed, as well as more heat resistant organisms which, if present, could produce spoilage under normal conditions of non-refrigerated canned food storage and distribution' (Tucker & Featherstone, 2011).

If the number of organisms in the product is excessive, recommended processes may not be adequate to prevent spoilage. Therefore, it is essential to exercise strict principles of sanitation while the raw commodities are prepared for canning.

The destruction by heat of the organisms naturally present in the sealed container is the fundamental operation of food preservation by canning and is known as processing to commercial sterility. The time and temperature combination at which the product is heated is known as 'the process'.

'The process' is determined from a study of the rate of heat penetration for the product and from a study of the heat resistance of significant spores. A theoretical process is then calculated and tested by inoculation of product with a known spore load.

An example is the determination of a process for canned corn. Because it is known that flat sour and sulphide thermophiles, as well as putrefactive anaerobic mesophiles, cause spoilage of corn, it is necessary to study the conditions under which these agents are destroyed. After preparing a spore crop of each test organism, a heat-resistance determination is made. By using thermocouples, the rate of heat penetration into canned corn is determined. Employing a mathematical correlation between heat resistance and heat penetration, what is known as a 'theoretical' process is determined. To test this theoretical process, containers of corn are inoculated with the test organisms. The containers are processed at various temperatures for varying periods, ranging from the 'theoretical process' to lower temperatures and then incubated to determine the spoilage levels.

The inoculated pack technique is valuable, especially for products such as spinach, which exhibit large variations in their rate of heat penetration. If the inoculated pack results confirm the mathematically derived theoretical process, the mathematical methods can usually be applied to the product in a variety of can sizes, thus precluding the need for studying the effects of the process on experimental packs in each can size. The process so determined will produce a 'commercially sterile' canned-food product with the greatest retention of quality.

'Commercial sterility' of equipment and containers used for aseptic processing and packaging of food means the condition achieved by application of heat, chemical sterilant(s), or other appropriate treatment which renders such equipment and containers free of viable forms of microorganisms having any public health significance, as well as any microorganisms of nonhealth significance capable of reproducing in the food under normal nonrefrigerated conditions of storage and distribution.

Some thermophilic (heat-loving) bacteria produce spores of such high resistance to heat that they cannot be destroyed in some products without processing to such a degree that the canned product would be unmarketable. Fortunately, these thermophilic bacteria are not infectious or poisonous and are, therefore, of no significance with respect to public health. When such thermophilic spores survive the process in canned foods, they are unable to germinate and cause spoilage at storage temperatures of 38 °C (100 °F) or lower. Prompt cooling of processed cans to an average temperature of 38 °C, avoiding high-temperature storage, safeguards against spoilage by thermophilic bacteria. Incubation of low-acid canned foods at 55 °C (131 °F) will obviously allow germination with recovery of vegetative cells.

For many years, laboratories connected with the canning industry and other laboratories have devoted much attention to *C. botulinum* and other spoilage organisms. A great amount of work has centred on heat-resistance studies, and processing recommendations for foods are based upon the results of this research. Concurrently, a study was made of the growth characteristics for *C. botulinum*, and it was found that it would not grow at a pH of 4.6 or below. This is a very important observation because, as a practical matter, it means that, in general, products of pH levels higher than pH 4.6 must be processed under pressure at temperatures above 100 °C (212 °F) to insure destruction of the spores, whereas products at pH 4.6 or lower may be safely processed at 100 °C. The experience of the industry over many years has established the validity of this product classification.

The sterilisation by pressure processing for some low-acid products, such as globe artichokes, pimientos, onions, and peppers, may result in a product of unsalable quality. Under proper control, such products may be processed in boiling water after they have been acidified to a pH level at which they are no longer a low-acid food. Experience has shown that careful supervision of all details is essential when acidification is included in the processing procedure. This procedure should not be followed without consulting a competent thermal processing authority.

Acid foods are not subjected to as much heat as low-acid foods; however, they are heated sufficiently to destroy all vegetative pathogenic and spoilage bacterial cells, yeasts, and essentially all moulds, which could, if not destroyed, cause spoilage.

11.3   Sterilisation metal (tinplate and or aluminium) cans in still, static, steam (discontinuous, non-agitating) retorts (summarised from NFPA Bulletin 26-L, 5th edition)

The importance of proper equipment and procedures cannot be overemphasised. Processes for canned foods are usually determined by tests made with containers in pure saturated steam (free from air) at precisely controlled times and temperatures. When these processes are applied in commercial practice, equivalent conditions must be met. Adequate venting of air from all parts of the retort and careful control and recording of temperatures and times are essential.

Close supervision of the cook room and close attention to details are extremely important to insure successful processing. Otherwise, irregularities may occur because of poor organisation of the cook room, carelessness, or failure to understand and apply the basic principles of safe practice (Bulletin 26L, 1984).

11.3.1　Type of retort

The equipment described in this chapter is for pressure processing of containers in pure saturated steam in discontinuous (batch type), nonagitating (still), vertical or horizontal retorts, including crateless retorts. For information on other types of processing equipment, consult a competent thermal processing authority and the equipment manufacturers.

11.3.2　Retort steam, equipment, and operation requirements

The following information provides a guide for minimum requirements necessary for installing and operating retorts for pressure processing low-acid foods in pure steam. Other equipment and operating procedures may be satisfactory, if proven adequate by appropriate tests.

1. *Steam supply.* Steam supply to the processing room must be sufficient to provide adequate steam to all retorts operating at any one time. Insufficient steam supply is indicated by inability to meet venting requirements, long come-up times, and temperature fluctuations in retorts in use when additional retorts are being brought up.
2. *Steam header.* The supply line delivering steam to a group of retorts should be large enough to provide sufficient steam for the greatest number of retorts that may be brought to retort temperature simultaneously. The following pipe sizes in Table 11.1 are suggested.
3. *Steam controller.* Each retort must be equipped with an automatic steam controller to maintain accurate retort temperature. This may be a self-actuated or air-actuated (air-to-open) type, responsive to either temperature or pressure. If the controller is smaller than the steam inlet pipe, the use of a steam bypass may be necessary during the come-up time.
4. *Steam bypass.* A steam bypass for the control valve is desirable to make hand operation of the retort possible in the event of failure of the control valve. In some installations, the steam bypass may be used regularly during the come-up time, if the steam demand is greater than the control valve provides. This is particularly true if a smaller control valve is used. Because controlled excessive pressure in the retort might lead to equipment damage and personnel injury, the operator should not leave the retort while the bypass is open.
5. *Steam inlet.* The steam inlet to individual retorts must be large enough to provide sufficient steam for venting the retort in a reasonable length of time. Steam may enter either the top or bottom of the retort but must enter the portion of the retort opposite the vent.

Table 11.1　**Steam header pipe size**

Number of retorts brought up simultaneously	Steam header pipe size	
	Vertical and horizontal retorts less than 15 ft long	**Horizontal retorts more than 15 ft long**
1	2 in	2–2½ in
2	2½ in	3–3½ in
3	3 in	3½–4 in
4	3½ in	4–5 in

Note: 1 in = 25.4 mm; 1 ft = 0.304 m.

6. *Steam spreader.* Steam spreaders are continuations of the steam line inside the retort, which may or may not be perforated.

Horizontal retorts must be equipped with steam spreaders that extend along the bottom of the entire length of the retort. The perforations should be along the top 90° of this pipe, that is, within 45° on either side of top centre. Horizontal retorts over 9 m (30 ft) long must have two steam inlets connected to the spreader at approximately equal divisions of its length.

In vertical retorts, bottom steam spreaders, if present, should be in the form of a cross or straight pipe with the perforations along the top or side of the pipes. In crateless retorts with top steam entry, steam should enter through a circular steam spreader.

To insure proper distribution of steam, the number of perforations in the steam spreaders should be such that the total cross-sectional area of the perforations is equal to 1½ to 2 times the cross-sectional area of the smallest restriction in the steam inlet line. Table 11.2 may be used as a guide.

7. *Vent.* Vents are large, valve-controlled openings in retorts, used for elimination of air during venting. They must be installed in such a way that all air can be removed from the retort before timing of the process is started. Vents must be controlled by gate or other suitable valves, which must be fully open to permit rapid discharge of air from the retort during venting. Other types of vent valves may be satisfactory, if tests have been conducted to establish an adequate venting schedule. All external vent lines, manifolds, etc. should be short and as free as possible from bends and other conditions which might retard rapid discharge of air. Such lines should discharge to the atmosphere as closely to the retort as possible. *They must not be connected directly to a closed-drain system.* If the overflow is used as a vent, there must be an atmospheric break in the line before it connects to the drain. This is required to prevent backpressure during venting and to meet plumbing codes. *The vent must be located in the opposite portion of the retort from which the steam is admitted.*

8. *Retort manifold and manifold header.* The best air removal is accomplished by venting directly to the atmosphere from the valve controlling the vent. However, it is often desirable to remove the steam vapour from the retort room by connecting the vent pipes to a suitably sized retort manifold or manifold header. A retort manifold connects several vent pipes from a single retort and is controlled by a gate or other suitable type valve. The manifold must be large enough so that the cross-sectional area of the pipe is greater than the total cross-sectional area of all connecting vents. *A manifold header* connects vents or manifolds from several retorts. It leads to the atmosphere within as short a distance as practical and

Table 11.2 **Number of holes in a steam spreader**

Hole sizes (in)	Smallest restriction in steam inlet line					
	¾ in	1 in	1¼ in	1½ in	2 in	2½ in
$3/16$	29–39	47–63	82–109	111–148	182–244	261–347
$7/32$	22–29	35–46	60–80	82–109	134–179	192–255
¼	12–22	27–36	46–61	63–83	103–137	147–196
$5/16$	11–14	17–23	30–40	40–54	66–88	94–125
$3/8$	–	12–16	21–28	28–37	46–61	66–86
½	–	–	12–16	16–21	26–35	37–49

Note: 1 in = 25.4 mm.

with as few bends as possible. It is NOT controlled by a valve. The manifold header must be sized so that the cross-sectional area is at least equal to the total cross-sectional area of all connecting manifold pipes from all retorts venting simultaneously. If the header is of excessive length, it should be increased at least one pipe size. The discharge of a retort manifold or manifold header must not be directly connected to a closed drain without an atmospheric break in the line.

9. *Venting procedures for systems utilising divider plates.* When perforated-divider plates are used between layers of cans, standard venting procedures and specifications may not be adequate. Consult a competent thermal processing authority for proper venting time and temperature determinations.

10. *Venting specifications.* No single venting schedule is applicable to all retorts. The choice of a satisfactory venting schedule is dependent upon a number of factors. These include: retort size, shape, and piping arrangement, as well as quantity and pressure of steam available, container size, method of stacking cans in the retort, desired length of come-up time, and resistance to outward flow of air from the retort, among others. *To assure complete air removal during venting*, it is necessary to meet both time and temperature requirements specified for a particular retort installation.

Some typical installations and operating procedures which have proven to be satisfactory are given in Figure 11.1(a)–(g). Although other installations and operating procedures may be satisfactory, these should be evaluated prior to use to establish specifications for adequate air removal. Any abnormal increase or decrease in retort come-up time could indicate a potential problem and should be investigated to determine the cause.

The venting schedules presented here are based upon the following conditions: the vent valve(s) and steam valve must be fully open and the water and air valves must be closed. Vent valves, other than gate valves, should be approximately equivalent in flow capacity to a fully opened gate valve of the specified size.

Timing of the venting period starts when steam is turned on. In crateless retorts with top steam entry in which steam is used to assist in cushion water removal, the method for timing the venting period may vary. For these retorts, the timing method should be specified in the venting schedule.

11. *Bleeders.* Do not confuse bleeders with vents! Bleeders are openings used to remove air entering the retort with the steam and to provide circulation of steam in the retort. Bleeders MUST be open and emit steam continuously and freely during the entire process, including come-up time. All bleeders must be arranged in such a way that the operator can observe steam escaping during the process. An 1/16-in (1.6 mm) or larger opening must be used to bleed wells for mercury thermometers and temperature recorder bulbs. Bleeders are necessary on all external wells. All other bleeders must be 1/8 in (3.2 mm) or larger.

Horizontal retorts must have one bleeder within approximately one foot of the outermost location of containers at each end along the top of the retort. Additional bleeders must be located not more than 8 ft (2.4 m) apart along the top. Vertical retorts must have one bleeder located in that portion of the retort opposite the steam inlet.

In retorts utilising top steam and bottom venting, an adequately sized bleeder must be installed in the bottom of the retort to indicate and assist in complete and continuous removal of condensate. Its discharge must be located so its operation can be observed. Bleeders may be installed in the bottom of any retort to remove condensate.

12. *Condensate bleeder.* For crateless retorts with top steam entry, there should be one or more ⅜-in (9.5 mm) or larger condensate bleeders. When a false bottom is employed in a crateless retort, it is useful to have a ⅛-in (3.2 mm) bleeder with its opening at a point higher than the condensate bleeder and just below the false bottom.

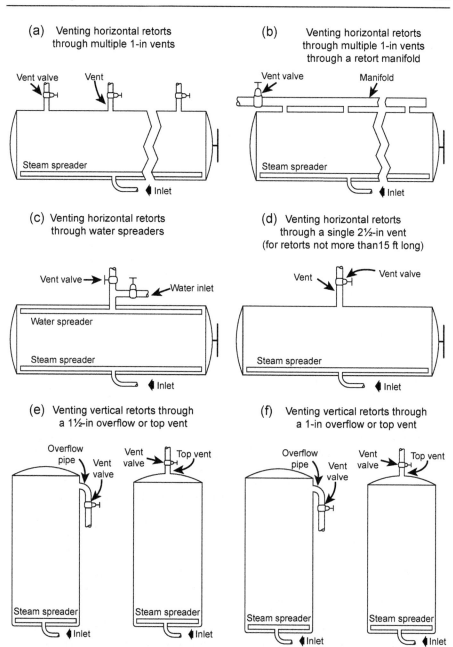

Figure 11.1 Vent arrangements in static, steam retorts with no agitation.
Source: From pages 314–319 in 13th Ed Book I.

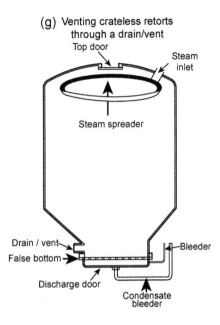

(g) Venting crateless retorts through a drain/vent

Figure 11.1   Cont'd

13. *Mufflers or noise suppressors.* Bleeders and vents may be equipped with mufflers or noise suppressors to reduce their noise level. If mufflers are used, evidence that they do not significantly impede air removal must be kept on file; this may be in the form of heat distribution data. A letter from the manufacturer, the designer, or a competent processing authority may also serve this function. Bleeder and vent mufflers should be periodically checked for proper operation; if clogged or in disrepair they should be repaired or replaced.

14. *Drain.* The drain should be large enough to permit rapid removal of water after cooling. Only when steam is admitted at the top should the drain be used as a vent. In this case, the drain line must be open to the atmosphere.

15. *Water line.* If containers are to be cooled in the retort, the line pressure and pipe size of the supply line and inlet should be adequate to allow for rapid filling of the retort. Water valves have replaceable seals and must be maintained in good condition to insure that water does not leak into the retort during processing; water leakage into the retort during processing may result in underprocessed containers

    Both top and bottom water inlets to the retort may be desirable to provide for the most efficient cooling procedure. The top inlet on vertical retorts should discharge water through a spray ring or several small openings around the shell.

    In horizontal retorts, water should enter at the top through a full-length spreader inside the shell and at the bottom through one or more openings. Three rows of holes in the top water spreader are recommended to insure that spray will strike all containers. A sufficient number of holes should be made in the water spreader to provide adequate water distribution for proper cooling of the containers. If the retort is vented through the water spreader, the number of holes should be such that their total cross-sectional area is at least equal to the cross-sectional area of the smallest restriction in the vent outlet. Table 11.3 may be used as a guide.

16. *Overflow line.* The overflow line should be located near the top of the retort above the top layer of cans. Gate or other suitable valves should be used to permit unrestricted flow. If the overflow is used as a vent, see Section 11.3.7.

Table 11.3 Minimum number of holes in water spreaders when used for venting

Hole size (in)	Smallest restriction in vent valve					
	1¼ in	1½ in	2 in	2½ in	3 in	3½ in
3/16	55	74	122	174	268	359
7/32	40	55	90	128	197	264
¼	31	42	69	98	151	202
5/16	20	27	44	63	97	129
3/8	14	19	31	44	67	90
½	–	11	18	25	38	51

17. *Vacuum breaker.* Retorts piped with water may be equipped with an adequately sized vacuum breaker to prevent vacuum formation which could result in retort and container damage. This condition can occur when water is rapidly admitted to the top of a retort while it is under steam pressure.

18. *Air line.* If retorts are equipped with an air supply for pressure cooling, the air valves should have replaceable seals; they must be maintained in good condition. Air leakage into the retort during processing must be prevented; air in the retort reduces process effectiveness and may result in understerilisation of the containers.

19. *Crate supports.* Some type of bottom crate support must be provided in vertical retorts, other than crateless retorts.

20. *Baffle plates.* Baffle plates must not be used in the bottom of vertical or horizontal still retorts because they tend to direct the flow of steam around the containers rather than up through them. The use of baffle plates in retorts may cause underprocessing due to the presence of cold spots or localised areas of air in contact with containers being processed.

21. *Safety valve.* Adequately sized safety valves are necessary to prevent excess pressure in retorts. These should comply with local safety codes for unfired pressure vessels. Because the relieving capacity of a given size safety valve varies with manufacturer, no specification as to required size can be given here. Such specifications for specific operating conditions should be obtained from the manufacturer.

22. *False bottom.* A false bottom is a perforated metal plate in the bottom of a retort which supports containers and allows steam to flow beneath them. Crateless retorts should be equipped with false bottoms to insure adequate steam circulation around containers in the bottom of the retort.

11.3.3 Instrumentation

All control and indicating equipment, such as thermometers, pressure gauges, automatic controllers, and recorders should be maintained in a clean and workable condition (see Table 11.4 for temperature–pressure relationship at different altitudes). Installation of this equipment must consider available light and suitable position so that they can be read easily.

Table 11.4 Gauge Pressure (in psi)[a] corresponding to specified process temperature at various altitudes

| Temperature (°F) | Sea level | Feet above sea level | | | | | | | Temperature (°C) |
		500	1000	2000	3000	4000	5000	6000	
200	–	–	–	–	–	–	–	–	93.3
205	–	–	–	–	–	–	0.5	0.9	96.1
210	–	–	–	0.4	0.9	1.4	1.8	2.3	98.9
212	0.0	0.2	0.5	1.0	1.5	2.0	2.4	2.9	100.0
215	0.9	1.1	1.4	1.9	2.4	2.9	3.3	3.8	101.7
220	2.5	2.7	3.0	3.4	3.9	4.4	4.9	5.3	104.4
225	4.2	4.5	4.7	5.2	5.7	6.2	6.6	7.1	107.2
230	6.1	6.3	6.6	7.1	7.6	8.0	8.5	9.0	110.0
235	8.1	8.3	8.6	9.1	9.6	10.0	10.5	11.0	112.8
240	10.3	10.5	10.8	11.3	11.7	12.2	12.7	13.1	115.6
242	11.2	11.4	11.7	12.2	12.7	13.1	13.6	14.1	116.7
245	12.6	12.9	13.1	13.6	14.1	14.6	15.0	15.5	118.3
248	14.1	14.3	14.6	15.1	15.6	16.0	16.5	17.0	120.0
250	15.1	15.4	15.6	16.1	16.6	17.1	17.5	18.0	121.1
252	16.2	16.4	16.7	17.2	17.7	18.1	18.6	19.1	122.2
255	17.8	18.1	18.3	18.8	19.3	19.8	20.2	20.7	122.9
260	20.7	21.0	21.2	21.7	22.2	2.7	23.1	23.6	126.7
265	23.8	24.0	24.3	24.8	25.3	25.8	26.3	26.8	129.4
270	27.3	25.5	27.8	28.3	28.8	29.3	29.8	30.3	132.2
275	30.9	31.2	31.5	32.0	32.5	33.0	33.5	34.0	135.0
	152	305	610	915	1220	1525	1829		
				Metres above sea level					

[a] 1 psi = 6.89 kPa.

The official temperature-indicating device of the retort no longer has to be a mercury-in-glass thermometer. As of March 2011, the US FDA ruled that any other type of temperature-measuring device could be used. These should be easily readable to 0.5 °C (half a degree Celsius) and should have a scale of at least 2 mm/°C (FDA 21, 2011).

1. *Indicating mercury-in-glass thermometer or suitable temperature-measuring device.* Each retort must be equipped with at least one mercury-in-glass thermometer (or equivalent) scale with no more than 3.5 °C/cm (17 °F/in) and divisions that are easily readable to 0.5 °C (1 °F). Temperature range should adequately encompass scheduled retort temperatures used.

   Indicating thermometer bulbs must be installed within the retort shell or in external wells attached to the retort. An external well or pipe must be connected to the retort through at least a 19 mm (¾ in) diameter opening and be equipped with a 1.6 mm (1/16 in) or larger bleeder, located to provide a full flow of steam past the entire length of the thermometer bulb. Thermometers with separable wells or sleeves should not be used. The bleeder must emit steam continuously and freely throughout the processing period.

   Thermometers should not be installed in the lid or door of a retort. Abrupt jarring from closing the retort may cause a divided mercury column. A thermometer with a divided mercury column must be repaired or replaced immediately.

   Thermometers must be tested for accuracy against a known accurate standard thermometer upon installation and at least once a year thereafter or any time their accuracy is questioned. Each thermometer should be properly identified, including the date on which it was last tested for accuracy. Records of these tests specifying date, standard used, and person performing the test should be maintained.

2. *Temperature recorder.* Each retort must be equipped with an accurate temperature-recording device to provide a permanent record of time and temperature of each retort load processed. Chart graduations must not exceed 1 °C (2 °F) within a range of 5 °C (10 °F) of the processing temperature. Each chart must have a working scale of not more than 12 °C/cm (55 °F/in) within a range of 11 °C (20 °F) of the processing temperature. Recorders may be combined with steam controllers and function as recording-controlling instruments. A means of preventing unauthorised changes in adjustment must be provided. A lock or suitably worded sign prominently displayed on or near the recorder is an acceptable method of prevention.

   The bulb may be installed within the retort shell or in an external well connected to the retort through at least a 19 mm (¾ in) opening. The well must have a 1.6 mm (1/16 in) or larger bleeder which is open at all times during the processing period. The temperature recorder must be checked and adjusted to agree as nearly as possible with, but in no event be higher than, the known accurate mercury-in-glass thermometer during the process time.

3. *Pressure gauge.* Each retort should be equipped with an easy-to-read pressure gauge. The gauge should have a range of 0–207 kPa (0–30 psi) and be graduated in divisions of 14 kPa (2 psi) or fewer.

11.3.4   Container loading equipment

1. *Crates, baskets, trays, and gondolas.* Equipment used to hold containers in retorts must be made of strap iron, adequately perforated sheet metal, or other suitable material. The number and size of perforations in the sides of this equipment are not critical. However, insufficiently perforated bottoms should not be used, because they may prevent adequate heat distribution in the retort. When perforated sheet metal is used for the bottoms, perforations should be approximately 1-in holes (25.4 mm) on 2-in (51 mm) centres (or their equivalent in percent open area), (⅜ in [9.5 mm] on ¾-in [19 mm] centres, ½ in [13 mm] on 1-in [25.4 mm]

centres, ¾ in [19 mm] on 1½-in [38 mm] centres, 1½ in [38 mm] on 3-in [76 mm] centres, or 1¾ in [44.5 mm] on 3½-in [89 mm] centres).

2. *Divider systems.* Retort loading systems employing divider plates between layers of containers generally require increased venting schedules. These divider plates should have at least the same perforations (1 in [25.4 mm] holes on 2-in [51 mm] centres) or their equivalent in percent open area, as used for crate bottoms. A competent thermal processing authority should be consulted for specific details to establish an adequate venting schedule. Poor heat distribution and possible underprocessing may result if a divider plate is placed on the crate bottom before container loading.

3. *Stacking of containers.* When stacked, containers should be arranged to permit free circulation of steam throughout the retort load. If it is necessary to separate two lots in one crate or tray, fish net or other material ½ in (13 mm) or larger mesh should be used. Do not use burlap sacks, boards, sugar sacks, towels, or other similar materials as separators because they may interfere with steam circulation, causing underprocessing.

4. *Container Nesting.* Unless the thermal process is designed to take into account container-nesting effects, step-sided containers or two-piece containers that nest must be placed in still retorts with an adequate divider between each layer to prevent container nesting. Other methods used to prevent container nesting should be validated by a competent thermal processing authority.

11.3.5   Process-timing equipment

A large, easily read clock or accurate timing device must be installed where it can be readily observed by the retort operator for accurate process timing. Pocket and wrist-watches and mobile phones are not considered satisfactory for timing purposes. Digital or other clocks not indicating seconds should not be used, unless the specified operating process and venting time have a 1 min or greater safety factor over the scheduled process.

11.3.6   Maintenance of equipment

Processing equipment should be maintained in satisfactory operating condition at all times. Safety valves should be tested frequently. Upon installation, instruments should be checked for accuracy. They should be checked annually, or whenever accuracy is questioned. Water and air valves should be checked frequently for leaks. Vent and bleeder mufflers, if used, should be checked often for proper operation.

Before each operating season and after any lengthy idle period, the entire retort hookup should be examined carefully. Each retort should be brought to processing temperature (without a load) to test for leaks, check vents, and test instruments and control equipment for proper operation and accurate recording. Never introduce water into an empty retort after testing until steam pressure has been released.

Holes in perforated steam spreaders and water spreaders used for venting should be periodically reamed or drilled to their original size if clogging with deposits is a problem.

11.3.7   Coding

All cans should be so closed and processed that their ends remain concave under normal commercial conditions of storage and distribution. An adequate code on each container is one of the key factors in retrieving merchandise in the event a recall or market withdrawal becomes necessary.

Each sealed container of canned food must be marked with an identifying code permanently visible to the naked eye. Consideration should be given to the coding system design to permit easy inspection by the consumer in the event of a recall. Experience has shown that the average retailer or consumer does not readily identify changes in position or design of symbols used in codes. This has resulted in returns of merchandise which were not actually part of the recalled code lot. This same situation has occurred when label perforations or scratch marks have been used for coding.

The container code must identify the establishment where packed, product contained therein, year packed, day packed, and period during which the product was packed, The packing period code must be changed with sufficient frequency to enable easy identification of lots during sale and distribution. Codes may be changed according to one of the following: intervals of every 4 to 5 h, personnel shift changes, or batches, provided the containers composing such batches do not extend over a period of more than one personnel shift.

11.3.8 Rapid handling and prompt retorting of filled containers

The time between closing and retorting should be kept to a minimum. It is desirable that this period be no longer than 1 h. If this is not practical, the proper procedure to be used should be obtained from a competent thermal processing authority and with appropriate product testing.

11.3.9 Initial temperature (IT)

The term 'initial temperature' designates the average temperature of the contents of the coldest container to be processed at the time the thermal processing cycle begins. Just prior to the start of the process, the contents of the container used to check initial temperature should be shaken or stirred and its average temperature determined. This container should be representative of the coldest cans in the retort load. It should have an initial temperature equal to or greater than the initial temperature specified for the process. Initial temperature is as important to process adequacy as retort temperature and process time. Initial temperature might not be the same as the closing temperature.

If a container is closed at a temperature higher than that of the canning-room atmosphere and is then held for some time in the room before processing, its contents will cool. However, the temperature at the container's centre may not be appreciably lower than the closing temperature. For this reason, initial temperature is taken on the contents of the container after it has been stirred or shaken.

In crateless retorts, initial temperature is the average temperature of one of the last few containers entering the top of the retort or the temperature of the cushion water, whichever is lower.

11.3.10 Preventing mistakes in the cook room

To minimise mistakes and prevent the occurrence of unprocessed or underprocessed cans, the following procedures are recommended.

1. Post all operating processes and retort vent schedules in a conspicuous place near the retorts or in a place readily accessible to operators.
2. Mark baskets, trucks, cars, or crates containing unprocessed material plainly and conspicuously with a suitable heat-indicating material that illustrates a change in appearance after exposure to processing temperatures. A visual check must be made to determine that the appropriate change has taken place.
3. Establish a system for product traffic control in the cook room to prevent unprocessed product from bypassing the retort process.
4. Identify containers of product requiring different processes which are being packed at the same time so that each will receive the proper process.
5. Hang a distinctive marker from the retort opening when it holds unprocessed cans. It should be placed so that the door or lid cannot be closed until the marker is removed.
6. Do not close the retort until ready to start the process.
7. Puncture and destroy all containers the process status of which is unknown (disposal must be under supervision and with numbers and reasons recorded).
8. Take adequate precautions to clear exhaust boxes and precookers of containers at the end of each day's operations.
9. Install a clean, new recorder chart of the correct type and specifications at the start of each day's operation. Set it at the correct time of day. Fill the recorder pen with ink and make sure the timing mechanism is working. Check that the recorder pen arc length is correct.

11.3.11   Steps in a retort process cycle (see Figure 11.2)

Starting a process cycle. Steam is turned on after all bleeders and vents have been fully opened. Bleeders must be left fully open during the entire processing period.

Venting. The vents must be left fully open for a sufficient time and temperature after steam is turned on to insure that all air is swept out of the retort. If the venting

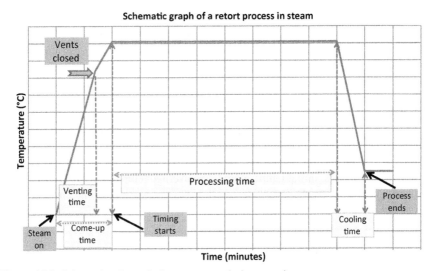

Figure 11.2 Schematic of a typical temperature during retorting.

schedule is only partially completed, air will remain trapped in the inner areas of the retort load. Air in the retort may cause underprocessing.

Sometimes a bottom drain, which is not being used as a vent, is opened to remove the initial condensate buildup or cooling-water residual. Unless the venting schedule specifies otherwise, this drain valve should be closed when the water is removed so that it does not interfere with venting adequacy.

The venting schedule must be timed with an accurate clock or timing device, not with the recorder chart. The venting temperature should be determined by reading the mercury-in-glass thermometer. After venting temperature and time have both been met or exceeded, the vent valve(s) may be closed. Both time and temperature at which the vent valve(s) is closed must be recorded.

Venting schedules and specifications for standard retort configurations were described earlier. All retorts must be vented before processing begins. Continuous cooker coolers must be vented at the beginning of the shift and again after any stoppages.

Come-up time. Steam is turned on until the appropriate process temperature is reached and maintained and process timing begins this period from the time. The venting schedule must be completed within this time. If the steam bypass is used to bring the retort up to processing temperature, it should be closed gradually to prevent a sudden temperature drop.

Check the agreement between the temperature recorder and the mercury-in-glass thermometer. These readings should be taken after the retort temperature has stabilised. They must be recorded.

Process timing. Timing of the process must not begin until the retort has been properly vented and the appropriate processing temperature has been reached and maintained, as indicated by the mercury-in-glass thermometer. Processing temperature is indicated by the mercury-in-glass, not the pressure gauge. Processes must be timed with an accurate clock or timing device, rather than a wristwatch or temperature recorder chart.

Ending a process. Before turning off the steam, make the following checks:

- Check that the appropriate process time has elapsed as indicated by a clock or accurate timing device.
- Check the temperature recorder chart to see that it shows that at least the scheduled process time has been recorded.
- Check the temperature recorder chart to see that no fluctuations below the required retort temperature have occurred.
- Check the mercury thermometer to see that it indicates the appropriate process temperature.

If the check of any of these items is unsatisfactory, appropriate steps should be taken to provide an adequate process. If the check of all of the above items is satisfactory, the steam may be turned off.

11.3.12 Cooling of cans/containers in the retort

The duration of the water-cooling period should be sufficient to bring the average temperature of the contents of the containers to approximately 38 °C (100 °F). However, water cooling should not be continued to the point at which external rusting of the container may occur. In humid locations, water cooling to a temperature of approximately 38 °C may lead to external rusting. Under these conditions, containers should be mechanically dried, or water cooling should be continued until container contents are approximately 48 °C (120 °F). Subsequently, the containers should be air cooled before they are put into storage. When containers the contents are at temperatures substantially above 38 °C are stacked closely together, or are cased, they cool so slowly that spoilage by thermophilic bacteria and heat damage to quality of the product (stack burning) may occur.

Various cooling methods are used, depending on container size and retort hookup, among other factors. If the cooling method involves recirculated cooling water or cooling canals, cooling water must be chlorinated or otherwise sanitised. There should be a measurable residual of the sanitiser at the water discharge point of the cooling system.

Some containers require cooling under pressure to prevent end distortion or buckling. For details on appropriate cooling procedures for particular containers, consult the container supplier.

When water is admitted before the retort overflow, drain, or vent valve is open and the pressure has been down to zero for a short time, a vacuum may be created in the retort when the water condenses from the steam. This condition may result in buckled (panelled) cans or damage to the retort.

Cooling in a canal: This procedure complements processing in vertical retorts. The primary advantage of using a cooling canal is that it frees retorts for immediate reuse. Cooling canals are not generally used in conjunction with horizontal retorts.

- Open the retort overflow, drain, or vent valve and permit steam to discharge until all pressure is relieved.
- Open the top water valve slowly and spray containers with water for about a minute to remove some of the heat.
- Remove baskets of containers from retorts and transfer to the cooling canal.

11.3.12.1   Cooling in retorts without pressure

- Open the retort overflow, drain, or vent valve and permit steam to discharge until all pressure is relieved.
- When pressure is down to zero, close the drain but not the overflow or vent valve. Open the top water valve slowly. Fill the retort with water.
- When retort is full, close the top water valve, and open the bottom water valve. This allows water to circulate from the bottom to the top and out through the overflow. Continue cooling for a few minutes. Then reverse the flow by closing the bottom water valve and opening the drain and the top water valves. If retort is not equipped with an overflow, admitting water at the top and out through the drain (with the retort kept full of water) will give a reasonably uniform cool. The length of time required to complete the cool may be variable, because it depends on such factors as water temperature, container size, and type of product.

Pressure cooling with steam and water. With this cooling method, steam admitted to the top of the retort is used to maintain pressure. Steam is prevented from condensing in the cooling water by admitting cold water slowly under a layer of hot water into the bottom of the retort. The layer of hot water may be obtained by connecting a steam line to the bottom water inlet pipe and heating the water as it is added for the first few minutes of the cool.

Proceed with the cool in the following manner:

- When process is completed, close all bleeders, close bottom steam valve, open top steam valve, and raise pressure in the retort about 14 kPa (2 psi) above that used during the process.
- Open steam line connected to bottom water line. Gradually open bottom water valve, thus running hot water into the retort.
- Continue admitting steam with water until retort is about one-quarter full. Then turn off steam and add cold water under layer of hot water.
- Maintain a constant pressure in the retort by gradually closing top steam valve as the retort fills with water.
- Maintain pressure in the retort at or slightly above that used for processing until the retort is nearly full of water and the containers are covered.
- It is advisable to install a petcock near the top of the retort to indicate when water has reached that level. When the water level nears the top, slightly open the overflow or drain valve and begin to close the water valve to maintain the pressure at the desired point. If the retort is allowed to fill completely with water under full line pressure without relief, pressure in the retort may rise rapidly to the pressure in the water line. This might result in panelling of the containers and retort damage.
- Maintain desired pressure in the retort by a proper balance of water entering at the bottom and discharging through the overflow. Retort pressure should be maintained until all containers are cooled sufficiently to avoid end distortion or buckling.
- Continue opening the overflow drain valve to release the pressure gradually.
- Continue cooling with water entering at the bottom and discharging through the overflow valve for a few minutes. Then reverse flow by bringing water in at the top and discharging through the drain while keeping the retort full of water. This flow reversal will provide a more uniform cool.

11.3.12.2 Pressure cooling with air and water

- When the process has been completed, close all bleeders, open the air valve, and increase the retort pressure about 14 kPa above that used during the process.
- Turn off steam.
- Gradually admit water into the top or bottom of the retort, holding the retort pressure by means of compressed air.
- It is advisable to install a petcock near the top of the retort to indicate when water has reached that level. When the water level nears the top, slightly open the overflow or drain valve, close air valve, and throttle the water valve to maintain the pressure at the desired level. If the retort is allowed to fill completely with water under full line pressure without relief, pressure in the retort may rise rapidly to the pressure in the water line. This might result in panelling of containers and retort damage.
- Maintain desired pressure in the retort by a proper balance of water entry and discharge. Retort pressure should be maintained until all containers are cooled sufficiently to avoid end distortion or buckling.
- Open the overflow or drain valve wider to release pressure gradually.
- Continue cooling by bringing water in through either the top or bottom inlet and discharging through the opposite end of the retort. Reversing the direction of flow helps provide uniform container cooling.

Air cooling. If the water cooling equipment's capacity is inadequate, or a shortage of water is experienced and air cooling is necessary, containers should be stacked in single rows, to allow free air circulation. Stacks should be arranged parallel to the cross ventilation of the warehouse. Careful attention to factors affecting air circulation will help prevent stack burn and safeguard against thermophilic spoilage.

When air cooling is used, care must be taken to protect containers from contact with contaminated surfaces.

See Appendix A for more information on pressure cooling (Bock, 1961).

11.3.13   Precautions for handling filled cans

The installation of some labour-saving devices for handling filled cans can introduce certain hazards, which, if not minimised, may result in some spoilage with the best possible double-seam construction. Before the cans are thoroughly cooled, the seams are slightly expanded and the compound lining is somewhat soft or plastic. In addition to the usual attention to good seam construction, precautions must be taken in handling the cans before they are thoroughly cooled to prevent even small dents on, or near, the double-seams. This involves elimination of fast runways with sharp turns and abrupt stops for handling filled cans, both before and after processing. Care should also be taken to avoid conditions which would strain the seams during the processing or cooling; e.g. excessive holding time of unprocessed cans, inadequate exhausting, or too rapid release of pressure during cooling. In cooling under pressure, particular attention must be paid to the magnitude of the pressure and the length of time it is maintained because the greater the differential pressure between the inside and the outside of the can, the greater the tendency towards forcing minute quantities of the cooling water into the can during this critical period. The microbiological content of water should be kept as low as possible, because the spoilage hazard is reduced in proportion to the number of microorganisms present.

11.4   Retort maintenance and testing

The efficiency of retort performance will deteriorate if retorts are not adequately maintained at regular intervals. In addition to the standardisation of thermometers and pressure gauges, gaskets, closure systems, valves, steam traps, air supply system bleeders, steam spreaders, and water spreaders must be examined to make sure that each piece of equipment is in good operating condition.

The following defects have been discovered in poorly maintained retorts:

1. Broken steam-spreader pipes.
2. Corrosion of holes in the steam spreader to the degree that openings are larger than specified.
3. Inoperative steam traps which allow water to accumulate in the bottom of the retort. This situation is especially serious in retorts with top steam entry.
4. Bottom bleeders filled with scale, resulting in a condensate volume which exceeds the capacity of the bleeder.
5. Inadequate airflow in retorts used for glass containers causing poor circulation of water and poor distribution of heat.

6. Retort-closure failures, which cause plant damage and endanger the lives of plant personnel.

7. Faulty manually operated water and air valves have caused questioning of the adequacy of processes for low-acid food.

11.4.1 Retort pressure safety testing

Retorts, like all pressure vessels, are subject to failure if subjected to excessive pressures. A periodic hydrostatic test of all retorts should be made to minimise the possibility of such failures and for the protection of the operators and other adjacent personnel. Such tests show up defects due to rusting and wear, and these faults may then be corrected. The hydrostatic test is simple and can easily be made either by maintenance personnel or under the supervision of insurance inspection. In some areas, such tests are mandatory and are made by state or local government agencies. In making such a test, all safety valves, pressure-release systems, etc. are removed and the openings thus made plugged or capped. A test gauge is installed, replacing the regular retort gauge. The retort is then completely filled with water, all overflows, etc. closed, and further water pressure is imposed on the entire system to the required testing pressure. This pressure is maintained for a time sufficient to allow for a visual inspection for any leaks.

It is suggested that the complete retort system be so checked once each year.

11.4.2 Valves

All manually operated valves should be carefully checked and repaired or replaced when found faulty. Valves which do not close properly or tend to stick in any position may be the cause of improper functioning in any of the retorting cycles. Improperly packed valve stems are very annoying and dangerous to the operator, and any leaking packing glands should be repaired immediately.

11.4.3 Testing mercury thermometers

Thermometers are the basic reference for the correct processing temperature. It is imperative that they be accurate, dependable, and easily seen by the operator. They should be checked for accuracy at least once a year against a standard instrument. The range should be 77–132 °C (170–270 °F), with scale divisions not greater than 1 °C (2 °F).

11.4.4 Maintenance of other retort equipment

Retort controllers should be checked for accuracy by the manufacturer of the specific controller. This service is usually rendered by the manufacturer through a contract at a fixed price, calling for two to three visits a year. In the event of an emergency on the part of the canner, he may contact the manufacturer who will make a special service call at a pre-established price. The controllers are usually checked by means of a standardised thermometer or hot-water bath of known temperature.

Safety valves should be tested frequently. Pressure gauges should be checked for accuracy at least once a year. Water valves and compressed-air valves, especially the latter, should be checked frequently for leaks.

11.5   Postprocessing can handling

Food preservation by heat processing in cans depends upon the fulfilment of two conditions:

1. The destruction, by heat, of bacteria capable of spoiling the food product.
2. The prevention of bacterial recontamination of the product by means of the sealed container.

Although heat processes have been developed that will insure the destruction of spoilage organisms normally present in the canned product, there remains the hazard of reentry of spoilage bacteria during postprocessing can-handling operations.

The use of down-gauged, double-reduced (DR) tinplate and aluminium cans calls for a fresh approach to can-handling methods, particularly because higher final vacuums in canned foods are preferable. Subsequent to the heat-processing operations, the canned product is usually cooled by water. During the cooling operations, filled cans go from pressure to vacuum and, in most instances, enter automatic or semiautomatic can-handling lines. When filled cans are handled on automatic equipment at high speeds, small deformations of the seams may be more significant as spoilage factors than they are under slow-speed, low-impact conditions.

11.5.1   Spoilage factors due to handling

1. The quality of the can's double seam(s).
2. The presence of bacterial contamination in cooling waters or on wet runways.
3. Excessive abuse, due to poor operation or adjustment of the filled-can-handling equipment.

11.5.2   Operating precautions in can handling

1. Inspect can seams frequently to insure that they are properly formed and that seamer adjustments have not exceeded tolerances.
2. Periodically inspect the can-handling system from the closing machine to the caser. When rough handling of the cans is apparent, smooth out the operation to minimise can seam and body damage. Roll-in casers must be adjusted carefully to prevent violent can-to-can contact or can-seam–can-body contact.
3. Do not allow cans to drop freely into crates from closing machine discharge tables.
4. Do not overfill the retort crates. This will eliminate protruding cans which could be crushed by the crate bales or by crates placed on top of them in the retort.
5. Prevent sharp impacts between filled crates or against protruding points.
6. Operate crate dumps smoothly to prevent impact denting.
7. Chlorinate all cooling waters to a point such that at least 1-ppm residual remains at the discharge end of the can cooler. If chlorination renders the water corrosive, use a suitable corrosion inhibitor.

8. Thoroughly scrub and sanitise all tracks and belts which are exposed to can seams at intervals frequent enough to prevent bacterial buildup.
9. Replace all worn and frayed belting, can retarders, cushions, etc. with new nonporous material.
10. Run cans through a can dryer immediately on leaving the cooling system or tip the full retort baskets to drain water trapped on can ends and allow the cans to dry in the retort crates before discharging into the can-handling unit to lessen the recontamination hazard.
11. High-velocity air blasts over the body and ends of cans remove excess water and maintain dry can tracks. Dry conditions do not encourage the development of bacterial contamination.
12. Microorganisms may develop on can-handling equipment in a film of water, lubricants, or other material. Microbial contamination of the contact point of the can body and double seam can be significant, and drying methods that permit contamination and then remove visible water, but leave contaminated water, at this point may not be sufficient.
13. Chlorinated cooling water will tend to depress microbial numbers in the cooling water.
14. Microbial growth on can runways should be minimised. Deliberate wet conditions can be tolerated by continuously running or spraying a sanitising solution of water containing 3–5 ppm of free residual chlorine on the can tracks. Adequate control of continuous drippage must be provided, and the cans must be dried before entering the labeller.
15. Can-transporting belts and elevators, unless completely dry, should be continuously sprayed at the beginning of the return with water containing 3–5 ppm of free residual chlorine.

11.5.3   Sanitary design of can handling equipment

Filled and sealed can-handling equipment must be designed to minimise pickup of microbial contamination around the double seams. The design must also prevent shock, strain, or even small denting of the cans, particularly on or near the seams, while the cans are still warm. This may cause aspiration of minute amounts of moisture, even through well-made seams. Spoilage may result if the moisture contains bacteria. The engineering and design objectives of postprocessing can lines should be designed to minimise contamination and prevent rough handling (i.e. shock, strain, and denting).

Recommendations to help accomplish these objectives are as follows:

1. Keep handling to a minimum. All switchbacks, quarter turns, elevators, and other changes of can direction or orientation should be engineered according to can speed, size, and weight to minimise strain. In general, sharp reversals of direction should be avoided. Quarter turns should have a long radius to handle the cans gently.
2. The need of 'bumpers' should be avoided, but when necessary, they should be of nonabsorbent, easily cleaned material. Fabrics, wood, or absorbent core belting are not satisfactory, as they will support high bacterial populations that cannot be eliminated. This source of contamination is at the point at which shock and seam strain occurs, and the harder the bump, the greater the hazard. Such bumping also increases the hazard from contamination picked up from subsequent equipment.
3. If possible, eliminate all rolling cans where can-to-can contact may occur. Each can should be positively controlled as far as practical, such as by flat belts or cables, and the drive mechanism automatically controlled so cans cannot slam into each other or have seams strained or damaged by belts and cables continuing to run after line stoppage. Plastic-covered cables are sanitary and considerably reduce friction and abrasion between cable and cans. Line flow control, if properly installed in the right location, will permit an even flow of cans without jamming and will shut off power-driven belts or cables when conveyors are full.

4. When cans are rolled, the slopes, can spacing, and can speeds should be engineered to prevent cans bumping each other. This is a common cause of seam strain and damage. Can track adjustments and can- and guide-clearance tolerances should be such as to assure that any unavoidable can-to-can contact will be seam-to-seam contact and not seam-to-body contact.

5. When cans are rolled, the double seams should not contact the runway surface, except in coolers. In angle-iron runways, the installation of metal half-rounds is one possibility to keep the can double seams from contacting wet or damp angle-iron guides.

6. Drill weep holes to prevent water accumulating in the tracks behind the half rounds. In all cases install properly, or use drip pans, to prevent water from one can track dripping onto another track or cans below. Consider safety, housekeeping, appearance, and comfort hazards caused by dripping water.

7. Where cans pass between belts or other retarders, and on some elevators, cut away the contacting material so that the can double seams ride free of contact. Do not slow the can to the extent the following cans will bump into it.

8. At palletisers and other take-off and transfer points, provide for a continuous and gentle deceleration of the cans. Usually this requires a take-off belt so the cans are moved out of the way of the following can, so that violent impact cannot occur. Do not run cans at high speed into a dead end where they are stopped suddenly by bumping into the can ahead.

9. Dirt and organic debris, as well as bacterial contamination, will accumulate on can-handling equipment. The equipment must be designed and installed so that it can be cleaned. This means it must be accessible for cleaning and because water, detergents, and sanitisers will be used, in a location that provides drainage. In some installations, drip pans or other shields will be necessary.

11.6   Precautions for safe canning operations

1. Comply with Good Manufacturing Practices and regulations as they apply to each operation. Become familiar with equipment requirements, process filing, and supervisor's training regulations.

2. Follow approved procedures. Understand the theory and use information from books and consultants who have practical experience.

3. Keep people informed. Irregularities in cook-room procedure, such as unprocessed cans getting into the warehouse, inadequate venting of retorts during coming-up period, and using a wrong process, will occur unless the cook room is properly supervised. Insist that all supervisory personnel and retort operators understand the significance of the tasks they perform.

4. Equip your retorts properly. Tested and properly installed retort instruments are essential. Instruments need regular testing.

5. Do not reduce processes.

6. Code your cans clearly and keep a record of all codes so that traceability is possible if required.

7. Keep your facility clean. Microbial contamination may lead to spoilage.

11.7   Important points for retort equipment operators

1. Be informed of the theory and practice of retort operation in order that you may understand the significance of the tasks you perform.

2. Do not allow crates of cans to stand around before processing. Spoilage may result.

3. Do not trust your memory on processes.
4. Keep retort instruments and valves in good working order. Instruments need regular testing.
5. Vent retorts thoroughly. Do not start timing process until temperature is up and pressure gauge and thermometer agree.
6. Keep the retort temperature constant and correct.
7. Insure timing is accurate.
8. Keep permanent time and temperature records of every retort load processed.
9. Cool cans properly. Too little cooling may cause either stack burn or spoilage; too much may cause rusting.

To minimise the possibility of mistakes and reduce the danger of unprocessed or underprocessed cans reaching the warehouse, it is recommended that the following cook-room operating procedures be used.

1. Post processes for all products being packed in a conspicuous place near the retorts.
2. Mark plainly and conspicuously all baskets, trucks, cars, or crates containing unretorted material. If several products are being packed at the same time, indicate each product plainly. The GMP regulations require that packers of low-acid foods place a heat-sensitive marker conspicuously in each retort basket, truck, car, or crate containing unretorted food product. Other effective visual indicators may be used instead of a heat-sensitive marker.
3. Hang a distinctive marker from the retort opening when the retort holds unprocessed cans. Place it so that the door or lid cannot be closed before the marker is removed.
4. Do not close a retort until the operator indicates that he/she is ready to start the process.
5. Pick up, puncture, and throw out cans of unknown status with regard to process found on the cook-room floor or any place in the cannery.
6. Take adequate precautions to clear exhaust boxes, precookers, and filled-can washers of all cans at the end of each day's operations.

11.8 Monitoring of sterilisation

Retort baskets, trucks, cars, or crates containing unretorted food products, or some of the containers on top of each basket, shall be plainly and conspicuously marked with a thermochromic (heat-sensitive) indicator or by other effective means, which will visually indicate to thermal processing personnel whether the unit has been retorted.

Thermochromic indicators are offered commercially in various forms, including paint, lacquers, crayons, pellets, tapes, labels, and ink. Some tapes and labels have an area which turns black upon reaching a certain temperature. Other heat-sensitive tapes are colour-coded and are available with stripes or other indicators, which appear after the tape has reached a specific temperature. Heat-sensitive indicators are used to provide assurance that a retort load has not bypassed the retorting operation.

ATI Cook–Chex retort-processing indicators are cardboard tags impregnated with a purple indicator ink that changes colour when exposed to various conditions of time and temperature in a pure steam atmosphere. They do not react to dry heat. They also serve as a permanent record of processing. The pigment is the key to the Cook–Chex reaction. It is a chromium compound which is purple in the anhydrous form. When exposed to an atmosphere of steam, the pigment picks up water and becomes hydrated. Further exposure to the steam causes a molecular shift and the pigment turns to a green colour. The Cook–Chex reaction can be altered to have a sterility value ($F_{121.1°C}$) range of 2 to 90 min while maintaining the same z value. This means that by selecting the proper Cook–Chex based on the time and temperature of the process, the processor can select a Cook–Chex which will not complete its colour change from purple to green until the product has been exposed to the required time and temperature in a saturated steam atmosphere.

11.9   Sterilising glass containers in still, static (discontinuous, non-agitating) retorts (NFPA Bulletin 30-L, 5th edition)

Foods packed in glass require proper processing and some special precautions. Low-acid foods in glass containers are processed under water with superimposed air pressure. For all glass containers to be commercially sterile, careful control and recording of temperatures, pressures, and adequate circulation of water to insure uniform temperatures are essential. The same sterilisation principles apply, and there are many similarities between processing cans and glass jars, but some added or different steps have to be taken when processing glass jars. These are discussed below (Bulletin 30L, 1982).

11.9.1   Retort equipment required for processing glass

The retort must be suitable for processing glass containers under water with superimposed air pressure. Construction and testing should be done in compliance with government and insurance regulations. Vertical or horizontal retorts are used to process glass containers, and the general details of these two types will be considered collectively and the specific details outlined separately.

Pressure control, recording and indicating equipment. The control of pressure when processing glass is critical, as even momentary loss of pressure can cause lifting and therefore leaking of the caps (lids) and too much pressure can damage the lining compound of the cap, which can also result in leakage. Use of adequate automatic controls is required.

Pressure controller. An automatic pressure-control unit should be installed in the overflow line. Its capacity should be sufficient to prevent undesirable increases in retort pressure when the water valve is wide open.

Indicating-pressure gauge. The pressure gauge should be graduated in 7 to 14 kPa (1–2 psig) divisions and should have a range of 0–310 kPa (0–45 psig) or 0–413 kPa (0–60 psig). The minimum diameter of the dial should be of such a size that it can be easily read and the gauge should preferably be of a type in which the operating mechanism is independent of the case for protection of the mechanism. The gauge should be connected to the retort by means of a gauge siphon or gooseneck. In a vertical retort, the gauge should, for reading convenience, be located in the thermometer pocket. In the horizontal retort, the gauge is installed near the mercury-in-glass thermometer or at the top of the retort.

Recording-pressure controller. The chart should be easily readable and should cover the normal operating range.

Water level indicator. A means of determining the water level in the retort shall be provided. This can be best accomplished by using an automatic device or sight gauge, although petcocks have been used but are not recommended. The top layer of glass containers shall be covered with about 150 mm (6 in) of water.

Overflow line. The overflow line should be located approximately 150 mm above the top layer of glass containers. The flow capacity of the overflow line should be larger than the flow capacity of the water supply line.

Drains, drain valves, and screens. The drain valve should be a suitable, nonclogging, water-tight valve. It is strongly recommended that screens be installed over drain openings to prevent foreign objects from clogging drain valves.

The presence of a piece of glass in the valve may cause a leak during a process. Screens should be installed to prevent debris from entering the drain line. A leak can result in the exposure of the upper layers of glass containers to the air/steam atmosphere above the processing water. Glass containers exposed in this manner would be underprocessed because of the lower temperature of the steam–air mixture. To prevent this occurrence, an automatic warning device, preferably activating a horn, should be installed in each retort to indicate the water has fallen below a safe level.

11.9.2   Glass container loading equipment

Crates, trays, gondolas, etc. for holding containers shall be made from suitable polymers, adequately perforated sheet metal, or other suitable material. When perforated sheet metal is used for the bottoms, the perforations should be, at a minimum, approximately the equivalent of 1 in (25.4 mm) holes on 2-in (51 mm) centres, ⅜ in (9.5 mm) on ¾-in (19 mm) centres, ½ in (12.7 mm) on 1-in (25.4 mm) centres, ¾ in (19 mm) on 1½-in (38 mm) centres, 1½ in (38 mm) on 3-in (76 mm) centres, or 1¾ in (44.5 mm) on 3½-in (89 mm) centres. The dividers between the layers of containers should be perforated as above. Dividers with a greater percentage of open area may offer some advantages by improving temperature distribution throughout the retort. Plastic dividers are recommended to reduce the potential for electrolytic reaction and abrasion with the closure during retorting. They should be checked periodically to prevent broken or brittle dividers from shedding fragments, which may cause valve problems or plug drains.

11.9.3   Installation of air supply and controls

A reliable supply of compressed air at the proper pressure and a means of introducing it into the retort at an adequate rate shall be provided. An automatic pressure control unit shall be provided for both vertical and horizontal retorts. If air is used to control circulation, the air shall be introduced with the steam at the bottom of the retort close to the entry of the steam line into the retort. Air circulation shall be maintained continuously during come-up and processing to insure uniform temperature distribution.

The amount of air pressure required depends upon the steam pressure in the retort line and the location of the point at which air is introduced into the steam line. It should be in the range of 350–480 kPa (50–70 psig). If air pressure at this point does not exceed steam pressure, air will not pass into the retort. This lack of airflow would seriously affect temperature distribution and increase 'chatter' during come-up. The amount of air required to prevent 'chatter' during come-up depends, to a considerable extent, on steam pressure in the line and backpressure in the steam distributor. The amount of air necessary may vary from 0.23 to 0.43 m³/min. The amount of air required during process and cool periods will vary with retort size. These suggestions assume that there is no leakage from the retort, either through the gasket or overflow valves. Airflow should be checked at regular intervals with a flow meter. Further details on the amount of air required during this period can be obtained from a qualified thermal processing authority equipment manufacturer.

A check valve shall be provided in the air supply line at the top of the retort to prevent water from the retort entering the system. A separate compressor is suggested for operating control instruments. If a separate compressor is not provided, the air line to the instruments should be connected to a separate air supply system. An adequate filter system to supply clean, dry, oil-free air is needed for the control instruments.

11.9.4   Steam supply and introduction

To achieve reasonable come-up times, steam pressure in the main line to the cook room should not be less than 620 kPa (90 psig) at all times during operation.

11.9.5   Retort details

Vertical Retorts. Several specific details of vertical retorts require special emphasis:

1. *Door or lid-securing devices.* The retort should be provided with eight wing nuts or equivalent lid-securing devices.
2. *Retort and crate diameters.* There should be a minimum of 3.8 cm (1½–2 in) clearance between the side wall of the crate and retort wall. This allows ample water circulation along the retort wall. The inside diameter of the retort should be a nominal 107 cm (42 in) and the outside diameter of the crates should be a nominal 97 cm (38 in). Not only should diametric clearance be provided, but centring guides should be installed to assure proper clearance between the sidewall of the crate and retort wall.
3. *Retort headspace.* A minimum of 100 mm (4 in) headspace should be maintained between the top water level and the top of the retort shell.
4. *Cooling water.* Cooling water should be introduced at the top of the retort below the water level but above the jar level, either through a minimum of four openings equally spaced around the circumference of the retort or through a perforated ring. This suggestion

assumes a feeder pipe of 2.54 cm (1 in) minimum size with a minimum of 344 kPa (50 psig) water pressure at the retort. When water pressures exceed 551 kPa (80 psig), the 2.54 cm (1 in) cooling-water line should be restricted or reduced to prevent excessive water-flow rates or pressures during cooling. A check valve should be included in the cooling-water line to prevent a pressure drop in the retort, if water pressure should fall below the retort pressure.

5. *Steam distributor.* Adequate results can be achieved by several methods. One means is an assembly of six pipes radiating from a centre coupling with fishtail nozzles at the end of each pipe to direct steam up the retort walls outside the crates. Another means is a four-legged cross in which each pipe leg is perforated with holes directed 15° below horizontal along one side only. The legs are arranged in opposing pairs to give alternate live and dead quadrants.

6. *Air supply.* The amount of air required during the process and cooling periods varies according to retort size. A flow rate of 0.084 m³ (3 ft³) per minute is suggested for three-crate vertical retorts and 0.113 m³ (4 ft³) per minute for four-crate vertical retorts. These suggestions assume no leakage from the retort, either through the gasket or through the overflow valves. Airflow should be checked at regular intervals with a flow meter.

Horizontal Retorts. The general arrangement of a suitable horizontal retort is given in a diagramme later in the text. Details of piping diagrammes and specifications may be obtained from a qualified thermal processing authority. Several specific details of horizontal retorts require special emphasis:

1. *Water circulating system.* A water circulating system is suggested to insure uniform heat distribution. This system should be installed so that water is drawn from the bottom of the retort through a suction manifold and discharged through a spreader, which extends the full length of the retort top. The pump for this system should have sufficient capacity to replace the retort water every 4–5 min. The size of the suction manifold, circulating system, pump, and water spreader depends on retort size. The holes in the water spreader should be uniformly distributed and have an aggregate area not greater than the cross-sectional area of the outlet line from the pump. The suction outlets should be protected with a screen to keep debris from entering the circulating system, because such debris may foul the pump and clog the water spreader holes. The pump should be equipped with a pilot light or other signalling device to warn the operator when it is not running. A bleeder should be provided to remove air when starting operations.

2. *Hot-water supply.* A hot-water supply for filling retorts to reduce come-up times and avoid thermal shock is advisable. A steam-heated storage tank equipped with a temperature controller and thermometer should be provided to save hot water discharged from cooling retorts. The hot-water outlet of the tank should be connected to the retort circulating system on the suction side; the circulating pump can then be used to pump water from the tank into the retort for the next cook.

3. *Retort headspace.* The amount of necessary headspace varies with the retort size. In general, water level should be above the bottom of the water spreader and far enough below the top of the retort to allow proper control of the overriding pressure.

4. *Steam distributor.* The steam distributor shall run the full length of the bottom of the retort with perforations distributed uniformly along the upper part of the pipe. For details on installation of a steam distribution system, consult a qualified thermal processing authority.

5. *Cooling water.* At the start of cooling, water should be introduced into the suction side of the pump so that cold water can be mixed with hot water to prevent breakage of glass containers from thermal shock. Near the end of the cool, water can be admitted rapidly through the spreader in the top of the retort.

11.9.6   Procedures for processing in glass containers

The glass containers described herein are sealed with vacuum-style metal closures, which are not intended to hold internal pressure (press-twist-type caps). Processing of closed glass containers is carried out in water-filled retorts, with a top cushion of air under pressure to hold the closures in hermetic seal throughout the retort cycle.

The headspace for most products should not be less than 6% of the container volume, measured at a sealing temperature of 50 °C (130 °F) for cooks in the range of 116–121 °C (240–250 °F). Unless product-sealing temperature is very high, smaller container headspaces do not allow sufficient space for expansion of the contents of the sealed container. As a result, the closure may be displaced. For most products, the temperature at sealing should be above 54 °C (130 °F). Lower sealing temperatures require larger headspaces to accommodate product expansion.

In general, retort operating pressure should be 138–240 kPa (20–35 psig) for cooks in the range of 116–121 °C (240–250 °F). This depends on the type of closure used and sealing conditions. Further information may be obtained from your closure manufacturer.

Container closure examinations: Glass closures should be inspected to insure that proper seals have been made. Nondestructive inspections should be conducted at intervals not to exceed 30 min and destructive inspections not to exceed every four h. In addition, nondestructive inspections are required after a jam in the capper, adjustment, or prolonged shutdown. If vacuum closures are used, capper efficiency must be measured by 'cold water vacuum'.

Position of glass containers in retort: In heat-processed foods which contain freely flowing liquids, heat penetrations are enhanced by convection currents. These currents move generally in a vertical direction; consequently, they seek channels which permit such motion. When their progress is impeded or baffled by solid material, the currents flow around the obstruction seeking channels of least resistance to flow. For this reason, alignment of certain foods in the glass container is important with regard to heat penetration.

Except in special cases, glass containers should be processed in an upright position. When packing or filling in glass containers results in stratification, these containers should be processed so the plane of stratification is vertical. In the case of asparagus, the spears are generally parallel and tightly packed in a vertical position, with channels containing liquid parallel to the spears. As a result, the rate of heat penetration is greater when glass containers are placed upright in the retort.

Stacking of glass containers: Glass containers should be stacked to permit free circulation of water throughout the retort load. Divider plates should be constructed as described earlier (Section 11.9.2). Retort crates should be made so they stack on each other, and in no case should the top layer of glass containers extend above the rim of the crate.

Rapid handling and prompt retorting of filled glass containers: A long holding period between filling and sealing or between sealing and retorting glass containers may result in souring, off-flavour, vacuum loss, and lower initial temperatures, depending on the nature of the product. Processing should follow promptly after sealing; it is desirable that this period be no longer than 1 h. If delays are encountered, partial loads should be processed.

Retort cycle: For processing most products, the highest quality is obtained when the retort is brought to processing temperature quickly and the process timed accurately, followed by prompt and rapid cooling. This procedure not only protects product quality but also shortens total time required for each processing cycle and results in efficient retort use.

During the retort cycle, a number of details must be observed for reliable processing:

1. *Water level.* In vertical retorts, the starting water level should be sufficient to cover the crates of filled glass containers as they are loaded into the retort, without excessive overflow. The water should be brought to the overflow pipe at the time the last crate is loaded. The water level should be at least 150 mm (6 in) above the top layer of glass containers. No extra glass containers should be placed on the top of the uppermost crate. The water level shall remain above the top layer of glass containers during the entire come-up time, cooking, and cooling. For efficient operation of horizontal retorts, a reserve-water storage tank is advised. As soon as the retort is closed, water should flow from the tank into the retort until the top layer of glass containers is covered by at least 6 in (15 cm) of water. Many water supplies have a mineral content that requires treatment to prevent unsightly deposits on the package and deleterious effects to the closure.

2. *Preheat water temperature.* The water temperature should approximate the water temperature of the product in the glass containers at the time of sealing. If the water temperature is more than 8 °C (15 °F) above the sealing temperature of the glass containers, closures may be displaced. Water at substantially lower temperatures may cause breakage and will reduce initial temperature. The temperature of the preheat water should be adjusted by the introduction of water or steam to approximate the temperature of the product in the filled glass containers to minimise thermal shock and promote rapid heating.

3. *Air pressure.* To insure an adequate air supply, air to the bottom of the retort should be turned on immediately after closing the retort and the overriding pressure built up promptly to operating range *before steam is turned on*. This pressure shall be maintained throughout come-up, processing, and cooling.

4. *Come-up time.* Timing of the process shall not begin until the specified retort temperature has been indicated by the official retort thermometer.

5. *Process timing.* The timing of the process shall not begin until the retort has reached the proper processing temperature.

6. *Cool.* At the end of the specified process time, steam is shut off and the necessary overriding pressure maintained with air. When a cooling canal is used, retort cooling should be continued under full retort pressure for a sufficient length of time to produce some vacuum in every container. The water-cooling period, either in the retort alone or in combination with cooling canals or sprays, should be sufficient to bring the temperature of the product to an average of 38 °C (100 °F). The product should not be cooled below ambient temperature, because condensation could accumulate on the container, promoting rusting of the container closure. Insufficient cooling can promote the growth of thermophilic bacteria or damage product quality (commonly known as stack burn).

All cooling canals or recirculated cooling water supplies shall be chlorinated or otherwise sanitised. There should be a measurable residual of the sanitiser at the container cooling water discharge point.

11.9.7   Precautions for handling glass containers

Glass containers can be damaged by excessive mechanical or thermal shocks. It is suggested that lines be planned in cooperation with package suppliers to assist in obtaining the best container cooling system and method of handling. With high-speed filling lines, particular attention should be directed towards eliminating

1. Bruising jars by abrupt stops or careless handling.
2. Scuffing on conveyors or transfer points.
3. Excess thermal shock during filling, processing, or cooling.

The damage done to filled containers by impacts usually is greater than that to empty containers because of the added weight of the contents. The hermetic seal of the friction closures applied to the jars can be maintained by careful attention to

1. Maintenance of maximum filling temperature of product.
2. Adequate headspace in the glass jar.
3. Correct application of the closure.
4. Proper temperature of water during retort loading.
5. Ample overriding pressure in retort during processing and particularly during cooling.
6. Proper heat distribution in the retort.
7. Provision for ample cooling after retorting.
8. Reasonable handling of jars during and after cooling.

Appendix A: Cooling cans under pressure in retorts

The following procedure for cooling cans under pressure is taken from the bulletin, Retort Installation, Equipment and Operating Procedures, by J.H. Bock (1961), Continental Can Company. See Tables 11.5–11.10 for useful information regarding the installation of vertical and horizontal steam retorts.

Pressure cooling hookup

Manually operated pressure cooling installations are relatively simple and are suitable for both vertical and horizontal retorts. There are two procedures for pressure cooling, one using steam and the other air, to maintain a desired pressure in the retort during cooling and thereby prevent buckling.

Pressure cooling under air pressure

The use of air for this imposed pressure is preferable over steam, due to the noncondensable properties of air. This makes the operation considerably more foolproof, as it eliminates the possibility of accidentally condensing the steam in the retort, thus producing a partial vacuum and increasing the buckling hazard. A pressure control system is essential to insure moderate accuracy in maintaining the desired air pressure inside the retort. A pressure-reducing valve serves as an inexpensive controller. One such valve placed in a header supplying a bank of retorts is sufficient for all the

Table 11.5 **Volume of air atmospheric pressure per volumetric unit of supply tank capacity in cubic feet/cubic feet (m³/m³)**

Cooling pressure (gauge)	Receiver pressure (psi)									
	60	**70**	**80**	**90**	**100**	**110**	**120**	**130**	**140**	**150**
10 psi	3.40	4.10	4.75	5.45	6.10	6.80	7.50	8.20	8.85	9.55
12.5 psi	3.20	3.90	4.60	5.30	5.95	6.65	7.30	8.00	8.70	9.35
15 psi	3.05	3.75	4.40	5.10	5.80	6.45	7.15	7.85	8.50	9.20
18 psi	2.85	3.55	4.20	4.90	5.55	6.25	6.95	7.65	8.30	9.00
21 psi	2.65	3.35	4.00	4.70	5.35	6.05	6.75	7.45	8.10	8.80

Table 11.6 **Retort volumes (some common sizes)**

Retort style	Dimensions (in)	Retort volume (cubic ft)
Vertical	42 × 71 in	58.0
Vertical	42 × 84 in	67.5
Vertical	42 × 96 in	77.2
Vertical	42 × 108 in	87.0
Horizontal	42 in diameter	9.65 cu ft per foot of length
Horizontal	54 in diameter	15.9 cu ft per foot of length
Horizontal	60 in diameter	19.7 cu ft per foot of length

retorts in that bank. Air lines to each retort should be ¾–1 in (19–25 mm) size, with the headers sized in accordance with the number of retorts which might be pressure cooled simultaneously.

The following formulae and the tables are designed to aid in the calculation of the amount of air required for pressure cooling, supply tank size, and compressor size required. It is suggested that all calculations be based on the volume of an empty retort to insure an ample air supply when only a partial load has been processed. Calculations should be made on a single retort and, if cooling should start simultaneously in more than one, the requirements would be governed accordingly.

Volume of air at atmospheric pressure required per retort:

$V1$ = Volume of retort.
$P1$ = Atmospheric pressure, 14.7 psi (101.3 kPa).
$V2$ = Volume of air at atmospheric pressure to obtain cooling pressure $P2$ in retort.
$P2$ = Pressure of cooling = $P1$ + manometric cooling pressure (gauge reading).
$V2 = V1 \times P2/P1$.

Supply tank capacity per retort = $V2$/(volume of air available per volumetric unit of supply tank).

Table 11.7 **Minimum requirements for 3- and 4-basket vertical retorts**

Steam pressure	90 psi
Steam inlet	1 in
Steam regulating valve	(consult manufacturer)
Steam spreader	1 in
Steam spreader – hole size	$^3/_{16}$–¼ in
Steam spreader – number of holes	47–62 ($^3/_{16}$ in)
Vent line	1¼ in
Vent valve	1¼ in gate
Bleeders	$^1/_8$ or ¼ in petcocks
Overflow	1¼ in (gate valve)
Drain (not critical)	1½–2 in (gate valve)
Safety valve	(meet ASME or local codes)
Water pressure	40 psi
Water inlet	1 in
Air for control instruments	20 psi, ¼ in tube, or $^1/_8$ in pipe
Air for pressure cooling	¼ in, 40 psi
Pressure relief valve	1¼ in (adjustable)
Temperature control instruments	Control to ± 1 °F

Compressor capacity = volume of air required per period of time = V2/(period of time between consecutive cooling operations).

Example: vertical retort

*   Size: 42×72 in (107×183 cm).
*   Cooling pressure (gauge reading): 10 psi (69 kPa).
*   Time between consecutive cooling operations: 5 min.
*   Supply tank pressure (gauge): 100 psi (689 kPa).

To determine the volume of air at atmospheric pressure necessary to obtain a gauge pressure of 10 psi (69 kPa) in the retort, solve for V2:

V2 = V1 × P2/Pl, in which
V1 = 58 cu. ft (1.64 ms) – from Table 11.6.
P1 = 14.7 psi (101.3 kPa).
P2 = 14.7 + 10 = 24.7 psi (170.3 kPa).
V2 = 58 cu. ft × 24.7 psi, 114.7 psi = 97.5 cu. ft (2.76 ms) per retort.

The supply tank capacity necessary for one retort is calculated with V2 and data from Table 11.5:
Supply tank capacity = 97.5 cu. ft/(6.10 cu. ft/cu. ft)
= 16 cu. ft (0.45 ms).
The compressor capacity is then calculated as:
Compressor capacity = V2/(period of time between cooling)
= 97.5 cu. ft/5 min = 19.5 cu. ft/min. (0.55 ms/min.).

Table 11.8 Minimum requirements horizontal retorts

	Up to 8 ft	8–15 ft	Larger than 15 ft
Steam pressure	90 psi	100 psi	125 psi
Steam inlet	1 in	1¼ in	2 in
Steam regulating valve	(consult manufacturer)		
Steam spreader	1 in	1 ¼ in	2 in
Steam spreader – hole size	³/₁₆ in	³/₁₆ in	³/₁₆ in
Steam spreader – number of holes	42–62 (³/₁₆ in)	81–108 (³/₁₆ in)	183–244 (³/₁₆ in)
Vent line	1¼ in	1½ in	2½ in
Vent valve	1¼ in gate	1½ in gate	2½ in gate
Bleeders	¹/₈ or ¼ in petcocks	¹/₈ or ¼ in petcocks	¹/₈ or ¼ in petcocks
Overflow	1¼ in (gate valve)	1½ in (gate valve)	2 ½ in (gate valve)
Drain (not critical)	1¼ in (gate valve)	1½ in (gate valve)	2 ½ in (gate valve)
Safety valve	(meet ASME or local codes)		
Water pressure	40 psi	50 psi	60 psi
Water inlet	1 in	1¼ in	2 in
Air for control instruments	20 psi, ¼ in tube, or ¹/₈ in pipe		
Air for pressure cooling	1 in, 40 psi	1¼ in, 50 psi	2 in, 60 psi
Pressure relief valve	1¼ in	1½ in	2 ½ in
Temperature control instruments	Control to ± 1 °F		

Table 11.9 **Venting schedule for vertical retorts**

		Time (minutes)		Temperature (°F)	
Steam inlet	Vent	Without divider plates	With divider plates	Without divider plates	With divider plates
1 in	1¼ in	4	6	220	225
1¼ in	1½ in	3	5	220	225
1½ in	2 in	3	5	220	225
2 in	2½ in	2	4	225	230

Pressure cooling under steam pressure

Pressure cooling with steam, of course, does not require a compressor system and air lines. However, such an operation requires careful and trained operators, because mishandling a few retort charges could easily incur losses sufficient to purchase and install an air system. If steam is used, remember that a baffle plate should be placed over the water inlet to prevent splashing, which might cause condensation of stream in the retort.

Water process with superimposed air pressure

It is often necessary, and usually desirable, for glass containers with press-on/twist-off closures, to process cans having large areas of flat surface in superheated water and with superimposed air pressure. Retort hookup for this procedure is practically the same as for pressure cooling with air. The water temperature is controlled by a temperature controller to admit steam for heating and a separate control system is used for controlling the superimposed air pressure. A source of hot water for filling the retort is advisable to reduce come-up time. Insulated storage tanks for hot water are often used, with pumps for transferring water to and from the retorts. A supply of compressed air capable of maintaining pressures up to 30 psi (207 kPa) in the retort is required.

Automatic operation

Hookups for automatic operation vary considerably, depending on the extent of automatic operation desired and the requirements determined by the type of the particular supplier's system selected. Such systems may vary from a simple timed-process cycle through computer systems which carry out all steps through the venting, processing, and pressure-cooling cycles. All such installations must be consistent with all of the fundamental requirements for efficient retort operation. The manufacturer of the automatic system selected should be consulted regarding proper installation of the control and operating equipment, and the final hookup should be checked by someone experienced in retort hookups and operation.

Table 11.10 Venting schedule for horizontal retorts

Steam inlet	Vent	Up to 8 ft				8–15 ft			
		Time (minutes)		Temperature (°F)		Time (minutes)		Temperature (°F)	
		Without divider plates	With divider plates	Without divider plates	With divider plates	Without divider plates	With divider plates	Without divider plates	With divider plates
1 in	1¼ in	4	6	220	225	–	–	–	–
1¼ in	1½ in	4	6	220	225	6	8	220	225
1½ in	2 in	4	6	220	225	6	8	220	225
2 in	2½ in	3	5	225	230	5	7	225	230

References

Bock, J. H. (1961). *Retort installation, equipment and operating procedures.* Continental Can Company.

Codex Alimentarius. (2011). *Code of hygienic practice for low and acidified low acid canned foods.* CAC/RCP 23–1979.

FDA 21 CFR Part 113 temperature-indicating devices; thermally processed low-acid foods packaged in hermetically sealed containers. *Federal Register, Final Rule, 76,* (March 3, 2011) Thursday, No. 42.

Thermal processes for low-acid foods in metal containers, Bulletin 26L (12th ed.) (June 1982). Washington, DC: National Food Processors Association.

Thermal processes for low-acid foods in glass containers, Bulletin 30-L (5th ed.) (September 1984). Washington, DC: National Food Processors Association.

Tucker, G., & Featherstone, S. (2011). *Essentials of thermal processing.* Wiley-Blackwell, United Kingdom (West Sussex).

Sterilisation systems

12

12.1 Introduction

Sterilised and pasteurised foods are shelf stable, without any preservatives, dehydration, or refrigerated storage by virtue of the heat treatment that they have received (and, in the case of pasteurised foods, by pH as well). Canned foods are sterilised inside hermetically sealed containers.

It is necessary to define what is implied by 'sterile'. From the standpoint of microbiology, sterile means the total absence of microorganisms in viable condition. In commercial practice, it is not desirable to sterilise foods to that point, due to the high heat resistance of the spores of certain thermophilic and thermoduric bacteria. The heat needed to destroy them would seriously impair the quality of food products. Therefore, foods which are thermally sterilised receive a process that leaves them in a condition called 'commercially sterile'. The concept of commercial sterility has been defined and discussed elsewhere in this book and will not be repeated here. It must be understood, however, that aseptically packaged low-acid foods, of pH higher than 4.6, must be commercially sterile. Otherwise, serious consequences could arise related to public health stemming from microbiological spoilage, if products are stored at ambient temperature. Commercially sterile foods are microbiologically stable and safe from the viewpoint of public health, if the container is appropriate and is hermetically sealed.

In the case of foods with high acidity, i.e. pH 4.6 or lower, it is much easier to obtain commercial sterility, because the lower pH value decreases the heat resistance of bacteria and their destruction can be effected in a relatively short time and at a temperature much lower than that needed for low-acid foods. With high-acid foods, the multiplication of certain species of bacteria that survive these relatively mild sterilisation processes is inhibited by the low pH. Therefore, it is very important to make a distinction between low-acid foods and high-acid foods, because by their nature they require very different thermal sterilisation procedures. In sterilising high-acid foods, like fruit juices and drinks, enzyme inactivation must be considered.

To destroy the heat-resistant spore-forming bacteria that can be found in low-acid foods, temperatures in the region of 110 to 135 °C are generally used. These are temperatures well above the boiling point of water under atmospheric conditions and can only be achieved in water by using steam, generated under pressure in special equipment. To pasteurise most acid foods, temperatures between 70 and 100 °C are generally used (depending on the product, pH, and other ingredients) (Tucker & Featherstone, 2011).

For aseptic packaging, thermal processes of 93 to 96 °C (200–205 °F) for 15–30 s are commercially used. Low-acid foods are generally heated to approximately 138–149 °C (280–300 °F) and are held for 1 to 30 s.

A Complete Course in Canning and Related Processes. http://dx.doi.org/10.1016/B978-0-85709-677-7.00012-8

The most basic and traditional type of retort is the static pure (saturated) steam or full water-immersion type that is used for metal cans and glass jars or bottles. These are still used very effectively by many canners, but retort manufacturers have produced a host of modifications and other designs to allow for the processing of foods in alternate types of packaging (e.g. pouches, bowls, cups, and trays) to achieve faster throughput, more convenient processing, and energy saving. The different designs are aimed at specific types or production scenarios and product type, so it is important to understand what is available so that an informed choice can be made if setting up a new cannery or replacing old equipment.

The methods used for sterilising low-acid foods are much more complex and must satisfy more stringent requirements from the standpoints of public health, food preservation, and quality control of food and container.

In aseptic processing systems, following sterilisation, the product is cooled to near ambient temperature and transported (pumped or gravity/pressure transferred) to a sterile container. Commercial sterility of the product must be maintained from the heat processing to the packaging operations. The product is a hermetically sealed container holding commercially sterile product. It may be stored at ambient temperatures for prolonged periods, up to several months or one or two years, depending on package characteristics and package handling.

In addition to the concept of commercial sterilisation, basic concepts related to determination and evaluation of thermal processes of sterilisation must be considered. Thorough knowledge of those concepts contributes to a better understanding of the procedures that are necessary to apply to aseptic processing and packaging to obtain safe, shelf-stable, high-quality products.

Thermal sterilisation has been successful in the case of many food products, because the equipment required is relatively simple and because of the characteristics of bacteria in relation to their heat resistance. Mechanically, heat can be easily measured and controlled. Much microbiological information has been developed during the last 100 years on the subject of thermal sterilisation of food products. Methods to determine the parameters of time and temperature needed to thermally sterilise foods are well known and in general use. Many products can be commercially sterilised prior to packaging by continuous processes so that their organoleptic and nutritional quality is not significantly affected. In the techniques applied to aseptic packaging, continuous heat exchangers can be designed so that any temperature profile may be applied. However, it is important to remember that the same basic principles that apply to conventional canning for achieving commercial sterility in foods also apply to aseptic processing and packaging. These concepts are discussed in the chapters, 'Microbiology' and 'Heat Penetration' in Book II.

12.2   Pure-steam static batch retorts

These can be horizontal or vertical. Still-steam retorts were one of the first types of retort systems used to process low-acid canned food. The heating medium is saturated steam, which is an excellent medium for heat transfer. As there is no agitation,

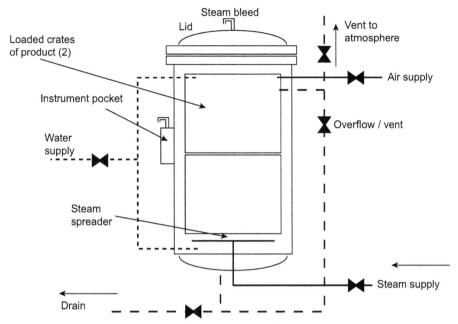

Figure 12.1 Schematic of a vertical steam static batch retort.

it is very important that, when using pure steam, still retorts have adequate venting to remove air before starting the process, as air pockets can create 'cold spots' (see Chapter 11).

Vertical Retorts are loaded and unloaded from the top, usually by overhead hoist into the retort, and usually hold 3–4 baskets (see Figure 12.1).

Horizontal Retorts have their retort baskets pushed or conveyed into the retort. They can vary from one small crate to large retorts holding 12 or more crates of product.

Solid pack products (e.g. corned meat or pet food), delicate products that get damaged from agitation during processing (e.g. fish in sauce), and products that would not benefit from any agitation during processing are generally processed in these kinds of retorts.

In summary, static-steam batch retorts are a relatively inexpensive, effective, and efficient method of processing batches of solid pack or delicate products and are commonly used. However, baskets have to be loaded and unloaded, delays before processing have to be controlled, and retorts must be vented for each batch.

12.3   Still-water immersion batch retorts

These can be horizontal or vertical and are similar in design to pure steam retorts; however, the product is processed in water rather than pure steam. The rate of heat transfer from water to product is significantly slower than from steam to product.

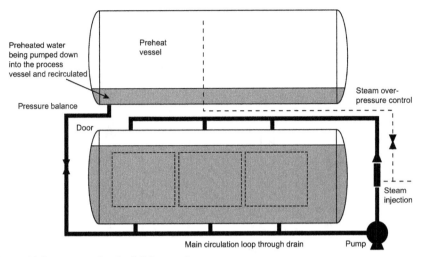

Figure 12.2 An example of a full immersion retort.

It is, however, easier to control the overpressure, and this is the method of choice for processing glass containers. The water level in the retort must be maintained to provide water above the top level of the containers at all times, as containers at the top might be underprocessed as they will be in a mixture of steam and air (see Figure 12.2).

12.4   Still-water cascade and water-spray batch retorts

A water-cascade retort is a retort in which a small amount of process water is drawn from the bottom of the retort by a high-capacity pump and distributed through a manifold or plate in the top of the retort. A water-spray retort is a retort in which a controlled amount of process water is drawn from the bottom of the retort by a high-capacity pump and distributed through spray nozzles located along the top and sides of the retort onto the product. Heating medium is therefore steam and water. The movement of the water aids heat transfer into the product (see Figures 12.3–12.5).

12.5   Crateless retorts

Crateless retorts are large vertical retort vessels with doors at either end. During filling, the top door is opened and cans are loaded in directly, their fall being cushioned by water. Once the retort is full, the water is flushed out by steam, and the cans are processed. After processing, the cans are water-cooled and then at the end of cooling, the bottom door is opened and the cans drop into a cooling canal (see Figure 12.6) (Mãlo Inc.).

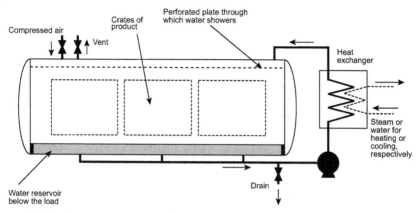

Figure 12.3 Schematic of a water cascade retort.

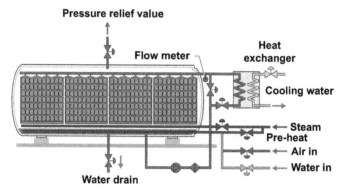

Figure 12.4 Schematic of a steam water spray (SWS) retort.
Courtesy of JBT Food Tech, www.jbtfoodtech.com.

Figure 12.5 Photograph of the inside of a water spray retort.
Courtesy of JBT Food Tech, www.jbtfoodtech.com.

Typical processing sequence

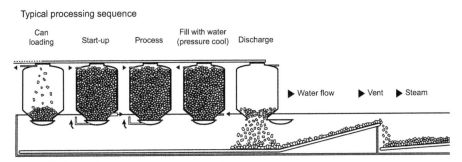

Figure 12.6 Crateless retort.
Courtesy Mãlo, Inc., www.maloinc.com.

These retorts are 2.4 m high and 1.8 m in diameter. Their capacity is four to five times greater than that of the conventional three-basket vertical retort. With an automatic conveyor system, one retort automatically top loads to the preset count for the particular can size, the conveyor gate closes, and the next retort in line starts to load. Before loading starts, the retort is filled to the top with water at the initial temperature desired. This water acts as a cushion for the cans falling through the 25–51 cm hole in the top. When the retort is loaded, the lid and the overflow valve are closed. Steam is admitted through a spreader at the top and forces the water out through the bottom. This water may be collected, reheated, and used in another retort. The venting time necessary before the bottom drain is closed has to be determined for individual installations. The bottom drain valve also acts as a vent valve for this type of retort.

At the end of the process time, there are two methods for unloading cans from the retort. In either method, pressure cooling may be performed in the retort at the end of the cook process when required.

- The first method requires that the water level in the cushion canal be maintained just below the bottom door of the retort. The hydraulic bottom door is opened and the cans are discharged into the cushion canal. A conveyor in the bottom of the canal moves cans to the cooling canal where they are cooled to the desired temperature. The cans are then conveyed to an unscrambler, which will orient the cans and feed the labelling machine. When all cans have been discharged from the retort, the bottom door is closed, the retort is refilled with warm cushion water, and the unit is ready to start a new cycle.
- The second method, the submerged or vacuum system of unloading cans, is strongly recommended, especially when running larger size cans. This method requires the water level in the cushion canal to be maintained above the bottom door of the retort. This requires a special double-chamber bottom door to keep condensate off the cans. When the cook process is completed, the sealed retort is refilled with cooling water or warm cushion water. The top door remains closed; when the bottom door is opened, cans gently float into the transfer tank, while a corresponding amount of warm water is drawn up into the vessel from the transfer tank because of the resulting vacuum effect. After all cans are discharged from the retort, the bottom door is closed and the unit is ready to start a new cycle.

12.5.1 Special characteristics of system

- Flexibility of Process and Can Size – This system will handle all standard size containers and any process at temperatures up to 134 °C (274 °F).
- Steam Savings – Steam usage is less than vertical still retorts, because there is substantially less metal mass to heat, less metal surface subject to heat radiation losses, and no extended venting requirements. Crates are eliminated.
- The system can be completely automatically operated.
- There has been some criticism regarding denting of cans in the unloading of crateless retorts. This can be kept to a minimum, once the operator is familiar with the process, to a point at which denting is no greater than in conventional still retorts. A moving belt under the discharge has been found helpful in this respect.

12.6 Rotating batch retorts

An effective way to increase the rate of heat transfer is to agitate or mix the product. Rotating batch retorts are horizontal, in which the crates are pushed or conveyed into the retort. The heating medium can be pure steam, air–steam mixtures, water–steam mixtures, or water. The kinds of products processed range from normal and vacuum-sealed cans, glass jars, aluminium and plastic trays and pouches, glass or plastic bottles, and retortable cartons (Stock America).

The operation is similar to other batch retorts. The agitation of rotation of a vertical can is end over end, which allows for efficient mixing, but headspace size is important.

The SuperAgi™ from JBT is a batch retort that can do end-over-end (EOE) rotation, rocking, or static processing. It comes as an empty vessel technology with Steam Water Spray (SWS) processing method or can be supplied as a multiprocess machine that can do immersion (full and partial) and SWS (see Figure 12.7).

12.6.1 End-over-end rotation

Heat transfer can be greatly enhanced for some containers and products by agitating the containers during cooking. Movement or agitation of the containers forces convection heating of the product inside the container. The level of convection is dictated by the headspace of the container, the viscosity of the product, the geometry and orientation of the container, the position inside the retort basket, and the type of movement applied to the container.

One of the more common types of forced-convection batch processing uses end-over-end (EOE) rotation of the containers. This is usually used for metal cans, plastic bowls, and glass jars.

In the EOE process mode, the rigid cylindrical containers are usually placed upright inside the retort baskets, and the containers are rotated as shown in Figure 12.8.

The mechanical action of the retort's rotation mechanism rotates the baskets EOE at speeds ranging from 2 to 25 rpm. This forces the headspace bubble inside the container to move from the container wall through the centre of the container, forcing hot and cooler product to thoroughly mix. The mixing greatly reduces the length of time required to produce a sterile product.

Figure 12.7 SuperAgi™ retort.
Courtesy of JBT Food Tech, www.jbtfoodtech.com.

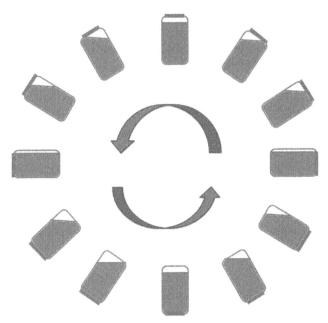

Figure 12.8 Headspace bubble movement in end-over-end rotation.

12.7   Horizontal circulating water retorts

Horizontal water retorts, with or without reels for EOE basket rotation, are in use worldwide from various manufacturers. The process vessel, or 'autoclave', is conventionally connected to a separate pressurised water supply tank from which heated water, preheated to a temperature of 8 to 19 °C (15–35 °F) above the intended process temperature, can be rapidly pumped into the process vessel to reduce come-up time.

Recent developments in these batch systems include the integration of multiple retorts into continuous container handling systems that automate all portions of the cycle, including cage loading and transfer operations. This type of solution provides significant labour savings for large output operations, while maintaining the flexibility to process diverse container and product types through the same system (see Figure 12.9).

12.7.1   Systems characteristics

Because these systems differ greatly from traditional steam retorts, it is advisable to conduct temperature-distribution and heat-penetration tests to properly validate production processes. With some of the available automatic controls, specific processes can and often are designed that optimise product quality and throughput by giving some 'come-up time' and 'cooling lethality' credit.

The temperature/pressure differential between the heated-water reservoir and the processing vessel (together with a steam overpressure, if needed) provides the necessary overpressure required for sterilising pouches, semirigid trays, buckle-prone cans having large areas of flat surface (half steam-table trays, meat cans), and glass jars.

Figure 12.9 Photograph of horizontal batch retorts with automatic loading.
Courtesy of JBT Food Tech, www.jbtfoodtech.com.

Overpressure during heating can contribute to heat-penetration efficiency in cans with large surfaces relative to product depth (and with conventional headspace volumes) by compressing the cover against the product surface. This can be a significant consideration in a half steam-table tray, for example, in which a 6 mm (¼ in) headspace represents about 22% of the shortest distance to the geometric centre.

For certain fluid products, agitation may enhance heat penetration and thereby permit a reduced process. A commercial agitating process should never be adopted for a basket-rotating steriliser until a comprehensive process development study has established the point of least lethality in the can and in the basket. The former study must take into account all reasonably possible variations of fill, liquid viscosity, and solid-to-liquid ratios, including possible shrinkage of product and the highly variable rotational path traversed by a container, depending on whether it is in the centre of the basket or adjacent to the basket wall.

If a reduced-time process, dependent on agitation, is employed and thickeners (starch, flour, tomato paste, pureed vegetables, etc.) are present in the liquid component, strict control of the headspace volume is mandatory. This is feasible in cans or jars in which 6 mm (¼ in) of hot product is the usual minimum but very difficult in semirigid or rigid trays, unless the container is intentionally designed for a generous headspace. In general, agitation to further reduce the process time is not recommended for tray-type containers in which a very significant reduction, relative to cylindrical containers of comparable capacity, is already justifiable on the basis of the wide/shallow geometry.

Generally, circulating retorts are operated at 6–21 rpm, depending on container, product viscosity, and a host of other factors. Specific products, such as cream-style corn in large (157 × 178 mm) cans, utilise 30–36 rpm to optimise cycle time and colour. Other products utilise a discontinuous or oscillating mode of rotation for best results.

Because heat distribution is dependent on a combination of forced water circulation throughout the baskets and mechanical movement of the containers through the water, dividers between the layers of containers must provide a minimum resistance to flow.

12.8   High-speed reciprocating motion (Shaka®)

A horizontal batch retort is loaded with baskets filled with packed food containers. The retort is brought up to sterilising temperature as a reciprocating actuator shakes the baskets and their contents back and forth vigorously. The vigorous agitation causes thorough mixing of the food in the container so the heat is rapidly transferred throughout the contents. All the containers in the retort undergo exactly the same degree of agitation. The thorough mixing and rapid heat transfer permits the use of higher sterilisation temperatures, without overcooking or burning on the wall of the container. The rapid agitation also speeds the cooling part of the cycle. The combination of the even, rapid heat transfer with the higher sterilisation temperature normally used in the Shaka® process leads to significantly shorter cooking times.

12.9   Continuous rotary pressure cookers and coolers (reel and spiral cooker coolers)

Still-static retorts were prevalent for low-acid foods until about 1950. Since then, the continuous agitating type retort has become more common in the industry, resulting in considerably reduced processing times made possible by a higher rate of heat penetration into the food and by the higher temperatures used. When processing relatively large packs, production costs are reduced through savings in labour as well as in steam, because there is no need for repeated venting procedures. In many instances, it is possible to obtain better product uniformity, higher quality, and improved nutrient retention due to the higher temperature–shorter processing times used and to product agitation during sterilisation and cooling. Can damage and product loss are also reduced.

12.9.1   Description and operation

Continuous rotary cookers use two to seven shells (chambers) that are shaped like horizontal cylinders. The cans travel through the cylinders in a long spiral path. The temperature in each shell can be set for effective cooking and cooling. The can-transfer mechanisms for the cylinders are linked so that the system is always synchronised and the speed of the transfer mechanism is the primary means of controlling the process. A feed device delivers the cans through the in-feed valve to the revolving reel of the cooker. The cans are fed into the retort individually, so there is no reason for any delay before processing. The reel carries the can through the cooking steam. There is generally some rolling and some carrying that takes place as the cans travel around the 360° of the cylinder (see Figure 12.10). The continuous spiralling motion and the rotation of the can through the cylinder give an even and faster cook to every can.

The cooler shell is approximately two-thirds full of water to provide flood cooling of cans as they progress through the shell. Water enters at the discharge end and exits at the feed end of the shell for a counterflow cooling effect. The reel in the cooler has a series of baffles that seals off the central portion of the reel to confine the movement of cooling water to that area occupied by cans.

The combination of reel baffles and counterflow movement of water insures efficient usage of cooling water and controlled, uniform cooling of cans. This line employs the operating principle of can agitation, which permits short-term, high-temperature cooking and rapid, efficient cooling in an automatic and continuous operation. Agitation is provided by the spiral-reel mechanism, each revolution of which produces a three-phase cycle.

Fixed Reel Travel – This phase takes place in the upper portion of the rotation cycle for a distance of approximately 220°C around the periphery of the shell, in which cans are carried around a central axis.

Free Rotation – This phase occurs in the lower portion of the shell in which cans roll on the spiral tees for a distance of approximately 100°. In this area, agitation of the product in the can takes place due to a free-rolling action of cans about their own axis.

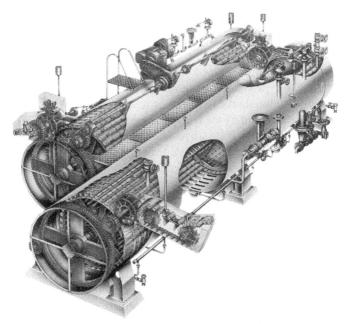

Figure 12.10 360° movement of the can as it travels around the shell.

Transitional Phase – This occurs on both sides of the shell's periphery, wherein free-rolling agitation commences, after fixed-reel travel, and where it stops after passing through the free-rotation phase.

The agitation of the product within the container caused by this three-phase rotation cycle in the cooker shell permits fast and even heat penetration throughout the container by means of induced convection. Conversely, in cooling operations, this agitation facilitates heat transfer. This principle increases the rate of heat penetration and/or transfer, assuring thorough, even cooking and cooling of every can, with all cans cooked and cooled uniformly and exactly alike. The rate of heat penetration and/or transfer depends on the consistency of product and operating speed of the machine.

The heating medium can be pure steam or water. These cookers are only used for cans and many different types of products are processed (e.g. fruit, vegetables, ready meals, and soups).

This kind of retort is commonly used and relatively energy efficient. As they are continuous, they only get vented at the start of production. The agitation is intermittent axial rotation (fixed reel, sliding rotation and free rotation). The headspace size is very important. Solid packs do not benefit from agitation. For validation of temperature, pressure, and can rotations in these systems, only wireless data loggers mounted inside the cans are able to be used. Typically, most products will be sterilised at about 121.1 °C, but depending on the products, temperatures can be higher or lower.

Cooling techniques employed for pressure-cooked products are no less important than the cooking and sterilising operation. In the cooking operation, pressure developed inside the can is counterbalanced by the steam pressure in the cooking vessel.

When the cans are transferred from the cooking to cooling operation, adequate external pressure must be maintained to avoid can bursting and/or buckling under certain conditions.

Pressure Cooler – With this arrangement, cans are introduced directly into the pressure cooler shell after travelling through the cooking shell. The cooler shell is designed to allow sufficient overriding air pressure to compensate for internal can pressure developed in the cooking cycle. In addition to this air under pressure, the can is subjected to counterflow water cooling for proper cooling of the contents.

Open (Atmospheric) Cooler – This system is one in which the cooler shell is not pressurised. Its principal application, however, is for cooling fruit. Maximum can centre temperatures should not exceed 104–107 °C (220–225 °F) upon completion of the cooking cycle.

Split Cool – This line provides a two-stage cooling operation in which the first stage is performed under pressure and the last stage under a nonpressurised unit. This arrangement requires two separate shells for the cooling function and is normally required for larger can sizes (103 × 119 mm) and over, and/or when higher cooking temperatures (118–135 °C (245–275 °F)) are employed. In these applications, the first phase of cooling requires a correspondingly overriding air pressure to prevent can buckling from internal pressure as the can is transferred from the cooker into the first cooler shell. The pressure in the pressure-cooling shell should be slightly less than the steam pressure in the processing shell. This relatively short pressure cooling reduces internal pressure in the can sufficiently to eliminate the possibility of can buckling when it is subsequently transferred into an open cooler for the balance of the cooling cycle.

12.9.2 Principal features of system

Centralised Panel Board – Each line is equipped with a centralised panel board that contains all control instruments.

Rotary Can Valves – Two rotary can valves are employed with each pressure shell. These serve the following purposes:

1. Feed cans continuously into the cooker shell without loss of steam and/or pressure from the shell.
2. Transfer cans from the cooker shell to the cooler shell without permitting steam to escape from the cooker shell into the cooler shell.

The rotor in each valve is constructed with equally spaced pockets on the valve's periphery. Each pocket is designed to hold one can and deposit it in synchronisation with the pocket in the reel. At the discharge end, a star wheel mounted within the reel gently ejects cans into the valve at the transfer and discharge points of the cooker and cooler, respectively.

Labour Savings – This equipment is often referred to as the 'one man cook room', because only one operator is required to maintain a battery of several lines in a given installation. In this continuous process, cans roll directly from closing machines through the pressure cooker and cooler on to any subsequent equipment, such as can dryer, labeller, and/or palletiser.

Improved Product Quality – Controlled agitation of cans permits more rapid heat penetration and reduced cooking time, thus preserving the natural flavour, nutrients, and appearance of the product.

Savings in Floor Space – Minimum auxiliary equipment is required – baskets, hoists, unscramblers, trucks, cooling canals, and air cooling space are not required with the line.

One line replaces many retort installations. This equipment may result in savings in space requirements and provides increased capacities in the space available.

Reduced Can Damage – Bulk handling of cans at feed and discharge is eliminated. All cans are handled positively and separately to minimise loss from can denting and damage.

Reduced Process Times – Achieved because of the faster heat penetration gained by can agitation and higher processing temperatures.

Venting Only at The Start of a Production Shift (or After Any Stoppages) – The following is a guideline for venting continuous cooker coolers:

Vent for a minimum of 7 min and to a minimum temperature of 105 °C. All vent lines should be fully open during the entire vent procedure. The drain is typically open until the vessel temperature reaches 95 °C and then turned down so that it is just partially open to facilitate the removal of the condensate.

Normal steam supply pressure to be minimum 5 bar – maximum 7 bar.

Both conditions must be met:

· If the temperature reaches 105 °C prior to 7 min, then the vent must be continued until the 7 min period is completed.
· If the 7 min period is completed before the vessel reaches 105 °C, then the venting must continue until the temperature (105 °C) is reached.

Sterilisation Temperatures and Capacity – Standard construction design allows sterilisation temperatures of up to 135 °C (275 °F); optional designs are available for processing at temperatures up to 146 °C (294 °F). The production capacity of the system is up to about 1000 cans per minute for the (81 × 111 mm) can size.

· Glass containers are loaded into metal cartridges automatically, passed through the cooker–cooler, and then automatically removed from the cartridge.
· Aluminium-end cans, aluminium-body cans, and lightweight steel cans are also being processed successfully in continuous rotary pressure sterilisers.

This steriliser line can be designed to accommodate various sizes and the changeover time for can size will range from a few minutes to a few hours.

The system uses approximately half the amount of water and steam per equivalent amount of product when compared with batch retorts.

12.9.3   Important factors to check when operating continuous rotary pressure sterilisers

(a) Cooker (reel) speed (rpm); (b) can headspace; (c) can diameter; (d) can geometry; (e) product consistency; (f) bubble mechanics; (g) product moisture absorption by starches; and (h) condensate accumulation and removal (condensate accumulation may slow down can rotation by up to 30%).

Continuous atmospheric cookers are the standard for high-acid foods, like tomatoes and fruits. The operation is followed by continuous atmospheric cooling.

12.10 Hydrostatic sterilisers

Hydrostatic sterilisers use a water lock to transfer cans into a pressurised steam chamber. The incoming water 'leg' can control the incoming product temperature (preheat) and the outgoing water 'leg' can serve as a precool. The processing chamber is usually filled with saturated steam, but air–steam mixtures can be used (see Figure 12.11). This sterilisation method is more commonly known as 'hydrostatic sterilisation', because steam pressure in these units is maintained by water pressure. The hydrostatic steriliser is similar to the continuous still retort through which the containers are conveyed by a continuous carrier chain, at a constant rate. The dimensions of the carrier bar dictate what size containers may be processed.

Hydrostatic cookers are made up of four chambers – a hydrostatic 'come-up' or 'in-feed' leg, a sterilising chamber, a hydrostatic 'comedown' or 'discharge' leg, and a cooling system. The temperature of water in the water chambers or legs varies from about 16 °C (60 °F) to about 102 °C (215 °F). The temperature of steam in the steam chamber is controlled by a steam controller and pressure produced by the water legs; it can be regulated by moving the overflow level up or down in the downtravelling leg. Steam temperatures between 116 and 129 °C (240–265 °F) are generally used.

12.10.1 Basic operation

The operation of hydrostatic cookers is as follows. Containers are conveyed through the machine by means of carriers connected to heavy-duty chains which produce positive can-travel control. Cans enter a water leg in which the temperature is about

Figure 12.11 Cutaway of a continuous rotary cooker–cooler.

82 °C (180 °F); this in-feed water leg is where can temperature begins to increase. As the can moves down this leg, it encounters progressively hotter water. In the lower part of this leg, the water temperature reaches about 102 °C (215 °F), and then, near the water-seal area next to the steam chamber, the water temperature is about 107 °C (225 °F). In the steam chamber, the can is exposed to the temperature range of 116–129 °C (240–265 °F), the steam temperature being set to suit the product undergoing sterilisation. Some hydrostatic cookers have the can make two passes, one up and one down the steam chamber; other models are designed for four, six, or eight passes. Upon leaving the steam chamber, the can again goes through a water seal into water at the temperature of about 107 °C (225 °F) in which the cooling cycle commences. It is then conveyed through progressively cooler water to the top of the discharge water leg in which the water temperature is between 88 and 93 °C (190–200 °F). In other models, the temperature of water in the discharge leg is as low as 16 °C (60 °F), can cooling then being completed within this leg. Some hydrostatic cookers utilise multiple chains, giving the added advantage of being able to process different can sizes at different process times simultaneously (see Figure 12.12).

In determining steam-dome residence time, the temperature of both inlet and outlet legs must be specified. There is no direct correlation between the water pressure and temperature in the legs.

12.10.2   Advantages and disadvantages

The following advantages for hydrostats are claimed by manufacturers:

1. Savings in floor space.
2. Large reduction in steam and water costs, because of regenerative heating and cooling. Savings of 50% steam and 70% water have been reported, compared to that of batch steam retorts.
3. High-capacity operation.
4. Capability of processing all sizes of cans, glass containers, and retort pouches.
5. Constant temperature operation. Steam flow is controlled by the water level in the water seal. A 10 cm difference in water level equals 0.06 °C (0.1 °F) temperature in the steam dome. Some units are also insulated.
6. Hydrostats use less cooling water.
7. The container is subjected to minimum thermal shock.
8. The containers are handled gently because of the low chain speed.
9. Low labour requirements.
10. The brines in vegetables, such as peas and beans, are clear because of the absence of agitation.

The main disadvantages of the hydrostatic cookers are the lack of can agitation and the very large capital investment required. Because of their very large size, there is a considerable amount of structural steel required in construction and installation costs are high.

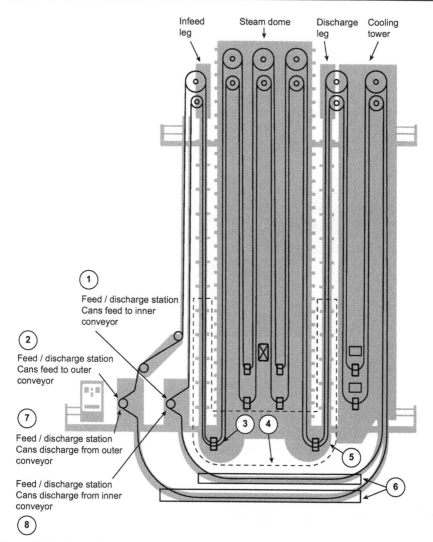

Figure 12.12 Operating principle of hydrostatic steriliser.
Courtesy of JBT Food Tech, www.jbtfoodtech.com.

Comparing hydrostatic sterilisers with continuous rotary cookers, the use of hydrostatic cookers is justified when the following conditions are encountered:

1. The unit is operated year round, preferably on a two-shift basis.
2. The product is sensitive to high temperatures of processing, for example, if the product discolours when the cooking temperature is higher than 245 °F (118 °C).
3. Filler speed is at least 400 cans per minute on consumer sizes.

Figure 12.13 Photograph of a hydrostatic steriliser.
Courtesy of JBT Food Tech, www.jbtfoodtech.com.

These retort systems are manufactured by various retort manufacturers, e.g. STORK and JBT Food Tech (FMC) but have the same basic principle of operation; they are very large and normally extend several stories into the air (see Figure 12.13).

12.11 Aseptic processing

Aseptic processing means the filling of a commercially sterile cooled product into presterilised containers, followed by aseptic hermetical sealing with a presterilised closure in an atmosphere free of microorganisms.

12.11.1 Advantages

The advantages of aseptic packaging of food products are the following:

1. A higher quality product is obtained by using high-temperature–short-time sterilisation. This reduces to a minimum the destruction by heat of the food components which are responsible for its typical organoleptic characteristics. It also results in better retention of nutrients when compared to conventional canning.
2. The advantages mentioned above are even more important when using large-capacity containers.

3. The basic principles and methods used in aseptic processing and packaging make possible the use of a wide variety of packaging materials and of many package sizes and shapes.
4. The sterilisation and packaging procedures that are used are continuous. Thus, handling of the containers during sterilisation is avoided, as well as the possible recontamination of the product during cooling.
5. Product heating and cooling equipment is designed with a high ratio of surface area to product volume, thereby providing very efficient heat transfer.
6. Because product and package are sterilised independently, the package need not be designed to withstand the high temperatures used for product sterilisation.

12.11.2   Limitations

Aseptic processing and packing has limitations and it does not offer advantages with all products. Some frequently cited limitations are

1. The large capital investment required to bring an aseptic system on line.
2. Because product transfer through the system is by means of pumps, the product must be a relatively homogeneous fluid that can be pumped.
3. A given system is designed for a limited range of product types.
4. Aseptic systems are quite complex and require sophisticated instrumentation to insure that an adequate process is delivered and that sterility is maintained. These systems require highly trained personnel.

12.11.3   Food sterilisation systems for aseptic processing

The key components, timing or metering pump, product heater, holding tube, cooler, and backpressure valve are characteristic of practically all aseptic processing systems (Chambers & Nelson, 1993).

With reference to holding-tube characteristics and maintenance, the following must be observed for low-acid foods and should be observed for acid foods:

1. The holding tube should always slant up at least 6.4 mm per 0.3 m (¼ in per linear foot).
2. When reassembling, the tube should always be put back the same way. Parts should not be left out or diameter changed.
3. The tube interior should be smooth with no gaskets protruding.
4. The tube should not be exposed to cold air drafts or condensate drops.
5. The product must not be heated between beginning and end of tube.

The timing pump must give a constant, uniform rate of product flow. The product must be of uniform composition; its characteristics must not vary. The type of aseptic processing equipment selected is dependent on pH, the viscosity or consistency of the product, and whether it contains particulates and their size.

12.11.4   Thermal sterilisation for aseptic processing

Generally, very high-temperature, short-time (HTST) thermal processes are used for commercial sterilisation of foods for aseptic packaging. HTST processes yield products with the best possible characteristics from the standpoints of flavour, aroma, colour, and nutritive value. Typical temperatures and times for commercial sterilisation of aseptically packaged low-acid foods are 138–149 °C (280–300 °F) for 1–30 s. For high-acid foods, these parameters typically are 93–96 °C (200–205 °F) for 15–30 s.

Temperatures used in aseptic sterilisation processes are always correlated with sterilisation time. When indirect heat exchanger systems are used, in addition to process lethality during product residence time at product sterilisation temperature, process lethality during come-up time and during product cooling time may also be considered. Table 12.1 illustrates an example.

It can be seen that the accumulated or total thermal process sterilising effect (Fo) value, when considering the total time that the process temperature was above 121.1 C, and calculated every second, is slightly greater than 40. If the total Fo value is figured strictly on the holding time of two s at 149 °C (300 °F), it would be equal to slightly more than 20, or approximately 50% of the total process Fo value.

Table 12.1 **Cumulative thermal process sterilising effect Fo values in an indirect heat exchange system**

	Time (minutes)	Temperature (°C)	Temperature (°F)	Contributed Fo
Heating section	1/60	126.7	260	0.060
	1/60	139.4	283	1.131
	1/60	144.4	292	3.835
	1/60	147.2	297	6.751
	1/60	148.3	299	8.380
Holding section	2/60	148.9	300	20.000
Cooling section	1/60	130.0	266	0.130
	1/60	124.4	256	0.040
Total	9/60 (9 s)			40.327

Fo is the accumulated lethality when reference temperature is 121.1 °C (250 °F) and z-value is 10 °C (18 °F).

Although it is possible to incorporate some of the come-up or heating section lethality in process design, in many cases the advantages of the reduced process may not justify the added expense of required extra monitoring and control devices, the burden of additional record-keeping requirements, and the increased possibility of process deviations. The biggest drawback to inclusion of this lethality or heating effect can be the fouling of the heat exchange surface, which will alter the temperature profile during heating in the course of a production run.

With direct systems of heating, lethality of come-up time and of cooling time are not considered in total process lethality calculations, because times during those portions of the sterilisation processes are so short that they do not add any significant lethality. The sterilisation value of the whole process is then equivalent to the Fo value of the holding value at the holding temperature.

The basic consideration for sterilising at high temperatures for a short time is that for each 10 °C (18 °F) increment increase in sterilisation temperature, the time required for destruction of bacteria is reduced by a factor of 10, whereas the rate of destruction of nutrients and of other chemical reactions affecting product colour and flavour increases by a factor of approximately three. Therefore, the higher the sterilisation temperature, the larger the difference between rate of destruction of microorganisms and rate of destruction of nutrients and other undesirable reactions. This is the concept upon which one of the most important advantages of aseptic processing and packaging systems is based. It is the reason why organoleptic and nutritive value changes in products processed by aseptic processing systems are less pronounced than in products processed by other systems, such as conventional canning.

The selection of the sterilisation system that is most advantageous for a specific food product depends upon many factors. Once the basic system, be it either direct or indirect, has been chosen, the details of the system must be determined so that the process as a whole produces commercially sterile products of high quality.

A list of factors that must be considered in the determination of times and temperatures of sterilisation that yield a product that is safe from the public health standpoint and that is of the highest quality attainable under the processing conditions considered follows.

12.11.5 Factors which affect scheduled sterilisation processes for aseptically packaged foods

1. Fo value required and existing z value.
2. Heat penetration rates (viscosity, particles).
3. Maximum temperatures product will withstand.
4. Allowable heating and cooling times in the lethal temperature range not detrimental to product.
5. Time–temperature relationship and lethality in the holding tube.
6. Lethality during heating and cooling.
7. Holding-tube size (length and diameter).
8. Product vapour pressure and required backpressure.
9. Product formulation and characteristics (pH).

The Fo value required for commercial sterilisation of low-acid foods and the z value associated with the product and potential spoilage microorganisms may be determined on the basis of previous research work, as long as the data are for the same product. Once that information is available, thermal sterilisation process times and temperatures may be determined using either the General or Graphic method of process calculations, or the Formula or Ball's Method.

Products which are especially heat labile must receive a sterilisation process so that the total heat received by the product is the minimum that will result in a product that is commercially sterile and of the highest attainable quality. The chapter on heat penetration in Book II can be consulted for further basic information on commercial sterilisation of low-acid foods. The basic principles and mathematical approach to process calculations also apply to aseptically processed and packaged foods (Tucker & Featherstone, 2011).

The commercial sterilisation of high-acid products, of pH 4.60 or lower, can be attained more readily than for low-acid foods. Normally, these products need only be heated to 93 to 96 °C (200°–205 °F) for 15–30 s to produce adequate sterilisation. All microorganisms that may cause spoilage in high-acid foods packed in hermetic containers are killed under those sterilisation conditions. Other microorganisms that are left viable cannot germinate at pH 4.60 or lower. To obtain aseptically packaged high-acid products with a shelf life longer than a few months, the thermal sterilisation process considerations discussed above must be incorporated into the overall aseptic processing of such high-acid foods, in addition to package and package-handling considerations.

12.11.6 Basic requirements of equipment for aseptic processing and packaging

The basic requirements that must be fulfilled by equipment for aseptic processing and packaging are the following:

1. It must be able to withstand initial sterilisation, i.e. it must withstand high internal pressures and temperatures.
2. It must be capable of being maintained under conditions of commercial sterility, i.e. it must be a hermetically closed system and must preferably operate under positive internal pressure.
3. It must produce a commercially sterile product.
4. It must be capable of cooling the product under sterile conditions and maintaining commercial sterility of the product during the cooling operation.
5. It must be capable of sterilising the containers or packages.
6. The filling and closing areas must be maintained under sterile conditions.
7. Container filling and closing must be handled in a very careful manner to avoid product recontamination.
8. Equipment must be designed so that it may be maintained under the sanitary conditions that are required for equipment used in food processing.
9. It must operate efficiently and with realistically low costs of operations and maintenance.
10. Equipment must be designed so that it conforms to local and national laws and regulations.

12.11.7 Aseptic processing systems

There are four different aseptic processing systems. Direct steam heating and tubular aseptic systems provide the shortest processing time. Certain products, like milk or ice cream mix, are highly susceptible to flavour changes when subjected to prolonged high temperatures and should be processed on a system that produces rapid uniform heating and cooling. Direct steam heating may be by steam injection or steam infusion.

The swept-surface heat-exchanger curve represents a gradual heating and cooling profile beneficial for heavy viscous products and products with particulates. Rapid heating and cooling is not as critical here, because slower heating allows penetration of particulates and produces uniform flavour (Richardson, 2004).

The plate heat-exchanger curve represents an even more gradual increase and decrease in product temperature. Juices and juice products are not as heat sensitive as dairy products and can be processed on a plate heat-exchange system providing the solids or pulp content are not sufficient to cause fouling. Normally, these products need only be heated to 93 °C (200 °F) for 30 s to produce adequate sterilisation.

Specific products and systems will dictate their own curve profile.

Direct heating: Steam sterilising is a direct heating/cooling method that utilises a steam chamber and vacuum vessel, sterilisation being accomplished by either steam injection or steam infusion. In steam injection, culinary steam is dispersed into the product stream by using an injector. With steam infusion, the product flows into a controlled atmosphere of culinary steam. The product may be sprayed into the steam or distributed as free-falling sheets, depending on the equipment manufacturer. In one arrangement, inverse distributor cups cause the product to flow in free falling, umbrella-shaped films which are immediately intermingled with the steam. The level of product plus condensate is controlled so that the infuser operates essentially empty, except for enough fluid to maintain a liquid seal at the outlet. The heater operates over an extended range of flow rates and temperatures. Operating pressure in the fusion heater is held at 34–69 kPa (5–10 psi) above the final product temperature saturation point. This allows instantaneous condensation of the steam, yet keeps a relatively small temperature difference between product and steam.

As an example, in the case of milk and milk products, the temperature is raised from approximately 77 °C (170 °F) to approximately 149 °C (300 °F) – so rapidly that flavour changes and product destabilisation are minimised. When limited to holding times of 2 to 4 s, the result is a clean pasteurised flavour and continuous operating runs of 20 h or longer.

Direct heaters are frequently combined with flash chambers, plate, or swept-surface heat exchangers to create a custom-designed sterilising system.

12.11.8 Advantages

Direct sterilising is a versatile aseptic processing method designed primarily to heat and cool fluid foods through their critical heat ranges in a matter of seconds. Direct heating produces the fastest heating of the four methods shown in the curve profiles. This rapid processing minimises flavour changes and product damage normally associated with high processing temperatures. It is especially important for low-acid products which require sterilising up to 149 °C (300 °F).

Direct sterilising systems are versatile. They may be prepiped package systems or field-assembled components arranged to meet specific plant-space requirements.

Acquisition costs for infusion heating systems are low when high flow rates are being processed. Because there are few moving parts, service costs are low.

12.11.9 Typical system operation

To sterilise the system, water from a supply tank is pumped through the system, heated to sterilising temperature, and circulated in a closed loop. When preset sterilising conditions have been met, the sterile side of the system is cooled down while running sterile water, then a signal is actuated to alert the operator.

Product from a supply tank is pumped through the regenerative heater section of a plate heat exchanger to the variable speed, product flow rotary pump. Product from this pump enters a plate preheater section and then enters the infusion heater approximately 77–82 °C (170–180 °F). In the infuser, the product is heated by direct contact with steam to the desired sterilising temperature. The temperature and product level are controlled automatically. The product is removed from the infusion heater by a variable speed rotary pump and passes through a short holder tube sized for 2 to 4 s.

After passing through a backpressure valve, the product enters an aseptic flash chamber. Here it immediately boils, evaporating off moisture. A shell and tube condenser uses cooling water in a closed loop to condense the vapours removed.

Generally, the product is cooled by evaporation to a few degrees above the temperature at which it entered the infuser heater. This restores in the sterilised product the same moisture content as the original raw product.

The product is removed from the flash chamber by means of an aseptic centrifugal pump and passes through an optional aseptic homogeniser. After homogenisation, the product flows to a heat exchange section in which it is cooled by a wire loop transferring heat to the plate regeneration section.

Downstream from the cooler section, an aseptic flow diversion valve controls routing of the sterilised product. In the event of improper temperature or holding time, routing is back to the product supply tank.

Indirect heating: Tubular aseptic sterilising is an indirect heating/cooling method that uses stainless steel coiled or trombone tubular heat exchangers. The tubing diameter is relatively small, compared to product flow. As a result, extremely high flow velocities within the tubing maximise turbulence. High turbulence induces very rapid heat transfer, allowing sterilisation to take place in a very short time, minimising harmful thermal effects.

Tubing diameter is suited to product flow and viscosity. Tubes are fabricated into coils or bundles and placed, along with special media baffles, into stainless steel jackets. Hot water, steam, or cold water pass through these jackets to heat or cool the product flowing within the tubes.

A high-pressure reciprocating pump is used to force product through the system. The pump can operate at pressures up to 5000 psi (34 MPa). The actual pressure required will vary according to capacity requirements, viscosity of the product, and homogenisation needed, if any.

12.11.10 *Advantages*

Tubular aseptic systems provide high heat-transfer rates and a scrubbing action that reduces 'burn off' or fouling in the tubes. This results in a very short processing time; generally, it takes less than 30 s for product to travel through the entire system. This helps to preserve the natural flavour of the product. This is especially important with low-acid products requiring sterilising treatment up to 149 °C (300 °F).

Length of runs varies with product nature and quality. With preconditioned product or deaeration and holding prior to final heating, runs of 14–16 h or more may be expected.

The systems have considerable flexibility in the range of products they can handle and the temperature range at which a specific product is processed. They are completely self-contained, requiring only product and utility hookups to be made during installation. Shipping and installation costs are minimised. There are no gaskets to replace on the high-temperature side.

Most systems are available with regeneration as an option. Regeneration may run as high as 85%, depending on flow rates, product characteristics, and the regeneration option used.

12.11.11 Typical system operation

To sterilise the system, water from a supply tank is pump circulated throughout and heated to sterilising temperature in a closed loop. When preset sterilising conditions have been met, a signal is actuated to alert the operator to switch to automatic product flow control.

Raw product is pumped from a supply tank through a regenerative plate heat exchanger and into an optional deaerating chamber. A centrifugal pump at the discharge of the deaerating chamber supplies product to a high-pressure timing pump. This pump is powered by a variable speed drive so the delivery rate of the product through the system can be adjusted to match the capacity requirement of the filler or fillers.

A series of horizontal tubular heat exchangers and a vertical holder tube heat and hold the product to the required sterilising temperature and time. For low-acid products, the temperature is generally around 149 °C (300 °F), with a holding time of 2 to 4 s. High-acid product would normally be heated to around 93 °C (200 °F) and held for approximately 30 s.

From the holding tube, product flows to another series of vertical tubular heat exchangers for cooling. In the first stage, the cooling medium is water tied in a closed loop with the plate heat exchange preheater.

If homogenisation is required, the product is routed from one of the initial cooling stages at the proper temperature to a remote homogenising valve. The homogenised product then continues through the final cooling stages.

Cool, sterilised product is then routed through a flow diversion valve and on to an aseptic tank or aseptic filler.

12.11.12 Indirect heating: scraped-surface sterilising

Scraped-surface sterilising is an indirect heating/cooling method incorporating a dasher with scraper blades that rotate inside a heat exchange cylinder. Product is pumped through the cylinder. Centrifugal force on the rotating dasher holds the scraper blades against the inside of the cylinder wall. The blades continuously scrape the cylinder wall clean of product, simultaneously producing a desirable level of product mixing. With the continual removal of product from the cylinder wall, product film is reduced to an absolute minimum, permitting long processing runs without product buildup on the heat exchanger wall.

Different dasher diameters may be selected to give the desired annular space between the dasher and cylinder wall. This space is dictated by the type of product or size of particulate to be processed.

12.11.13 Advantages

Scraped-surface heat exchangers are capable of processing heat-sensitive products that cannot be handled by tubular or plate-type systems. Flexibility in the choice of cylinder material, dasher diameter, scraper blade positioning, shaft seals, and dasher mounting positions make the surface heat exchanger a versatile choice for aseptic processing.

Scraped-surface heat exchangers process product over a broad temperature range. Product viscosity may vary from light fluid to heavy viscous pastes, with or without particulates.

Various horizontal and vertical configurations allow this form of heat exchanger to be adapted to specific systems or plant requirements.

12.11.14   Typical system operation

The system is initially sterilised by circulating water slightly above the sterilising temperature through the heat exchangers and interconnecting piping. Because temperatures required are usually above boiling, a backpressure valve or a holdback pump is used to maintain pressure on the system. A condenser cools the sterilising water below atmospheric boiling. This arrangement is also used to maintain a constant pressure to the filler and to bypass product back to a surge tank in case of temporary filler stoppage.

Once the system is properly sterilised, raw product is pumped to the heating cylinder using a rotary pump. When products contain particulates, these can be blended in the product makeup tank. Alternately, particulates may be added with an inline ingredient feeder. This is positioned just ahead of the first heating cylinder.

Product is held at the sterilising temperature in a holder tube between the heating cylinder and the first cooling cylinder. High-acid products may only need to be heated to 90 °C (200 °F) and held for 30 s. When large particulates are present, the product may need to be held as long as 3 min; this allows time for uniform heat penetration of each particle.

Low-acid products are heated up to 149 °C (300 °F) and held for only a few seconds. The product is then cooled in water-cooled, heat-exchange cylinders.

12.11.15   Indirect heating: plate sterilising

Plate heat exchange is an indirect heating/cooling method. It incorporates a number of stainless steel-gasketed plates, which are compressed and locked together in a rugged frame or press. Narrow spacing between the plates accommodates the product and the heating or cooling media in adjacent streams.

As product is pumped through the plate heat exchanger, the flow is distributed as a thin film which moves over the irregular plate surface, producing a level of turbulence desirable for uniform heating and maximum run lengths. Porting within the plates directs the medium to one side and the product to the other. Heat exchange takes place through the stainless steel plate.

The addition of terminals allows several independent sections to be located in one press or frame. These sections can be used for preheating, regeneration, sterilising, holding, and cooling.

During regeneration, the hot sterilised product may be used to preheat incoming cool product. This is accomplished by substituting the hot product for the heating medium in that particular section of the press.

12.11.16   Advantages

Plate heat exchangers are exceptionally efficient in processing fluid products (see Table 12.2 for a summary of advantages of aseptic food sterilisation systems). Product-to-product regeneration is possible; as much as 90% of the heating and cooling requirement can be done through regeneration.

Plate heat exchangers provide a maximum amount of heat-exchange surface in a minimum amount of floor space. They are also the least expensive of the four methods discussed. This major advantage, however, should be weighed against high residence volumes and exposure to gasket surface.

12.11.17   Typical system operation

Juices prepared from concentrate require either two blending tanks or an on-the-run metering system for supplying a continuous flow of product to the plate heat exchanger. The system shown here uses the two-tank method.

Concentrate is pumped by rotary pump from drums or bulk storage tanks to either of two blending tanks. These tanks are equipped with agitators which quickly disperse concentrate into a measured amount of water. Formulation can be done either with a volumetric metering system, or the tanks can be placed on a load cell system that weighs each ingredient.

The blended juice is pumped through an optional deaerating vessel. From the deaerator, juice is pumped through the plate heat exchanger in which it is heated up to 93 °C (200 °F). Hot water is normally used as the heating medium. Usually, 30 s is required at sterilising temperatures to kill yeast and moulds and to inactivate enzymes. In some cases, a holder tube may be used for this purpose, whereas in other cases, the time for the juice to flow to the filler, be filled, and then the container to get to the cooling tunnel may be sufficient to satisfy the holding requirement. Hot juice sterilises the container.

An alternate method of processing juice from concentrate is to heat and cool the blended juice rapidly before it enters the container. This enables the processor to take advantage of product-to-product regeneration. Juice is preheated in the generator

Table 12.2 Comparative advantages of food sterilisation systems for aseptic packaging

Method	Advantages
Infusion into steam	Fastest heat exchange system; handles heat-sensitive products best; can remove volatile flavour compounds
Tubular heat exchanger	Simpler maintenance; faster heat exchange than systems mentioned below
Scraped surface heat exchanger	Capable of handling viscous or solid containing product
Plate heat exchanger	The least expensive; has high throughput; is characterised by low-pressure operation

section of the plate heat exchanger; as much as 90% of the heating can be done in this manner. The temperature of the juice is raised to approximately 200 °F (93 °C), held for a sufficient time, and then cooled in the regenerator section, in which up to 90% of the cooling is done. The product may then pass through a final cooler section and on to an aseptic surge tank or aseptic filler.

12.11.8 Surge tank

An aseptic surge tank may be used to hold sterilised product after processing and before filling and packaging. Product can be delivered to a surge tank and then removed at the proper rate for the filler.

Aseptic surge tanks are sterilised before processing. Proper precautions must be taken to insure that all parts of the tank are exposed to the sterilisation medium for the proper length of time and that the air supply to the tank is sterile. A sterile air or sterile food-grade gas supply to the tank is necessary to maintain a protective positive pressure and to displace the contents. Protective steam seals or other types of barriers are provided at such places as valve stems or agitator shafts. Steam and sterile air piping kits can be preassembled for easy installation.

Aseptic surge tanks are generally available in sizes from 570 L (150 gallon) to 22, 700 L (6000 gallon) capacity. Surge tanks provide a buffer between processing and packaging systems for holding aseptic product prior to packaging.

The advantages of an aseptic surge tank are

1. An increase in flexibility so that processing and packaging equipment can operate at varying production speeds independent of one another.
2. Production efficiency can be increased, because one part of the system can be operating while the other is down.
3. In some cases, the surge tank is actually used to maintain backpressure on the processing system.

Surge tanks greatly increase the complexity of the system, because they must be designed and operated in a manner which will insure that asepsis is maintained.

12.12   Alternate thermal technologies

12.12.1 Ohmic heating

In ohmic heating processes, an alternating current flows though the food, causing heat to be generated due to the electrical resistance of the foods. In a liquid–particulate food mixture, if the electrical conductivity of the two phases is comparable, heat could be generated at the same or comparable rate in both phases; if they are not, heat will be generated faster in the one compared with the other. Electrical conductivity is influenced by ionic content; therefore, it is possible to adjust the electrical conductivity of a product by adding ion levels (e.g. salts) to achieve effective ohmic heating. Ohmic heating is a way of processing particulate food at the rate of HTST processes but without the limitation of conventional HTST on heat transfer to particulates (Rice, 1995).

12.12.2 Radio frequency (RF) heating and microwave processing

Microwave and radio frequency heating refers to the use of electromagnetic waves of certain frequencies to generate heat in food. Typically, microwave food processing uses the frequencies of 5800, 2450, 2375, and 915 MHz. Of these, 2450 MHz frequency is used for home microwave ovens. Radio frequency heating can be performed at various designated frequencies in the radio frequency range (e.g. 13.56 MHz ±6.68 kHz; 27.12 MHz ±16.00 kHz; 40.68 MHz ±20.00 kHz) (Richardson, 2001).

Heating with microwave and RF involves primarily two mechanisms: dielectric and ionic. Water in the food is often the primary component responsible for dielectric heating. Because of their dipolar nature, water molecules try to follow the electric field associated with electromagnetic radiation as it oscillates at the very high frequencies. Such oscillations of the water molecules produce heat. The second major mechanism of heating with microwaves and RF is through the oscillatory migration of ions in the food that generates heat under the influence of the oscillating electric field.

Microwave and RF heating for pasteurisation and sterilisation are rapid and therefore require less time to come up to the desired process temperature compared with conventional heating. Microwave and RF heating may be relatively more uniform than conventional heating. Other advantages of microwave and RF heating systems are that they can be turned on or off instantly, and the product can be pasteurised after being packaged. Microwave and RF processing systems also can be more energy efficient.

Microwave and RF heating can be used for baking and cooking, tempering and defrosting, drying, and pasteurisation and sterilisation.

References

Chambers, J. V., & Nelson, P. E. (Eds.). (1993). *Principles of aseptic processing and packaging* (2nd ed.) Washington, DC: The Food Processors Institute.

John Bean Technologies Corp., www.jbtcorporation.com, www.jbtfootech.com (formerly FMC).

Mālo, Inc., www.maloinc.com.

Rice, J. (1995). Ohmic adventures. *Food Processing*, 87–91.

Richardson, P. (2001). *Thermal technologies in food processing*. Woodhead Publishing Ltd, England (Cambridge).

Stock America, Inc., www.stockamerica.com.

Tucker, G., & Featherstone, S. (2011). *Essentials of Thermal processing*. Wiley-Blackwell, United Kingdom (West Sussex).

Computer-integrated manufacturing

13

13.1 Introduction

Few businesses in the world today neglect to utilise computers in their daily transactions. Increased power, decreased cost, and general usefulness account for their popularity. The comfort level of nonexpert users of computers has increased greatly over the last decade. Computers are better suited than humans for many activities. Tasks which are complicated and slow for the human mind are done much more quickly and accurately with a computer. The rapidly expanding and changing field of computer technology allows only a brief mention of several concepts. Some are very well known and others perhaps less familiar.

13.1.1 Hardware

Hardware is the machinery or equipment. It consists of items such as the central processing unit (CPU), disk drives, circuit boards, memory chips, monitors, and printers. In recent years, CPUs have become very powerful and fast. Memory capacity has expanded greatly. Work that could previously be done only on large mainframe computers can now be accomplished with a small personal computer.

13.1.2 Software

Software refers to sets of codes or computer instructions used to accomplish certain purposes. They are usually referred to as programs. Common programs include word processors, spreadsheets and databases. A wide variety of these are available depending upon the amount and size of the work to be accomplished and the size and type of computer.

13.1.3 Networks

With the large increase in numbers of personal computers in the workplace, systems to connect them so that information can be shared have become widespread. These are referred to as networks. Networks provide many advantages within a company and even between companies. Information exchange can be made quickly and economically. One copy of software can be installed on a system called a server. By paying a licensing fee to the software provider, other computers can more economically share this single copy.

A Complete Course in Canning and Related Processes. http://dx.doi.org/10.1016/B978-0-85709-677-7.00013-X

13.1.4 Vision systems

Specialised devices have been developed which can visually input images without having to enter the information indirectly through a keyboard or some file-transfer mechanism. These include scanners, barcode readers, and image analysers. Barcode readers are the simplest devices and have found widespread use in inventory control.

Scanners and image analysers must convert an essentially analogue image to a digital format. The computer is then able to store, analyse, and manipulate the data. Image analysers can be used for online inspection purposes. Hardware improvements in recent years have made this approach more practical. High-speed image analysis systems have been developed for vegetable grading and sorting.

13.2 Intelligent systems

One of the most significant developments in computer applications over the last decade has been intelligent or knowledge-based systems. Although this deals mainly with software, powerful computers are usually required to efficiently operate this software. Intelligent systems allow the computer more freedom to find relationships and solve problems than traditional programs. Terms for this software include neural networks, fuzzy logic, and expert systems. Expert systems are rule-based systems in which sets of rules are assembled to find logical conclusions from various input data. These have often been used to capture the knowledge of an expert in the form of these rules. Although it can be an arduous task to assemble expert systems, once established they can serve as advisors solving problems such as tracing defects in product quality to their source, scheduling routes, or giving advice on equipment repair. Fuzzy logic allows computer analysis of incomplete or noisy data. Neural networks are programs which create patterns between input and output data. To develop these patterns, the programs proceed through a 'learning' process using previous data. Sometimes referred to as Artificial Neural Networks (ANN), these programs are very useful at discriminating essential from redundant data (Hourigan, 1994). Fuzzy logic and neural networks are used for modelling and control of food processes ranging from fermentations and extrusion cooking to drying cereal grains (Eerikainen, Linko, Linko, Siimes, & Zhu, 1993).

13.3 Use of computers in the food industry

In many areas of the food industry, competitiveness has required cost-cutting measures resulting in reductions in employees and physical resources. As savings in these areas have been maximised, automation and the intelligent use of information play a key role in continued competitiveness (Hollingsworth, 1994). Various industry terms have been used to refer to computer incorporation. They all generally refer to computer-based control over the production process and demonstrate the interest and intent of the industry to optimise computer use. Two basic categories of computer usage in manufacturing are information management and process control. However, these categories are becoming less distinct with increased integration within the plant and between the plant, suppliers, and customers. An ultimate objective for a company wanting to take advantage of computer resources is to integrate decision making across the supply chain from supplier to consumer (Schutt, 1993).

13.3.1 Information systems

Information was initially a front office function. Data such as inventories, billings, payrolls, and production records were some of the first subjects of computerisation. Improved methods of data collection in the processing area, receiving, warehouse and storage area, and laboratory have both allowed and required better integration of increasing amounts of data. Automated data collection, using barcode scanners and data logging devices in many formats, has contributed to the data proliferation.

Computerised maintenance management systems (CMMS) have proved to be cost effective in food industries. They handle the paperwork associated with maintenance including work order preparation and tracking, labour scheduling and allocation, spare parts inventories and purchasing, report generation, and preventive maintenance scheduling. A large US processor reported a $2.5 million reduction in maintenance parts inventory over 18 months by using CMMS. The system detected inactive and/or obsolete parts, defined adequate stocking levels for required parts, and helped perform regular physical inventory. In general, the use of CMMS could cut inventory costs by 10–30% and improve equipment productivity by 20–30% and workforce productivity by 10–20%, due to reduced machine breakdowns and maintenance costs (Mans, 1994).

13.3.2 Purchasing, sales, and distribution

With computerised storage, supplier files can contain a great deal of information on product specifications and costs (Blythe, 1994). This is valuable in formulation and costing work. Computer systems can even monitor farm production costs and climatic parameters and predict potential attack by pathogens and probable pesticide residues.

Sales and marketing can maintain information on customer activities with much greater accuracy. Consumer trends can be tracked using information obtained immediately from sales outlets. This is the basis for Efficient Consumer Response (ECR) in which a plant's production schedule is tied directly to the customer's purchasing pattern. Computers are an integral part of this process.

A major US processor is using electronic data interchange (EDI) through a wide-area network to carry information between the company and its customers. About 70% of the purchase orders and 50% of the invoices are transmitted through EDI. The company analyses the sales volume, inventory information, local market intelligence, and any other information provided by the buyer and initiates purchase orders on its own, resulting in a continuous replenishment program. A 12-month pilot program produced a 50% reduction in inventory, 50% fresher on-shelf product, and a 40% drop in unsalable product (Demetrakakes, 1994).

Handheld units for distributors can print bills and receipts on site. They can also display instructions and give geographical directions. Upon returning to the distribution outlet, the handheld computer is downloaded and a report printed.

Packaging material waste can be saved by using programs which maximise use of package size and product arrangement, space on pallets, space in trucks, design of containers, and overwraps.

Product recalls are greatly assisted with proper product information storage and retrieval systems.

13.3.3 Production control

Supervisory control and data acquisition (SCADA) systems are becoming well established. These involve communication with control systems, communication with various databases, formulation controls, and various reporting and program interfaces. Computerised production control depends upon the type of food-processing operation. Typically, a process with a high degree of automation has been a more likely candidate for computer interfaces. Processes requiring more flexibility have tended to be more manual in nature. Processes with little automation can still take advantage of computer technology using specific formulation programs and data-logging devices.

Automation and computer control can be seen along most of the manufacturing chain, from raw materials to finished product. Automatic computer-based fruit and vegetable sorters and graders are gaining in speed and accuracy, while the costs are decreasing (Demetrakakes, 1995). Computerised control of basic processing steps, such as weighing, can be performed using machines equipped with programmable logic controllers that weigh and record different ingredients.

Computer control in thermal processing of canned foods provides for greater accuracy. Control operations such as venting, cooking, and cooling are sequenced automatically. Predictions of the extent of heat treatment can be made. Methods have been developed using optimisation techniques to predict temperature changes in a can during sterilisation. The model determines a lethality value which is used in the heating turn-off decision (Ryniecki & Jayas, 1993). Boiler efficiency can also improve with the use of microprocessor controls to optimise the air and fuel ratios. Large hydrostatic retorts' processors can benefit from expert systems developed to determine and correct problems inherent to these sterilisation systems.

Vision systems have been developed to check cans online (Robe, 1992). Capabilities of digital cameras to capture images of food products during the process and to evaluate and store essential parts of the digitised image are a rapidly developing and expanding field (Onyango and Marchant, 1994). High-speed video motion analysis can be used to solve production problems caused by can making and food-processing machines (Larsen, 1989). In the food industry, image processing has been applied to detection of slicing defects, colour of baked goods, dough thickness, surface texture and size, missing or damaged labels, faulty closures, defective pieces, and extraneous matter such as glass or bone fragments (Mans, 1988).

In the packaging area, computer-controlled printers can help meet legal requirements for date and production details. Programs can store variable text, logos, designs,
or barcodes.

Warehouse automation offers reduced product cost and improved customer service through either the use of minor software automation or multimillion dollar ren-
ovation/construction, depending on production volume and physical restrictions. The
simplest approach is to implement a warehouse management system (WMS) involving forklift operators interfacing with a remote computer. Reported improvements by
WMS included 20% in daily throughput and 40% in lift-truck productivity (Berne,
1994).

Another area in automation is the use of robotics. Today, robots are smarter, faster,
and capable of operating at speeds similar to or better than human operators in some
applications. The main advantage is that they do not suffer from fatigue or develop
repetitive-motion illnesses. The robots can be equipped with machine vision and interchangeable grippers to handle various shapes and different layer patterns.

13.3.4   Quality assurance

Statistical Process Control (SPC) and Hazard Analysis Critical Control Points
(HACCP) programs are commonly monitored using computer systems. The large
amounts of data generated can more easily be manipulated and schedules maintained
using electronic media. A promising area of development is the use of neural networks
to correlate processing or raw material information with final product analysis and
quality. Feedback can be rapid with high accuracy.

13.3.5   Product development

Least-cost formulations using linear regression are facilitated and in many cases made
possible by the use of computer programs. Software is available to assist with experimental design.

Computer modelling can help to formulate beverages. Models consider potential ingredients which may contribute to a typical type of beverage. Rather than
one-variable-at-a-time experimentation, computed multidimensional modelling
and optimisation combined with experimental test design can identify essential
variables. A trained sensory panel is essential in this case (Bomio, 1994).

13.4   Application considerations

Before any efforts to computerise a process or information flow, the state of the current
business must be examined. This is particularly important anticipating the need for
most companies to network computers for data and information exchange. Computers
cannot efficiently help a poorly organized business. However, if a company is suffer-
ing from disjointed activities, computerisation is an opportunity to take an integrated

approach to the product/process design cycle (Bruin, 1992). Regarding computers, consider that the more natural tasks that humans do almost without thinking tend to be very difficult for a computer. Highly analytical jobs involving complex number manipulations are much better handled by computers.

An individual or committee needs to be responsible for assessing the needs of the company. Input should be obtained from all individuals involved. Once needs have been established internally, a vendor can be contacted. These responsibilities would belong to the Information Systems department in most medium and large companies. It is still valuable to involve as many interested parties as possible. Features to consider include conversations and reports from other users, interfaces, vendor support, and upgrades.

Once a system is selected and installed, training is essential for a smooth transition. Effective education helps to teach basic principles and helps the user to have confidence in the system. The full range of tasks for which the computer upgrade is designed should be evaluated.

Acknowledgements

The 13th Ed author wishes to thank Robert L. Olsen, Schreiber Foods, Inc., Green Bay, WI and Olga Padilla-Zakour, Cornell University/NYSAES, Geneva, NY for their contributions to this chapter.

References

Berne, S. (1994). The ins and outs of automated warehousing. *Prepared Foods* (4), 81–85.

Blythe, K. (1994). Computerized monitoring of suppliers and supplies. *Food Technology International Europe*, 155–157.

Bomio, M. (1994). Magic ingredients. *International Food Ingredients* (3), 37–42.

Bruin, S. (1992). Integrated process design: issues and opportunities. *Food and Bioproducts Processing*, 70(C3), 126–130.

Demetrakakes, P. (1994). Casting a wide net. *Food Processing*, 55(10), 69–70.

Demetrakakes, P. (1995). Graders appeal to fruit, vegetable market. *Food Processing*, 56(2), 49–51.

Eerikainen, T., Linko, P., Linko, S., Siimes, T., & Zhu, Y. H. (1993). *Trends in Food Science & Technology*, 4(8), 237–242.

Hollingsworth, P. (1994). Computerized manufacturing trims production costs. *Food Technology*, 48(12), 43–45.

Hourigan, J. A. (1994). Artificial neural networks in the dairy industry. *Australian Journal of Dairy Technology*, 49, 110–133.

Larsen, M. (1989). Slow down motion to speed up line. *Packaging (Newton)*, 34(16), 89.

Mans, J. (1988). Sensors: windows into the process. *Prepared Foods*, 157(3), 81.

Mans, J. (1994). Software for the toolbox. *Prepared Foods* (6), 95–96.

Onyango, C. M., & Marchant, J. A. (1994). Transputers for high speed grading. *Food Control*, 5(1), 29–32.

Robe, K. (1992). On-the-fly can line checking. *Food Processing, 53*(8), 128–130.

Ryniecki, A., & Jayas, D. S. (1993). Automatic determination of model parameters for computer control of canned food sterilization. *Journal of Food Engineering, 19*(1), 75–94.

Schutt, J. H. (1993). Integrated planning makes inroads at food companies. *Food Processing, 54*(11), 112–114.

Warehousing of canned foods 14

14.1 Introduction

Almost every conceivable kind of warehouse and system of handling products is to be found among canneries.

Certain essentials should be incorporated in every canned-foods warehouse (see Figure 14.1 for an example of a well-designed and -constructed cannery warehouse).

- There should be ample strength to carry the heavy loads which will be stacked upon the floors.
- Every precaution should be exercised to make the warehouse dry to prevent rusting of the cans and to have the area reasonably cool.
- There should be adequate climate control systems and lighting.
- Convenience of location with respect to production and shipping.

Most canned-food warehouses are one story, thus avoiding costs of elevating loads. The warehouse usually has a flat metal roof deck, with a weatherproof roofing system on the outside and adequate insulation on the inside. If space is available, this form of construction is preferred, due to ease of maintenance and straightline warehousing flows. Precast concrete or wood covered with a built-up tar compound roof is also possible. Multi-storey facilities require less land area but may be more difficult to keep clean.

Site selection criteria, such as parking, access for incoming and outgoing materials, zoning ordinances, transportation routes, rail sidings, and proximity to customers should receive careful consideration.

Climate control is a major concern in a canned-foods warehouse. Initial design or retrofits must include adequate heating, ventilation, and air circulation to provide a stable, dry environment. If a hot water or steam heating system is used in a warehouse, more space should intervene between the stacks, and increased air movement using fans should be considered.

14.2 Stacking and cased

The reliability of canning procedures makes it possible to label and carton the finished product at the end of the canning line or after a relatively short can-cooling period. However, depending upon the product being packed, the grade may only be determined after quality control checks. Most canned foods packed are cased in shipping cartons after processing, cooling, and labelling and placed in storage in the cartons. However, some canners prefer to store the cans in the warehouse uncased (bright

A Complete Course in Canning and Related Processes. http://dx.doi.org/10.1016/B978-0-85709-677-7.00014-1

Figure 14.1 An example of a typical warehouse.
Source: Photograph courtesy of Tiger Brands Western Cape Warehouse.

stacked) on pallets and perform the casing operation at the time of labelling or shipping. Some store cans cased and unlabelled.

14.3   Warehousing of uncased cans versus cased cans

Systems of storing uncased cans allow the use of some types of automatic filled-can handling equipment, such as magnetic depalletisers. The appearance of cases which have been used for storing cans in large stacks is perhaps not as good at the time of shipment as that of new cartons used for cans from uncased storage, however; this is offset by the probability that cans which are stored in cases (cartons) will be cleaner than those stored uncased and unprotected from dust. There is no practical advantage, with respect to saving of space, ease of inspection, or lack of corrugated carton insulation in uncased storage. In the event of leaking or bursting of cans during warehousing, uncased storage saves cartons but exposes a larger number of cans to contamination and the possibility of external corrosion and secondary spoilage through pitting. The practice of storing filled-cans in case, in bulk, is costly as it requires more labour.

14.4 Palletising

No cans or cases should be stacked directly onto a cement floor, but upon wooden strips, at least 25 mm thick, or pallets. This protects against moisture condensation caused by changes in temperature and/or humidity.

A pallet is a flat transport structure that supports goods in a stable fashion so constructed that the lifting arms or 'forks' of a forklift, pallet jack, and front loader lift truck may be inserted between its faces and load transported or stored as a single unit. A pallet is the structural foundation of a unit load which allows handling and storage efficiencies. Goods or shipping containers are often placed on a pallet secured with strapping, stretch wrap, or shrink-wrap and shipped.

Although most pallets are wooden, pallets can also be made of plastic, metal, paper, and recycled materials. Each material has advantages and disadvantages for different application relative to the others. Pallets used in handling canned foods are most commonly made of wood. Soft wood pallets are lighter and cheaper than those made of hard wood or heavy plywood, but they splinter and wear quickly, require more maintenance, have a shorter useful life, and cannot support as much weight.

Significant labour savings over methods of individual case handling have been documented for even small-volume receiving, storing, loading, and shipping of a wide variety of materials. These economies are because it is cheaper to handle a single load of many units in one movement than it is to handle each of the units singly. Nevertheless, when this principle is applied to the warehousing of canned foods, other circumstances bear upon the effectiveness of the system and the economies that may be gained by its installation.

Because pallets are handling aids, they must be used repetitively if benefit is to be gained from unit load handling. If they are used to hold canned foods in long storage, they will be less cost-effective. Palletised handling is less economical for product in storage that is dormant for long periods. Seasonal canners will gain less by the use of pallets for storing all of the season's pack, whereas year-round packers, who do not store products in the warehouse for long periods, benefit more. Pallet handling makes it easier to utilise space in high-roof one-story warehouses. In multi-storey warehouses where ceilings are lower and between-floor transport is required for some of the goods in storage, the system is less practical. Time studies of warehousing operations, or even simpler step-by-step analyses and determination of labour requirements for the various warehousing tasks of transport, stacking, picking, labelling, pallet loading, empty pallet handling, and storage, will reveal what savings are possible by adoption of unit load handling systems. Volume of goods handled, methods of labelling and casing, equipment investment necessary, and methods of shipping all influence the effectiveness of the unit load method. Frequently, combinations of continuous flow, bulk storage, and unit load handling will be found best suited to an operation.

There are other uses for pallets in cannery warehouse operation, besides handling cased cans. Storage and transport of materials, such as sugar, salt, starch, dry beans, or other bagged or packaged goods can be economically handled by a pallet system. Pallets are also used for efficient transport in receiving and dock handling of raw produce in lug boxes, crates, tote boxes, hampers, and for loading empty containers for reuse.

The unit load can be bound by cross stacking of alternate layers in the same fashion used in making blocks in bulk piling of cased cans. In palletised canned-food handling, the unit load weighs usually between 500 and 1400 kg (1100–3100 lbs). For shipment of unit loads, it is advisable to strap, glue, or stretch wrap the cases rather than to ship loose on pallets. The danger of loss by shifting of the load and resulting damage to cartons and cans is thereby minimised.

14.5   Bright stacking

Storing of uncased cans without labels is commonly done in palletised form, with cans placed on pallets after sterilisation and cooling; this is called 'bright stacking'. Forklift trucks move the palletised cans to the warehouse. In other instances, sterilised and cooled cans may be mechanically conveyed to the warehouse, where they are palletised and the pallets positioned in the warehouse by means of a forklift truck.

For palletising unlabelled cans, the cans are packed directly on the pallets with paperboard dividers between every layer to prevent slipping of the cans and shifting of the load; a fibreboard or plastic dust cover is sometimes added, helping to stabilise the load. Many processors use full pallet stretch wrap to minimise shifting. Regardless, vent holes must be provided for excess moisture to escape.

Palletised cans may be safely stacked as high as cased canned foods on pallets. An occasional can suffers damage by dropping from outside rows, but the system effects a significant saving by eliminating the necessity of handling cartons when the cans are subsequently labelled. Loaded pallets are usually tiered three high and sometimes four. When the bright stacked pallet is tiered up on the storage block, the same precautions for even piling and correct placement of the load apply as for bulk piling of cased cans. Training lifttruck drivers in the proper handling of bright stack pallet loads is a critical factor in a safe and successful use of this system.

14.6   Effect of warehouse temperature on quality

Temperature is the most important variable of storage in the warehousing of canned foods. Qualitatively, the beneficial effects of reduced temperatures have been recognised for a long time, but only a limited amount of quantitative information is available.

The chemical reactions which bring about changes in canned foods during storage are extremely complicated, but as the temperature rises, the speed of these reactions increases. It is generally assumed that the rate at which chemical reactions occur doubles with each 10 °C rise in temperature. The importance of this assumption is evident when one considers the many ways by which chemical reactions may alter flavour, texture, and nutritive value of canned foods. The effect of storage temperature is economically significant. The variation in storage temperature between 10 and 30 °C (50–86 °F) may mean a fourfold difference in the rate of formation of hydrogen swells and therefore in the shelf life of cans containing products prone to this type of internal corrosion.

Internal corrosion results primarily from the action of the product on the metal of the can. Certain high-acid fruits and vegetables are prone to liberate hydrogen and perforate the container by their corrosive action on the container walls. The effect of temperature is apparent in its influence on the storage life of such foods in cans. Many other products attack the plate of the can less severely, but storage temperature is important here, because it accelerates or retards the action and the resulting loss of vacuum. In some foods, the action of product on cans may result in the appearance of unsightly discolouration; this reaction, likewise, is accelerated by high temperatures or retarded by low temperatures.

Microbiological changes are also influenced by temperature. In the growth of thermophilic (heat-loving organisms), temperature is a governing factor. For the thermophilic organisms which most commonly cause spoilage in the canning industry, this range is between 49 and 66 °C (120–155 °F). The necessity for control of storage temperature, at least in avoidance of this high thermophilic range, has often been demonstrated by thermophilic bacteria spoilage loss. Contamination of plants by thermophilic bacteria is difficult to avoid in the canning of some low-acid vegetables; it is possible for thermophilic spore-forming bacteria to survive in cans which have been processed according to commercially acceptable schedules. If storage temperatures are held below the thermophilic growth range, these bacteria do not develop and loss can be avoided.

The quality of canned foods is perhaps not as sensitive to storage temperatures as are the reactions in which internal corrosion occurs. However, desirable texture, flavour, and colour are retained over longer periods of storage at optimum temperatures. Excessively high temperatures may cause softening of texture, loss of fresh flavour, and darkening of colour in some canned products.

The nutritive value of canned foods is maintained satisfactorily, provided storage temperatures are not excessively high. In a very general way, research in laboratory and plant has demonstrated that storage temperatures between 0 and 10 °C (32–50 °F) are optimum for preserving the vitamin content of canned foods, whereas temperatures above 32 °C (90 °F) permit destruction of some of the vitamins originally in the product. Consumer interest in nutritional values of food makes consideration of methods for preserving these values in manufacture and storage increasingly important to canners.

To retard all chemical and bacteriological changes which are detrimental to the canned product, it is apparent that temperatures lower than those usually maintained in canned food warehouses are preferable. Although it is not practical to maintain warehouses of the type commonly used for canned food storage at 10 °C temperature year around, extremes of high and low temperatures should be avoided. Records of temperatures measured in surveys show extremes ranging from several degrees over 38 °C in summer months to temperatures well below 0 °C in northern areas in winter. Warehouse temperature is seldom the temperature of the canned food, because changes in warehouse temperature are very slowly reflected to the canned food in stacks, except for the first few outside rows.

Because the influence of adverse storage temperatures may be evidenced in a variety of undesirable changes, accurate and consistent temperature control in cannery warehousing deserves serious consideration.

14.7   Effect of freezing on canned food

Cool storage is desirable for all canned foods, but freezing should not be allowed. Although there is no change in the nutritive value, colour, or flavour caused by freezing, the texture of certain products, such as green beans and tomatoes, is softened. Canned products that contain starch (e.g. cream-style corn, squash, soups) show a marked change in physical consistency if frozen. Although some canned foods show no effect upon freezing, few products return to their original appearance and consistency on being heated.

Light freezing has very little, if any, effect on cans. After thawing, they gradually return to their normal shape with no appreciable loss of vacuum. Heavy freezing will likely burst can seams; it may cause distortion of the can seams without bursting them. Distorted seams may eventually produce leaks and spoilage.

If foods packed in glass containers are allowed to freeze solid, two things may occur: the jar may break, or the lid may come off, due to the expansion of the contents. Although processed foods do not freeze quickly, all glass containers should be protected from freezing temperatures. Fruits packed in syrup freeze at a lower temperature than vegetables packed in brine. All freeze at temperatures below 0 °C. In all instances, the internal quality of products should be examined whenever containers may have been exposed to freezing conditions.

14.8   External can corrosion in the warehouse

External corrosion of cans in storage may cause large losses to the warehouser. In slight degree, rusting means unsightly appearance of the can end or staining of the label. In severe cases, a totally nonmerchantable package may result. Spoilage of the product, because of rusting through the can from the outside, may also occur. Several factors must be controlled within proper limits to avoid external rusting, not the least of which is conditions prevailing in the warehouse while cans are in storage.

The presence of moisture on the surface of the can is essential to rust formation; the common phenomenon by which it forms during storage is condensation from surrounding air. In the industry, this is called 'sweating'. It occurs when the temperature of the can surface is at or below the dew point, that is, sufficiently colder than the surrounding air to condense moisture from it. When relative humidity is high and the temperature of the cans is considerably lower than that of the surrounding air, conditions are optimum for sweating. The importance of dry warehouse atmosphere and constant, even temperature is important. Sudden increases in temperature and humidity, such as may occur on a warm spring day when a cool warehouse is opened to outside air, almost surely result in sweating. Proper air circulation around the stacks, adequate temperature control, and the minimisation of unnecessary door openings decrease this danger.

The first step in the control of sweating is the measurement of the temperature of the cans and of the dew point. The determination of the dew point is made with a psychrometre, which is an instrument utilising the principle of cooling by evaporation to indicate relative humidity.

Cans cased or bright stacked at 35 °C (95 °F) and stored in the centre of large solid blocks may retain most of their heat for periods of months, even long after the warehouse temperature has become considerably lower. To avoid external corrosion, containers received from the cannery must be at temperatures high enough to permit rapid drying; a minimum of 35 °C is recommended. In this respect, palletised handling is advantageous, because the large block of many pallet loads is better ventilated than the solid block of cased cans stored in bulk. Thorough ventilation and air circulation around stacks is necessary when cans are being stored in the warehouse.

Condensation of moisture on cans in storage can result from excessive humidity in the warehouse atmosphere. If the can-storage areas are above or near improperly ventilated cannery preparation or processing rooms, these areas must be exhausted in order that moisture-laden air is removed away from the warehouse.

Another common cause of external corrosion is the error of casing or palletising wet cans at temperatures too low to create drying conditions. At temperatures much below 35 °C, cans stay wet for a considerable length of time, because the heat retained by the cans is not sufficient to evaporate the moisture from the surface of the cans. This effect is aggravated when cans are piled in the inner portions of large stacks, particularly if moisture-repellent solid-fibre cartons are used. Rusting can take place quickly under these circumstances. Mechanical removal of excess water from cans after cooling by air blasts, tipping of crates, or other methods of draining help avoid excess amounts of water in the cases. Nevertheless, the surest method is to control cooling so that cans are at the proper temperature when they are palletised or cased. The usual recommendation is that the average temperature of the contents of the can fall in the range of 35 to 38 °C (95–100 °F) at the time of casing.

Rusting of cans may result from the use of damp fibreboard or wooden cases made of green lumber. Several cases of external corrosion have occurred when wood having moisture content higher than 15% was used for making up the cases.

Sweating may occur during shipment of canned foods from cool warehouses into warmer climates or storage space. Warming of cans before shipping and use of insulated trucks and freight cars may help to avoid this.

Several other less common conditions of storage may cause external corrosion. Seaside storage requires extra precaution to avoid rusting, because of the extremely corrosive action of salt spray and the high humidity found in coastal areas. Cans should be stored in tight dry cases and warehouses built and ventilated to avoid as completely as possible having air currents from the ocean enter the storage space.

Severe corrosion can also result from contact of cans with corrosive dusts coincidental with high humidity (examples of such materials are cement, lime, and any chlorides, sulphates, or acid salts). Particularly dangerous are hygroscopic dusts, which are dusts of material having the ability to absorb moisture from the air even at temperatures well above the dew point. Many chlorides have hygroscopic properties, and if such dusts accumulate on cans, corrosion can be expected, even when sweating has not occurred.

14.9   Corrosion resulting from cannery operations

External corrosion frequently occurs during storage, not directly because of poor storage conditions, but because of bad practices in various canning operations. These factors are mentioned here because the corrosion they create often does not appear until after the cans have been warehoused and, even when not severe, they increase the possibility of extensive loss if the mistreated cans encounter adverse conditions during storage.

Retort operation can be the source of external corrosion of cans. Improper venting of steam retorts, carryover from boilers of corrosive condensates in wet steam through improperly trapped lines, and contact of cans with rusty retort crates, especially in the presence of alkaline waters, all cause rusting, pitting, tin removal, or etching of the cans. The use of naturally corrosive water for processing baths is also a cause of corrosion. Waters excessively acid, or alkaline, or high in chlorides, sulphates, or iron are dangerous for such use.

Corrosive residues may be deposited on cans at a number of points in the canning process. Failure to wash product from the outside of cans, failure to rinse cans thoroughly after washing with alkaline detergents, buildup of acid and salt dissolved in processing baths from the outside of cans, and the use of cooling waters containing high concentrations of hygroscopic salts all result in the deposition of corrosive residues on the cans. Rusting may follow within a short time after the cans are warehoused.

Corrosive cooling waters often cause rusting of cans. Such waters may attack tin plate exposed by mishandling in previous operations or may initiate corrosion themselves. When such corrosive water must be used in cannery operation, precautions can be taken to minimise corrosion. Proper retort operation, the use of approved chemicals as a corrosion inhibitor in processing baths and cooling waters, spraying or dipping cans in soft or noncorrosive water after cooling, and control of cooling so cans will be cased at safe temperatures are all helpful in avoiding loss.

Abrasion of the tin coating through rough handling anywhere in the cannery invites rapid external rusting by exposing the base steel of the tin plate. Scratching or abrasion may occur in can runways and cable conveyors, unscramblers, casers, gravity drops, dividers, seamers, fillers, or at any place where the cans are handled.

Acid fumes can contribute to rusting during storage. Their effect on stored cans should be considered in connection with the canning of sauerkraut, preparation of pickle brines, packaging of vinegar, and manufacture of condiments containing vinegar.

14.10 Secondary spoilage

Frequent inspection of canned foods in storage should be a routine part of warehouse operations. The early detection of any kind of spoilage frequently prevents large losses resulting from secondary spoilage, which occurs when sound cans rust externally after being contaminated by leaking or burst cans. It has often been found in the investigation of spoilage discovered in goods stacked in warehouses and not immediately reconditioned that the loss from secondary spoilage was far greater than that which resulted directly from the original spoilage itself. A few burst or leaking cans may eventually cause the loss of hundreds of cases of canned foods. Leaking cans wet sound cans which rust through and leak onto many other cans, the damage progressing almost geometrically. The development of this type of secondary damage can be especially rapid where leaking occurs in stored cans of high-acid products. Storing of cans in small stacks in accessible locations, frequent inspection, and prompt reconditioning when spoilage is found will keep potential loss to a minimum.

In summary, the probability of loss through corrosion resulting from mishandling in the processing and warehousing operations can be minimised by

1. Insuring that cooled cans have as much surface water removed as possible.
2. Insuring that cooled cans contain enough residual heat to completely evaporate any remaining moisture.
3. Frequent inspection for, and removal of, spoiled and leaking cans from the warehouse.
4. Maintaining close control of conditions prevailing during the storage period. Constant temperature and low humidity are the most important factors for safe, rust-free storage. Temperature and relative humidity recordings should be taken at least daily and management notified if the dew point is approached.
5. Training warehouse operators to implement emergency procedures based on dew point observations.

Acknowledgements

The 13th Ed author wishes to acknowledge the assistance of Dale Ford, Comstock Michigan Fruit Division of Curtice, Burns Foods in reviewing this chapter.

Appendix

Table 1 **Temperature conversion**

The numbers in bold face type refer to the temperature either in degrees Centigrade or Fahrenheit which it is desired to convert into the other scale. If converting from degrees Fahrenheit (°F) to degrees Centigrade (°C) the equivalent temperature will be found in the left column, while if converting from degrees Centigrade to degrees Fahrenheit the answer will be found in the column on the right.

°C		°F	°C		°F	°C		°F
-100 to 0			-8.89	**16**	60.8	7.78	**46**	114.8
			-8.33	**17**	62.6	8.33	**47**	116.6
-73.3	**-100**	-148	-7.78	**18**	64.4	8.89	**48**	118.4
-67.8	**-90**	-130	-7.22	**19**	66.2	9.44	**49**	120.2
-62.2	**-80**	-112	-6.67	**20**	68.0	10.00	**50**	122.0
-56.7	**-70**	-94						
-51.1	**-60**	-76	-6.11	**21**	69.8	10.6	**51**	123.8
			-5.56	**22**	71.6	11.1	**52**	125.6
-45.6	**-50**	-58	-5.00	**23**	73.4	11.7	**53**	127.4
-40.0	**-40**	-40	-4.44	**24**	75.2	12.2	**54**	129.2
-34.4	**-30**	-22	-3.89	**25**	77.0	12.8	**55**	131.0
-28.9	**-20**	-4						
-23.3	**-10**	14	-3.33	**26**	78.8	13.3	**56**	132.8
-17.8	**0**	32	-2.78	**27**	80.6	13.9	**57**	134.6
			-2.22	**28**	82.4	14.4	**58**	136.4
0 to 100			-1.67	**29**	84.2	15.0	**59**	138.2
			-1.11	**30**	86.0	15.6	**60**	140.0
-17.8	**0**	32						
-17.2	**1**	33.8	-0.56	**31**	87.8	16.1	**61**	141.8
-16.7	**2**	35.6	0.00	**32**	89.6	16.7	**62**	143.6
-16.1	**3**	37.4	0.56	**33**	91.4	17.2	**63**	145.4
-15.6	**4**	39.2	1.11	**34**	93.2	17.8	**64**	147.2
-15.0	**5**	41.0	1.67	**35**	95.0	18.3	**65**	149.0
-14.4	**6**	42.8	2.22	**36**	96.8	18.9	**66**	150.8
-13.9	**7**	44.6	2.78	**37**	98.6	19.4	**67**	152.6
-13.3	**8**	46.4	3.33	**38**	100.4	20.0	**68**	154.4
-12.8	**9**	48.2	3.89	**39**	102.2	20.6	**69**	156.2
-12.2	**10**	50.0	4.44	**40**	104.0	21.1	**70**	158.0
-11.7	**11**	51.8	5.00	**41**	105.8	21.7	**71**	159.8
-11.1	**12**	53.6	5.56	**42**	107.6	22.2	**72**	161.6

Continued

Table 1 **Temperature conversion – cont'd**

The numbers in bold face type refer to the temperature either in degrees Centigrade or Fahrenheit which it is desired to convert into the other scale. If converting from degrees Fahrenheit (°F) to degrees Centrigrade (°C) the equivalent temperature will be found in the left column, while if converting from degrees Centigrade to degrees Fahrenheit the answer will be found in the column on the right.

°C		°F	°C		°F	°C		°F
−10.6	13	55.4	6.11	43	109.4	22.8	73	163.4
−10.0	14	57.2	6.67	44	111.2	23.3	74	165.2
−9.44	15	59.0	7.22	45	113.0	23.9	75	167.0
24.4	76	168.8	41.11	106	222.8	60.56	141	285.8
25.0	77	170.6	41.67	107	224.6	61.11	142	287.6
25.6	78	172.4	42.22	108	226.4	61.67	143	289.4
26.1	79	174.2	42.78	109	228.2	62.22	144	291.2
26.7	80	176.0	43.33	110	230.0	62.78	145	293.0
27.2	81	177.8	43.89	111	231.8	63.33	146	294.8
27.8	82	179.6	44.44	112	233.6	63.89	147	296.6
28.3	83	181.4	45.00	113	235.4	64.44	148	298.4
28.9	84	183.2	45.56	114	237.2	65.00	149	300.2
29.4	85	185.0	46.11	115	239.0	65.56	150	302.0
30.0	86	186.8	46.67	116	240.8	66.11	151	303.8
30.6	87	188.6	47.22	117	242.6	66.67	152	305.6
31.1	88	190.4	47.78	118	244.4	67.22	153	307.4
31.7	89	192.2	48.33	119	246.2	67.78	154	309.2
32.2	90	194.0	48.89	120	248.0	68.33	155	311.0
32.8	91	195.8	49.44	121	249.8	68.89	156	312.8
33.3	92	197.6	50.00	122	251.6	69.44	157	314.6
33.9	93	199.4	50.56	123	253.4	70.00	158	316.4
34.4	94	201.2	51.11	124	255.2	70.56	159	318.2
35.0	95	203.0	51.67	125	257.0	71.11	160	320.0
35.6	96	204.8	52.22	126	258.8	71.67	161	321.8
36.1	97	206.6	52.78	127	260.6	72.22	162	323.6
36.7	98	208.4	53.33	128	262.4	72.78	163	325.4
37.2	99	210.2	53.89	129	264.2	73.33	164	327.2
37.8	100	212.0	54.44	130	266.0	73.89	165	329.0
			55.00	131	267.8	74.44	166	330.8
			55.56	132	269.6	75.00	167	332.6
	100 to 400		56.11	133	271.4	75.56	168	334.4
			56.67	134	273.2	76.11	169	336.2
			57.22	135	275.0	76.67	170	338.0

Table 1 Temperature conversion – cont'd

The numbers in bold face type refer to the temperature either in degrees Centigrade or Fahrenheit which it is desired to convert into the other scale. If converting from degrees Fahrenheit (°F) to degrees Centigrade (°C) the equivalent temperature will be found in the left column, while if converting from degrees Centigrade to degrees Fahrenheit the answer will be found in the column on the right.

°C		°F	°C		°F	°C		°F
37.78	**100**	212.0						
38.33	**101**	213.8	57.78	**136**	276.8	77.22	**171**	339.8
38.89	**102**	215.6	58.33	**137**	278.6	77.78	**172**	341.6
39.44	**103**	217.4	58.89	**138**	280.4	78.33	**173**	343.4
40.00	**104**	219.2	59.44	**139**	282.2	78.89	**174**	345.2
40.56	**105**	221.0	60.00	**140**	284.0	79.44	**175**	347.0
80.00	**176**	348.8	99.44	**211**	411.8	133.33	**272**	521.6
80.56	**177**	350.6	100.00	**212**	413.6	134.44	**274**	525.2
81.11	**178**	352.4	100.56	**213**	415.4	135.56	**276**	528.8
81.67	**179**	354.2	101.11	**214**	417.2	136.67	**278**	532.4
82.22	**180**	356.0	101.67	**215**	419.0	137.78	**280**	536.0
82.78	**181**	357.8	102.22	**216**	420.8	138.89	**282**	539.6
83.33	**182**	359.6	102.78	**217**	422.6	140.00	**284**	543.2
83.89	**183**	361.4	103.33	**218**	424.4	141.11	**286**	546.8
84.44	**184**	363.2	103.89	**219**	426.2	142.22	**288**	550.4
85.00	**185**	365.0	104.44	**220**	428.0	143.33	**290**	554.0
85.56	**186**	366.8	105.56	**222**	431.6	144.44	**292**	557.6
86.11	**187**	368.6	106.67	**224**	435.2	145.56	**294**	561.2
86.67	**188**	370.4	107.78	**226**	438.8	146.67	**296**	564.8
87.22	**189**	372.2	108.89	**228**	442.4	147.78	**298**	568.4
87.78	**190**	374.0	110.00	**230**	446.0	148.89	**300**	572.0
88.33	**191**	375.8	111.11	**232**	449.6	150.00	**392**	575.6
88.89	**192**	377.6	112.22	**234**	453.2	151.11	**304**	579.2
89.44	**193**	379.4	113.33	**236**	456.8	152.22	**306**	582.8
90.00	**194**	381.2	114.44	**238**	460.4	153.33	**308**	586.4
90.56	**195**	383.0	115.56	**240**	464.0	154.44	**310**	590.0
91.11	**196**	384.8	116.67	**242**	467.6	155.56	**312**	593.6
91.67	**197**	386.6	117.78	**244**	471.2	156.67	**314**	597.2
92.22	**198**	388.4	118.89	**246**	474.8	157.78	**316**	600.8
92.78	**199**	390.2	120.00	**248**	478.4	158.89	**318**	604.4
93.33	**200**	392.0	121.11	**250**	482.0	160.00	**320**	608.0
93.89	**201**	393.8	122.22	**252**	485.6	161.11	**322**	611.6
94.44	**202**	395.6	123.33	**254**	489.2	162.22	**324**	615.2
95.00	**203**	397.4	124.44	**256**	492.8	163.33	**326**	618.8

Continued

Table 1 Temperature conversion – cont'd

The numbers in bold face type refer to the temperature either in degrees Centigrade or Fahrenheit which it is desired to convert into the other scale. If converting from degrees Fahrenheit (°F) to degrees Centigrade (°C) the equivalent temperature will be found in the left column, while if converting from degrees Centigrade to degrees Fahrenheit the answer will be found in the column on the right.

°C		°F	°C		°F	°C		°F
95.56	204	399.2	125.56	258	496.4	164.44	328	622.4
96.11	205	401.1	126.67	260	500.0	165.56	330	626.0
96.67	206	402.8	127.78	262	503.6	166.67	332	629.6
97.22	207	404.6	128.89	264	507.2	167.78	334	633.2
97.78	208	406.4	130.00	266	510.8	168.89	336	636.8
98.33	209	408.2	131.11	268	514.4	170.00	338	640.4
98.89	210	410.0	132.22	270	518.0	171.11	340	644.0
172.22	342	647.6	188.89	372	701.6		400 to 500	
173.33	344	651.2	190.00	374	705.2			
174.44	346	654.8	191.11	376	708.8	204	400	752
175.56	348	658.4	192.22	378	712.4	210	410	770
176.67	350	662.0	193.33	380	716.0	216	420	788
						221	430	806
177.78	352	665.6	194.44	382	719.6	227	440	824
178.89	354	669.2	195.56	384	723.2	232	450	842
180.00	356	672.8	196.67	386	726.8	238	460	860
181.11	358	676.4	197.78	388	730.4	243	470	878
182.22	360	680.0	198.89	390	734.0	249	480	896
						254	490	914
183.33	362	683.6	200.00	392	737.6	260	500	932
184.44	364	687.2	201.11	394	741.2			
185.56	366	690.8	202.22	396	744.8			
186.67	368	694.4	203.33	398	748.4			
187.78	370	698.0	204.44	400	752.0			

Interpolation Factors					
°C		°F	°C		°F
0.56	1	1.8	3.33	6	10.8
1.11	2	3.6	3.89	7	12.6
1.67	3	5.4	4.44	8	14.4
2.22	4	7.2	5.0	9	16.2
2.78	5	9.0	5.56	10	18.0

Table 2 **Conversion factors, English to metric**

Mass

1 lb. (avoir) = 453.6 gram (g)
= 16 oz. (avoir)
= 7000 grain

1 kilogram (kg) = 1000 g
= 1,000,000 milligram (mg)
= 2.2 lb. (avoir)

1 ton (short) = 2000 lb.
= 907.2 kg

1 ton (long) = 2240 lb.
= 1016 kg

Length

1 ft. = 30.48 centimeters (cm)
= 0.3048 meters

1 in. = 25.4 millimeters (mm)
= 2.54 centimeters
= 0.025 meters (m)

1 mm = 0.0394 in.

1 cm = 0.394 in.
= 0.0328 ft.

1 meter = 39.37 in.
= 100 cm
= 1000 mm

Viscosity

1 poise (absolute viscosity) = 100 centipoises
= 1.00 gram per (sec) (cm)
= 0.06721b./(sec) (ft)

kinematic viscosity = Viscosity in poises;
Density in grams per cc

Volume

1 gal. (US) = 128 fl. oz. (US)
= 231 cu. in.
= 3.785 liter (l)
= 3785 milliliter (ml)
= 0.833 gal. (Brit.)

1 cu. ft. = 7.48 gal. (US)
= 28.32 (l)
= 0.0283 cu. meter

1 liter = 0.264 gal. (US)
= 1,000 ml
= 33.8 oz. (US fl.)

1 fl. oz. (US) = 29.57 ml
1 ml = 1.000027 cu.
centimeter (cc)

Continued

Table 2 Conversion factors, English to metric – cont'd

Specific gravity

Baume Hydrometers for liquids lighter than water....

$$\text{sp. gr.} = \frac{140}{130 + °\text{Baume}}$$

$$°\text{Baume} = \frac{140}{\text{sp. gr}} - 130$$

Baume Hydrometers for liquids heavier than water...

$$\text{sp. gr.} = \frac{145}{145 - °\text{Baume}}$$

$$°\text{Baume} = 145 - \frac{145}{\text{sp. gr.}}$$

A.P.I. Hydrometers...

$$\text{sp. gr.} = \frac{141.5}{131.5 + °\text{A.P.I.}}$$

$$°\text{A.P.I.} = \frac{141.5}{\text{sp. gr.}} - 131.5$$

Brix or Balling Hydrometers...

Scales read direct in percent (%) pure sucrose in water solutions, when measurement is taken at temperature indicated on hydrometer.

Salometer or Salimometer...

Salometer deg. = deg. Baume × 4.0

$$\text{sp. gr.} = \frac{\text{lbs./gal. at } 60°F}{8.337}$$

$$= \frac{\text{lbs./cu. ft. at } 60°F}{62.36}$$

Temperature

$$°F = \left(°C \times \tfrac{9}{5}\right) + 32$$

$$= (°C \times 1.8) + 32$$

$$°C = (°F - 32) \times \tfrac{5}{9}$$

$$= \frac{°F - 32}{1.8}$$

Continued

Liquid weight

1 gal. (US) = 8.34 lb. × sp. gr.
= 3.78 kg × sp. gr.

1 cu. ft. = 62.4 lb. × sp. gr.
= 28.3 kg × sp. gr.

1 lb. = 0.12 US gal. ÷ sp. gr.
= 0.016 cu. ft. ÷ sp. gr.
= 0.454 liter ÷ sp. gr.

Flow rate

1 gpm = 0.134 cu. ft./min.
= 500 lbs./hr. × sp. gr.

500 lbs./hr. = 1 gpm ÷ sp. gr.

1 cu. ft./min.
(cfm) = 448.8 gal./hr. (gph)
= 472 cc/sec

Work

1 Btu = 251.9 g cal (mean)
(mean)* = 778 ft. lb.
= 0.293 watt hr.
= $\frac{1}{180}$ of heat
required to change
temp. of 1 lb. water
from 32°F to 212°F

1 gram cal = $\frac{1}{100}$ of heat
(mean)
required to change
temp. of 1 gram water
from 0°C to 100°C

1 hp-hr = 2545 Btu (mean)
= 0.746 kw hr
= 641.3 kg cal (mean)

1 kw hr = 3413 Btu (mean)
= 860 kg cal (mean)
= 1.34 hp-hr

Table 2 Conversion factors, English to metric – cont'd

Power	Pressure
1 Btu/hr. = 0.293 watt = 12.96 ft. lb./min. = 0.252 kg cal/hr. = 0.00039 hp	1 lb./sq. in. = 2.31 ft. water at 60°F = 0.068 atmosphere = 0.07 kg/sq. cm = 2.04 in. hg at 60°F
1 ton refrigeration (US) = 288,000 Btu/24 hrs. = 12,000 Btu/hr. = 200 Btu/min. = 3024 kg cal/hr. = 83.33 lbs. ice melted/ hr. from & at 32°F = 2000 lbs. ice melted/ 24 hrs. from & at 32°F	1 ft. water = 0.433 lbs./sq. in. at 60°F = 0.884 in. hg at 60°F = 0.0295 atmosphere = 0.03 kg/sq. cm
	1 in. hg = 0.49 lbs./sq. in. at 60°F = 1.13 ft. water at 60°F = 0.033 atmosphere = 0.034 kg/sq. cm
1 hp = 550 ft. lb./sec. = 746 watt = 2545 Btu/hr.	1 atmosphere = 14,696 lbs./sq. in. (or 14.7) = 33.9 ft. water at 60°F = 29.921 in. hg at 32°F = 1.033 kg/sq. cm
1 boiler hp = 33,480 Btu/hr. = 34.5 lbs. water evaporated /hr. from and at 212°F = 9.8 kw	lbs./sq. in. = lbs./sq. in. gauge absolute (psig) + 14.7 (psia)
1 kw = 3413 Btu/hr.	

Courtesy of Votator Division, Chemetron Corporation.

Table 3 **Metric conversion table**

Millimeters	x	0.03937	=	Inches
Millimeters	=	25.400	x	Inches
Meters	x	3.2809	=	Feet
Meters	=	0.3048	x	Feet
Kilometers	x	0.621377	=	Miles
Kilometers	=	1.6093	x	Miles
Square Centimeters	x	0.15500	=	Square inches
Square Centimeters	=	6.4515	x	Square inches
Square Meters	x	10.76410	=	Square feet
Square Meters	=	0.09290	x	Square feet
Square Kilometers	x	247.1098	=	Acres
Square Kilometers	=	0.00405	x	Acres
Hectares	x	2.471	=	Acres
Hectares	=	0.4047	x	Acres
Cubic Centimeters	x	0.061025	=	Cubic inches
Cubic Centimeters	=	16.3866	x	Cubic inches
Cubic Meters	x	35.3156	=	Cubic feet
Cubic Meters	=	0.02832	x	Cubic feet
Cubic Meters	x	1.308	=	Cubic Yards
Cubic Meters	=	0.765	x	Cubic Yards
Liters	x	61.023	=	Cubic inches
Liters	=	0.01639	x	Cubic inches
Liters	x	0.26418	=	US Gallons
Liters	=	3.7854	x	US Gallons
Grams	x	15.4324	=	Grains
Grams	=	0.0648	x	Grains
Grams	x	0.03527	=	Ounces, Avoirdupois
Grams	=	28.3495	x	Ounces, Avoirdupois
Kilograms	x	2.2046	=	Pounds
Kilograms	=	0.4536	x	Pounds
Kilograms per square centimeter	x	14.2231	=	Pounds per sq. inch
Kilograms per square centimeter	=	0.0703	x	Pounds per sq. inch
Kilograms per cubic meter	x	0.06243	=	Pounds per cubic ft.
Kilograms per cubic meter	=	16.01890	x	Pounds per cubic ft.
Metric tons (1000 Kilograms)	x	1.1023	=	Tons (2000 pounds)
Metric tons (1000 Kilograms)	=	0.9072	x	Tons (2000 pounds)
Kilowatts	x	0.746	=	Horse-power
Kilowatts	=	1.3405	x	Horse-power
Calories	x	3.9683	=	British Thermal units
Calories	=	0.2520	x	British Thermal units

Table 4 **Decimal equivalents, hundreths of a millimeter into inches**

Millimeters	Inches	Millimeters	Inches	Millimeters	Inches
0.01	0.0004	0.34	0.0134	0.67	0.0264
0.02	0.0008	0.35	0.0138	0.68	0.0268
0.03	0.0012	0.36	0.0142	0.69	0.0272
0.04	0.0016	0.37	0.0146	0.70	0.0276
0.05	0.0020	0.38	0.0150	0.71	0.0280
0.06	0.0024	0.39	0.0154	0.72	0.0283
0.07	0.0028	0.40	0.0157	0.73	0.0287
0.08	0.0031	0.41	0.0161	0.74	0.0291
0.09	0.0035	0.42	0.0165	0.75	0.0295
0.10	0.0039	0.43	0.0169	0.76	0.0299
0.11	0.0043	0.44	0.0173	0.77	0.0303
0.12	0.0047	0.45	0.0177	0.78	0.0307
0.13	0.0051	0.46	0.0181	0.79	0.0311
0.14	0.0055	0.47	0.0185	0.80	0.0315
0.15	0.0059	0.48	0.0189	0.81	0.319
0.16	0.0063	0.49	0.0193	0.82	0.0323
0.17	0.0067	0.50	0.0197	0.83	0.0327
0.18	0.0071	0.51	0.0201	0.84	0.0331
0.19	0.0075	0.52	0.0205	0.85	0.0335
0.20	0.0079	0.53	0.0209	0.86	0.0339
0.21	0.0083	0.54	0.0213	0.87	0.0343
0.22	0.0087	0.55	0.0217	0.88	0.0346
0.23	0.0091	0.56	0.0220	0.89	0.0350
0.24	0.0094	0.57	0.0224	0.90	0.0354
0.25	0.0098	0.58	0.0228	0.91	0.0358
0.26	0.0102	0.59	0.0232	0.92	0.0362
0.27	0.0106	0.60	0.0236	0.93	0.0366
0.28	0.0110	0.61	0.0240	0.94	0.0370
0.29	0.0114	0.62	0.0244	0.95	0.0374
0.30	0.0118	0.63	0.0248	0.96	0.0378
0.31	0.0122	0.64	0.0252	0.97	0.0382
0.32	0.0126	0.65	0.0256	0.98	0.0386
0.33	0.0130	0.66	0.0260	0.99	0.0390
				1.00	0.0394

Table 5 **Decimal equivalents**

64ths	Decimals	Millimeters	64ths	Decimals	Millimeters
1/64	0.0156	0.397	33/64	0.5156	13.097
1/32	0.0313	0.794	17/32	0.5313	13.494
3/64	0.0469	1.191	35/64	0.5469	13.891
1/16	0.0625	1.588	9/16	0.5625	14.288
5/64	0.0781	1.984	37/64	0.5781	14.684
3/32	0.0938	2.381	19/32	0.5938	15.081
7/64	0.1094	2.778	39/64	0.6094	15.478
1/8	0.1250	3.175	5/8	0.6250	15.875
9/64	0.1406	3.572	41/64	0.6406	16.272
5/32	0.1563	3.969	21/32	0.6563	16.669
11/64	0.1719	4.366	43/64	0.6719	17.066
3/16	0.1875	4.763	11/16	0.6875	17.463
13/64	0.2301	5.519	45/64	0.7031	17.859
7/32	0.2188	5.556	23/32	0.7188	18.256
15/64	0.2344	5.953	47/64	0.7344	18.653
1/4	0.2500	6.350	3/4	0.7500	19.050
17/64	0.2656	6.747	49/64	0.7656	19.447
9/32	0.2813	7.144	25/32	0.7813	19.844
19/64	0.2969	7.541	51/64	0.7969	20.241
5/16	0.3125	7.938	13/16	0.8125	20.638
21/64	0.3281	8.334	53/64	0.8281	21.034
11/32	0.3438	8.731	27/32	0.8438	21.431
23/64	0.3594	9.128	55/64	0.8594	21.828
3/8	0.3750	9.525	7/8	0.8750	22.225
25/64	0.3906	9.922	57/64	0.8906	22.622
13/32	0.4063	10.319	29/32	0.9063	23.019
27/64	0.4219	10.716	59/64	0.9219	23.416
7/16	0.4375	11.113	15/16	0.9375	23.813
29/64	0.4531	11.509	61/64	0.9531	24.209
15/32	0.4688	11.906	31/32	0.9688	24.606
31/64	0.4844	12.303	63/64	0.9844	25.003
1/2	0.5000	12.700	1	1.0000	25.400

Table 6 **Metric weights and measures**

Linear measure

10 millimeters (mm) = 1 centimeter (cm)
    10 centimeters = 1 decimeter (dm) = 100 millimeters
    10 decimeters = 1 meter (m) = 1,000 millimeters
    1,000 meters = 1 kilometer (km)

Area measure

100 square millimeters (mm^2) = 1 square centimeter (cm^2)
    10,000 square centimeters = 1 sq. meter (m^2) = 1,000,000 square millimeters
    10,000 square meters = 1 hectar (ha)
    1,000,000 square meters = 1 square kilometer (km^2)

Volume measure

1,000 millileters (ml) = 1 liter (L)
    1,000 liters (L) = 1 cubic meter (m^3)

Weight

10 milligrams (mg) = 1 centigram (eg)
    10 centigrams = 1 decigram (dg) = 100 milligrams
    10 decigrams = 1 gram (g) = 1,000 milligrams
    1,000 grams = 1 kilogram (kg)
    1,000 kilograms = 1 metric ton (t)

Table 7 **Tin plate basis weights**

Weight		
Lbs./Base box (basis weight)	**Lbs./Sq. Foot**	**Thickness (inches)**
45	0.2066	0.0050
50	0.2296	0.0055
55	0.2525	0.0061
60	0.2755	0.0066
65	0.2985	0.0072
70	0.3214	0.0077
75	0.3444	0.0083
80	0.3673	0.0088
85	0.3903	0.0094
90	0.4133	0.0099
95	0.4362	0.0105
100	0.4592	0.0110
107	0.4913	0.0118
112	0.5143	0.0123
118	0.5418	0.0130
128	0.5877	0.0141
135	0.6199	0.0149
139	0.6383	0.0153
148	0.6796	0.0163
155	0.7117	0.0171
168	0.7714	0.0185
175	0.8036	0.0193
180	0.8265	0.0198
188	0.8633	0.0207
195	0.8954	0.0215
208	0.9551	0.0229
210	0.9643	0.0231
215	0.9872	0.0237
228	1.0469	0.0251
210	1.0791	0.0259
240	1.1020	0.0264
248	1.1388	0.0273
255	1.1709	0.0281
268	1.2306	0.0295
270	1.2398	0.0297
275	1.2627	0.0303

The base box is the unit area of 112 sheets, 14 × 20 inches or 31.360 square inches (217.78 sq. ft). Basis weights, which determine the approximate thickness of the plates, are customarily expressed in pounds per base box. Tin Plate is also produced for special uses in weights other than those shown in the above table.

Table 8 Case equivalents

Equivalent cases				
Case of...	No. 303's	No. 2's	No. 2 ½'s	No. 10's
48 6Z	0.7122	0.5854	0.4034	0.4386
72 8Z short	1.4064	1.1562	0.7965	0.8661
24 8Z tall	0.5134	0.4220	0.2908	0.3161
48#1 flat	1.0552	0.8674	0.5976	0.6498
48#1 picnic	1.2938	1.0634	0.7328	0.7967
24#211 cyl	0.8042	0.6610	0.4555	0.4952
24#2 vac (12Z)	0.8724	0.7171	0.4941	0.5372
24#300	0.9021	0.7415	0.5109	0.5555
24#1 tall	0.9852	0.8098	0.5580	0.6067
24#303	1.0000	0.8220	0.5664	0.6158
24#300 cyl	1.1513	0.9463	0.6521	0.7089
24#2	1.2166	1.0000	0.6891	0.7492
24#3 vac	1.4154	1.1634	0.8017	0.8716
24#2½	1.7656	1.4512	1.0000	1.0873
12#29Z	0.9644	0.7927	0.5462	0.5939
12#32Z quart	1.0534	0.8659	0.5967	0.6487
12#3 cyl	1.5341	1.2610	0.8689	0.9447
6#5 squat	1.0111	0.8311	0.5727	0.6227
6#10	1.6239	1.3348	0.9198	1.0000

The capacity of a 16 oz. and No. $2^{1}/_{2}$ glass jar is approximately the same as the No. 303 and No. $2^{1}/_{2}$ can, respectively.

The above table gives the equivalent in cases of 24/303's, 24/2's, $24/2\frac{1}{2}$'s and 6/10's of the more commonly used cans.

Source: The *ALMANAC of the Canning, Freezing and Preserving Industries*, E. E. Judge & Sons, Inc., Pub., Westminster, MD.

Table 9 Can dimensions and conversions, from US systems to metric

Dimensions (diameter × height)			
Inches	Millimeters	Inches	Millimeters
202 × 204	54.0 × 57.2	307 × 113	87.3 × 46.0
202 × 214	54.0 × 73.0	307 × 200.25	87.3 × 51.2
211 × 109	68.3 × 39.7	307 × 306	87.3 × 85.7
211 × 212	68.3 × 69.9	307 × 400	87.3 × 101.6
211 × 304	68.3 × 82.6	307 × 409	87.3 × 115.9
211 × 400	68.3 × 101.6	307 × 510	87.3 × 142.9
300 × 109	76.2 × 39.7	307 × 512	87.3 × 146.1
300 × 400	76.2 × 101.6	401 × 411	103.2 × 119.1
300 × 407	76.2 × 112.7	404 × 307	108.0 × 87.3
301 × 106	77.8 × 34.9	404 × 700	108.0 × 177.8
301 × 408	77.8 × 114.3	603 × 405	157.2 × 109.5
303 × 406	81.0 × 111.1	603 × 408	157.2 × 114.3
303 × 509	81.0 × 141.3	603 × 700	157.2 × 177.8

Can sizes are given in the nomenclature usually employed in the industry, which avoids the confusion incident to conflicting local names of cans. In this system the cans are identified by a statement of their dimensions (overall diameter and overall height). Each dimension is expressed as a number of three digits. The first digit gives the number of whole inches, while the next two digits give the additional fraction of the dimension expressed as sixteenths of an inch. The first number given in the size of each can is the diameter, and the second number is the height. For example, a can designated as 303 × 406, is $3^3/_{16}$ inches in diameter and $4^6/_{16}$ inches high, within manufacturing tolerances. The dimensions are "overall," the diameter being measured to the outside of the double seam, and the height including the entire seam at each end of the can. The following table lists the dimensions of common can sizes used in the canning industry and their approximate metric equivalents. The metric equivalents were calculated on the basis of 1 inch = 25.40 mm.
Source: National Food Processors Association Bulletin 26-L, 12th Ed., 1982.

Table 10 Container dimension conversion chart, from metric (mm) to US system (inches and sixteenths)
Example: a container dimension of 77 mm would convert to 0301 ($3^1/_{16}$ inches)

Dimension mm	Inches + 16ths of In.	mm	Inches + 16ths of In.	mm	Inches + 16ths of In.	mm	Inches + 16ths of In.
1	0001	41	0110	81	0303	121	0412
2	0001	42	0110	82	0304	122	0413
3	0002	43	0111	83	0304	123	0413
4	0003	44	0112	84	0305	124	0414
5	0003	45	0112	85	0306	125	0415
6	0004	46	0113	86	0306	126	0415
7	0004	47	0114	87	0307	127	0500
8	0005	48	0114	88	0307	128	0501
9	0006	49	0114	89	0308	129	0501
10	0006	50	0115	90	0309	130	0502
11	0007	51	0200	91	0309	131	0503
12	0008	52	0201	92	0310	132	0503
13	0008	53	0201	93	0311	133	0504
14	0009	54	0202	94	0311	134	0504
15	0009	55	0203	95	0312	135	0505
16	0010	56	0203	96	0312	136	0506
17	0010	57	0204	97	0313	137	0506
18	0011	58	0205	98	0314	138	0507
19	0012	59	0205	99	0314	139	0508
20	0013	60	0206	100	0315	140	0508

Continued

21	0013	61	0206	101	0400	141	0509
22	0014	62	0207	102	0400	142	0509
23	0014	63	0208	103	0401	143	0510
24	0015	64	0208	104	0402	144	0511
25	0100	65	0209	105	0402	145	0511
26	0100	66	0210	106	0403	146	0512
27	0101	67	0210	107	0403	147	0513
28	0102	68	0211	108	0404	148	0513
29	0102	69	0211	109	0405	149	0514
30	0103	70	0212	110	0405	150	0514
31	0104	71	0213	111	0406	151	0515
32	0104	72	0213	112	0407	152	0600
33	0105	73	0214	113	0407	153	0600
34	0105	74	0215	114	0408	154	0601
35	0106	75	0215	115	0408	155	0602
36	0107	76	0300	116	0409	156	0602
37	0107	77	0301	117	0410	157	0603
38	0108	78	0301	118	0410	158	0604
39	0109	79	0302	119	0411	159	0604
40	0109	80	0302	120	0412	160	0605
161	0605	197	0712	233	0903	269	1009
162	0606	198	0713	234	0903	270	1010
163	0607	199	0713	235	0904	271	1011
164	0607	200	0714	236	0905	272	1011
165	0608	201	0715	237	0905	273	1012
166	0609	202	0715	238	0906	274	1013

Table 10 Container dimension conversion chart, from metric (mm) to US system (inches and sixteenths) Example: a container dimension of 77 mm would convert to 0301 ($3^{11}/_{16}$ inches) – cont'd

Dimension mm	Inches + 16ths of In.	mm	Inches + 16ths of In.	mm	Inches + 16ths of In.	mm	Inches + 16ths of In.
167	0609	203	0800	239	0907	275	1013
168	0610	204	0801	240	0907	276	1014
169	0610	205	0801	241	0908	277	1014
170	0611	206	0802	242	0908	278	1015
171	0612	207	0802	243	0909	279	1100
172	0612	208	0803	244	0910	280	1100
173	0613	209	0804	245	0910	281	1101
174	0614	210	0804	246	0911	282	1102
175	0614	211	0805	247	0912	283	1102
176	0615	212	0806	248	0912	284	1103
177	0615	213	0806	249	0913	285	1104
178	0700	214	0807	250	0913	286	1104
179	0701	215	0807	251	0914	287	1105
180	0701	216	0808	252	0915	288	1105
181	0702	217	0809	253	0915	289	1106
182	0703	218	0809	254	1000	290	1107
183	0703	219	0810	255	1001	291	1107
184	0704	220	0811	256	1001	292	1108
185	0705	221	0811	257	1002	293	1109
186	0705	222	0812	258	1003	294	1109
187	0706	223	0812	259	1003	295	1110

188	0706	224	0813	260	1004	296	1110
189	0707	225	0814	261	1004	297	1111
190	0708	226	0814	262	1005	298	1112
191	0708	227	0815	263	1006	299	1112
192	0709	228	0900	264	1006	300	1113
193	0710	229	0900	265	1007	301	1114
194	0710	230	0901	266	1008	302	1114
195	0711	231	0902	267	1008	303	1115
196	0711	232	0902	268	1009	304	1200

Table 11 Sodium chloride brine tables for brine at 60 °F

Salometer degrees	Specific gravity	Baume degrees	% Sodium chloride by wt.	Lbs. per gal. brine		Grams per Liter brine		freezing Pt. °Fa
				NaCl	Water	NaCl	Water	Pt. °Fa
0	1.000	0.0	0.000	0.000	8.328	0.0	998.0	+32.0
2	1.004	0.6	0.528	0.044	8.318	5.28	996.5	+31.5
4	1.007	1.1	1.056	0.089	8.297	10.61	994.0	+31.1
6	1.011	1.6	1.584	0.133	8.287	15.97	992.8	+30.5
8	1.015	2.1	2.112	0.178	8.275	21.38	991.3	+30.0
10	1.019	2.7	2.640	0.224	8.262	26.84	989.9	+29.3
12	1.023	3.3	3.167	0.270	8.250	32.35	988.3	+28.8
14	1.026	3.7	3.695	0.316	8.229	37.86	985.9	+28.2
16	1.030	4.2	4.223	0.362	8.216	43.43	984.5	+27.6
18	1.034	4.8	4.751	0.409	8.202	49.03	982.7	+27.0
20	1.038	5.3	5.279	0.456	8.188	54.65	981.0	+26.4

Continued

Table 11 Sodium chloride brine tables for brine at 60 °F – cont'd

Salometer degrees	Specific gravity	Baume degrees	% Sodium chloride by wt.	Lbs. per gal. brine		Grams per Liter brine		freezing
				NaCl	Water	NaCl	Water	Pt. °Fa
22	1.042	5.8	5.807	0.503	8.175	60.32	979.4	+25.7
24	1.046	6.4	6.335	0.552	8.159	66.13	977.4	+25.1
26	1.050	6.9	6.863	0.600	8.144	71.88	975.7	+24.4
28	1.054	7.4	7.391	0.649	8.129	77.27	973.9	+23.7
30	1.058	7.9	7.919	0.698	8.113	83.56	971.8	+23.0
32	1.062	8.5	8.446	0.747	8.097	89.49	969.9	+22.3
34	1.066	9.0	8.974	0.797	8.081	95.42	968.0	+21.6
36	1.070	9.5	9.502	0.847	8.064	101.4	965.8	+20.9
38	1.074	10.0	10.030	0.897	8.047	107.5	963.8	+20.2
40	1.078	10.5	10.558	0.948	8.030	113.5	961.8	+19.4
42	1.082	11.0	11.086	0.999	8.012	119.6	959.6	+18.7
44	1.086	11.5	11.614	1.050	7.994	125.8	957.2	+17.9
46	1.090	12.0	12.142	1.102	7.976	132.0	955.3	+17.1
48	1.094	12.5	12.670	1.154	7.957	138.2	953.0	+16.2
50	1.098	12.9	13.198	1.207	7.937	144.5	950.6	+15.4
52	1.102	13.4	13.725	1.260	7.918	150.9	948.5	+14.5
54	1.106	13.9	14.253	1.313	7.898	157.2	946.2	+13.7
56	1.110	14.4	14.781	1.366	7.878	163.6	943.7	+12.8
58	1.114	14.8	15.309	1.420	7.858	170.1	941.3	+11.8
60	1.118	15.3	15.837	1.475	7.836	176.7	938.8	+10.9
62	1.122	15.8	16.365	1.529	7.815	183.2	936.2	+9.9
64	1.126	16.2	16.893	1.584	7.794	189.8	933.7	+8.9

66	1.130	16.7	17.421	1.639	7.772	196.5	931.2	+7.9
68	1.135	17.2	17.949	1.697	7.755	203.2	928.9	+6.8
70	1.139	17.7	18.477	1.753	7.733	209.9	926.3	+5.7
72	1.143	18.1	19.004	1.809	7.710	216.7	923.7	+4.6
74	1.147	18.6	19.532	1.866	7.686	223.5	921.0	+3.4
76	1.152	19.1	20.060	1.925	7.669	230.6	918.6	+2.2
78	1.156	19.6	20.588	1.982	7.645	237.4	915.9	+1.0
80	1.160	20.0	21.116	2.040	7.620	244.4	912.9	−0.4
82	1.164	20.4	21.644	2.098	7.596	251.5	910.5	−1.6
84	1.169	21.0	22.172	2.158	7.577	258.5	907.7	−3.0
86	1.173	21.4	22.700	2.218	7.551	265.7	904.6	−4.4
88	1.178	21.9	23.228	2.279	7.531	272.9	902.1	−5.8
[b]88.3	1.179	22.0	23.310	2.288	7.528	274.1	901.6	−6.0[b]
90	1.182	22.3	23.755	2.338	7.506	280.1	899.1	−1.1
92	1.186	22.7	24.283	2.398	7.479	287.4	896.0	+4.8
94	1.191	23.3	24.811	2.459	7.460	294.7	893.5	+11.1
95	1.193	23.5	25.075	2.491	7.444	298.4	891.7	+14.4
96	1.195	23.7	25.339	2.522	7.430	302.1	890.1	+18.0
97	1.197	23.9	25.603	2.552	7.417	305.8	888.6	+21.6
98	1.200	24.2	25.867	2.585	7.409	309.6	887.4	+25.5
99	1.202	24.4	26.131	2.616	7.394	313.4	885.7	+29.8
99.6	1.203	24.5	26.285	2.634	7.386	315.4	884.8	+32.3[c]
100	1.204	24.6	26.395	2.647	7.380	317.1	884.2	+60.0[d]

The above table applies to brine tested at the temperature of 60 °F.

[a]Temperature at which freezing begins. Ice forms, brine concentrates, and the freezing point lowers to eutecic.

[b]Eutecic point. For brines stronger than eutecic, the temperatures shown are the saturation temperature for sodium chloride dihydrate. Brines stronger than eutecic deposit excess sodium chloride as dihydrate when cooled, and freeze at eutecic.

[c]Transition temperature from anhydrous salt to dihydrate.

[d]Saturated brine at 60 °F.

Source: Morton Salt Company.

Table 12 Normal pH ranges of selected commercially canned foods

Kind of food	pH range, approximate
Apples, whole	3.4–3.5
Apple juice	3.3–3.5
Asparagus, green	5.0–5.8
Beans	
Baked	4.8–5.5
Green	4.9–5.5
Lima	5.4–6.3
Beans, with pork	5.1–5.8
Beef, corned, hash	5.5–6.0
Beets, whole	4.9–5.8
Blackberries	3.0–4.2
Blueberries	3.2–3.6
Boysenberries	3.0–3.3
Broccoli	5.2–6.0
Carrots, chopped	5.3–5.6
Carrot juice	5.2–5.8
Cherry juice	3.4–3.6
Chicken	6.2–6.4
Chicken with noodles	6.2–6.7
Chop Suey	5.4–5.6
Cider	2.9–3.3
Clams	5.9–7.1
Codfish	6.0–6.1
Corn	
On-the-cob	6.1–6.8
Cream-style	5.9–6.5
Whole grain	
Brine packed	5.8–6.5
Vacuum packed	6.0–6.4
Crab apples, spiced	3.3–3.7
Cranberry	
Juice	2.5–2.7
Sauce	2.3
Currant juice	3.0
Fruit cocktail	3.6–4.0
Gooseberries	2.8–3.1
Grapefruit	
Juice	2.9–3.4

Table 12 Normal pH ranges of selected commercially canned foods – cont'd

Kind of food	pH range, approximate
Pulp	3.4
Sections	3.0–3.5
Grapes	3.5-4.5
Ham, spiced	6.0–6.3
Hominy, lye	6.9–7.9
Huckleberries	2.8–2.9
Jam, fruit	3.5–4.0
Jellies, fruit	3.0–3.5
Lemons	2.2–2.4
Juice	2.2–2.6
Lime juice	2.2–2.4
Loganberries	2.7–3.5
Mackerel	5.9–6.2
Molasses	5.0–5.4
Mushrooms	6.0–6.5
Olives, ripe	5.9–7.3
Orange juice	3.0–4.0
Oysters	6.3–6.7
Peaches	3.4–4.2
Pears (Bartlett)	3.8–4.6
Peas	5.6–6.5
Pickles	
Dill	2.6–3.8
Sour	3.0–3.5
Sweet	2.5–3.0
Pimentos	4.3–4.9
Pineapple	
Crushed	3.2–4.0
Sliced	3.5–4.1
Juice	3.4–3.7
Plums	2.8–3.0
Potatoes	
White	5.4–5.9
Mashed	5.1
Potato salad	3.9–4.6
Prune juice	3.7–4.3
Pumpkin	5.2–5.5
Raspberries	2.9–3.7

Continued

Table 12 Normal pH ranges of selected commercially canned foods – cont'd

Kind of food	pH range, approximate
Rhubarb	2.9–3.3
Salmon	6.1–6.5
Sardines	5.7–6.6
Sauerkraut	3.1–3.7
Juice	3.3–3.4
Shrimp	6.8–7.0
Soups	
Bean	5.7–5.8
Beef broth	6.0–6.2
Chicken noodle	5.5–6.5
Clam chowder	5.6–5.9
Duck	5.0–5.7
Mushroom	6.3–6.7
Noodle	5.6–5.8
Oyster	6.5–6.9
Pea	5.7–6.2
Tomato	4.2–5.2
Turtle	5.2 5.3
Vegetable	4.7–5.6
Spinach	4.8–5.8
Squash	5.0–5.3
Strawberries	3.0–3.9
Sweet potatoes	5.3–5.6
Tomatoes	4.1–4.4
Juice	3.9–4.4
Tuna	5.9–6.1
Turnip greens	5.4–5.6
Vegetable	
Juice	3.9–4.3
Mixed	5.4–5.6
Vinegar	2.4–3.4
Youngberries	3.0–3.7

Table 13 **Calculated sterilizing values (F_o) for some current commercial processes**

Product	Can size	Approximate calculated sterlizing value, F_o
Asparagus	All	2 to 4
Green beans, brine packed	#2	3.5
Green beans, brine packed	#10	6
Chicken boned	All	6 to 8
Corn, whole kernel, brine packed	#2	9
Corn, whole kernel, brine packed	#10	15
Cream style corn	#2	5 to 6
Cream style corn	#10	2.3
Dog food	#2	12
Dog food	#10	6
Mackerel in brine	301 × 411	2.9 to 3.6
Meat loaf	#2	6
Peas, brine packed	#2	7
Peas, brine packed	#10	11
Sausage, vienna, in brine	Various	5
Chili con carne	Various	6

Courtesy of former American Can Co.

Glossary

AAS Atomic absorption spectrophotometry. Method used to quantitatively analyse for mineral elements, like sodium, phosphorus, chromium, cobalt.

Absolute humidity Actual weight of water vapour contained in a unit volume or weight of air. See Relative humidity.

Absorbent A substance having the ability to soak up or retain other substances, such as sugar or salt, absorbing water when exposed to high relative humidity atmospheres.

Acid A substance which increases the concentration of hydrogen ions (H+) in water and reacts with a base to form a salt. See Hydrogen-ion concentration.

Acid food A food that has a natural pH of 4.6 or below (in Europe it is common to use pH 4.5 as the upper limit).

Acid number Number or amount of KOH required to neutralise the free fatty acids in 1 g of fat, wax or resin.

Acidified food A low-acid food to which acid(s) or acid food(s) are added and which has a finished equilibrium pH of 4.6 or below and a water activity (aw) greater than 0.85.

Acidified low-acid food A food which has been treated so as to attain an equilibrium pH of 4.6 (or 4.5 in Europe) or lower after heat processing.

Acidulant An acidifying agent, such as acetic acid or vinegar.

Aciduric Microorganisms that can grow in high acid foods, i.e. with a pH value below 3.0. Generally are of low heat resistance.

Activated sludge Sludge floc produced in raw or settled wastewater by the growth of bacteria or other organisms in the presence of dissolved oxygen.

Activated sludge process A biological wastewater treatment process in which a mixture of wastewater and activated sludge is agitated and aerated.

Additive Any substance, the intended use of which results or may reasonably be expected to result, directly or indirectly, in its becoming a component or otherwise affecting the characteristics of any food.

Additive, food See Food additive.

Adsorbent Material on whose surface adsorption takes place.

Adsorption Adhesion of a substance to the surface of a solid or liquid.

Adulterant (adulteration) Foreign material in food, especially substances which are aesthetically objectionable, hazardous to health, or which indicate that unsanitary handling or manufacturing practices have been employed.

Aeration The bringing about of intimate contact between air and a liquid by bubbling air through the liquid or by agitation of the liquid to promote surface absorption of air.

Aeration tank A tank in which sludge, sewage, or other liquid waste is aerated.

Aerobes Microorganisms that need oxygen for their growth. Obligate aerobes cannot survive in the absence of oxygen.

Aerobic Living or active only in the presence of free oxygen.

Aerator A device used to promote aeration.

Aerosol Colloidal suspension in which gas is the dispersant. Dispersion or suspension of extremely fine particles of liquid or solid in a gaseous medium.

Aflatoxins Highly toxic substances produced by certain moulds on moist peanuts, corn, pecans, and other foodstuffs during the growing and postharvest period.

Agar Dried, purified stems of seaweed. Partly soluble and swells with water to form a gel. Used in soups, jellies, ice cream, meat and fish pastes, in bacteriological media, and as a stabiliser for emulsions. Also called agar–agar.

Agglomerate To gather, form, or grow into a rounded mass, or to cluster densely.

Aging See Maturation.

Air flotation Synonymous with flotation.

Albedo The white inner layer of citrus fruit peel. Consists of sugars, cellulose, and pectin; used as a source of pectin for commercial manufacture.

Albumen The white of an egg, composed principally of the protein albumin.

Albumin Any of a group of plant and animal proteins which are soluble in water, dilute salt solutions, and 50% saturated ammonium sulphate.

Aldehydes A class of highly reactive organic chemical compounds obtained by oxidation of a primary alcohol.

Algae Major group of lower plants, single- and multicelled, usually aquatic, and capable of synthesising their foodstuff by photosynthesis.

Alginates Salts of alginic acid found in many seaweeds. Used as thickeners and stabilisers in ice cream and synthetic cream, in artificial cherries, and as alginate sausage casings.

Allergen Any substance capable of producing allergy.

Allergy A hypersensitivity to a specific substance or condition which in similar amounts is harmless to most people.

Allspice Dried fruits of the evergreen *Pimenta officinalis*. Also known as pimento or Jamaica pepper.

Almond, Bitter Ripe seed of *Prunus amygdalus var. amars* (almond tree).

Almond, Sweet Ripe seeds of *Prunus amygdalus var. dulcis.*

Alpha-Tocopherol See Vitamin E.

Alum (Aluminium and potassium sulphate). Used in foods as a buffer, a neutralising agent, and a firming agent; in baking powders to help generate carbon dioxide; and in water purification as a flocculating agent.

Amino acid Proteins are composed of about 23 amino acids. Eight of them must be provided in the human diet, the essential amino acids. The remaining 15 can be synthesised in the body. Many amino acids are manufactured synthetically, and, lysine and methionine in particular, can be added to food and feeds to increase their nutritive value.

Amylopectin A branched polysaccharide which, together with amylose, makes up starch.

Amylose Straight chain polysaccharide which, together with amylopectin, makes up starch.

Anaerobes Microorganisms that grow in the absence of oxygen. Obligate anaerobes cannot survive in the presence of oxygen. Facultative anaerobes normally grow in oxygen but can also grow in its absence.

Anaerobic Living or active in the absence of free oxygen.

Analogs See Food analogs.

Anhydroglucose units The basic $C_6H_{10}O_5$ unit that occurs repeatedly in all starch molecules.

Anion Any negatively charged atom or group of atoms, such as hydroxide (OH^-), carboxylate ($COOH^-$) and phosphate ($PO_4^=$).

Anionic surfactants Ionic surface-active agents in which the portion that associates with the internal phase is the anion; they include carboxylic acids, sulphuric acid esters, and sulphonic acids.

Annealing Heating process used in tin plate manufacture to soften the steel strip after cold rolling and to impart the required hardness; the process can either be continuous (continuous annealing or CA) or in batches (batch annealing or BA).

Anthocyanins Violet, red, and blue colouring matter of many fruits, flowers, and leaves. Depolarisers in electrochemical reactions; as such they cause trouble in canned foods by accelerating internal can corrosion.

Anthoxanthins Yellow to orange-red pigments present in plant materials.

Antibiotic A substance that inhibits the growth of microorganisms usually produced by other organisms such as penicillin.

Anticaking agent Substance used in many salts and powders to keep them free-flowing. Anticaking agents are used in such products as table salt, garlic and onion salts and powders, powdered sugar, and malted milk powders.

Antifoamer Liquid of low intrinsic surface tension that prevents formation of foam.

Antimicrobial A compound which inhibits the growth of a microbe.

Antimycotic agent A substance which destroys or inhibits the growth of moulds and other fungi.

Antioxidants Substances that retard the oxidative rancidity of fats or the oxidation of other substances.

Antiseptic A substance that prevents or inhibits the growth of microorganisms on animate surfaces, such as skin.

AOAC Association of Official Analytical Chemists.

Apparent viscosity See viscosity. Viscosity of a complex (non-Newtonian) fluid under given conditions.

Aqueous Containing water.

Ascorbic acid See Vitamin C.

Asepsis Clean and free of microorganisms.

Aseptic Processing and Packaging The filling of a commercially sterilized cooled product into presterilized containers, followed by aseptic hermetical sealing, with a presterilized closure, in an atmosphere free of microorganisms.

Ash Ash is one of the components in the proximate analysis of biological materials. It is the name given to the residue that remains after a sample is heated at high temperature (usually 450 °C–550 °C) and consists mostly of metal oxides.

ATP (Adenosine triphosphate). The prosthetic group of the enzyme hexokinase, which is involved in the fermentation of sugars such as glucose.

Autoclave A vessel in which high temperatures can be reached by using high steam pressure. Bacteria are destroyed more readily at elevated temperatures, and autoclaves are used to sterilise food, for example in cans.

A_w A symbol for 'water activity'. See Water activity.

B vitamins See Vitamin B Complex.

B-Carotene Pro-Vitamin A. A compound found naturally in many foods and also synthesised, which is converted by the human body into Vitamin A. See Vitamin A.

Bacillus A rod-shaped bacterium. Some *Bacillus* produce spores.

Bacillus cereus Spore-forming, rod-shaped bacterium, aerobic to facultative aerobic, proteolytic. It produces gastroenteritis caused by the release of an exo-enterotoxin during lysis of *B. cereus* in the intestinal tract.

Bactericide Any substance that destroys bacteria, although not necessarily the spores of bacteria.

Bacteriostatic Preventing the growth of bacteria without killing them.

Baffle Partition or plate that changes the direction or restricts the cross section of a fluid, thus increasing velocity or turbulence.

Baking powder Leavening agent which acts through the release of carbon dioxide (CO_2) during the baking process. Baking powder consists of sodium bicarbonate (baking soda), an acid or an acid salt which reacts with the bicarbonate prior to and during baking to release the carbon dioxide, and starch to absorb moisture during storage.

Baking soda (Sodium bicarbonate). Produces carbon dioxide when heated or when reacting with an acid.

Barrier, grease-resistant A material that prevents or retards the transmission of grease or oils.

Barrier, water-resistant A material that retards the transmission of water vapour.

Base Alkaline substances (pH greater than 7.0) which yield hydroxyl ions (OH^-) in solution. See Hydrogen-ion concentration.

Base box A unit of area of tin plate equivalent to $20.23\,m^2$ (31, 360 sq. in or 217.78 sq. ft). The term '90# plate' means tin plate of such thickness that the above area weighs 40.8 kg (90 lb). A unit of sale under the imperial system.

Base plate pressure The force of the base plate holding the can body and end against the chuck during the seaming operation. In general, has the following effect on the seaming formation: Low Pressure – short body hook. High Pressure – long body hook.

Baume The name of one of the many hydrometer scales used for determining the relative density of liquids as compared to a standard liquid. There are two Baume scales: one for liquids lighter than water, the other for liquids heavier than water.

Bead A rounded depression around the surface of a container or end: used to stiffen or improve its appearance.

Beaded can A can reinforced by bead indentations in the body.

Bentonite Colloidal clay used as an absorbent. Also used in model systems for determining the rate of heat penetration.

Beriberi A disease caused by the deficiency of B-complex vitamins in the diet.

BHA Butylated Hydroxyanisole. An antioxidant.

BHT Butylated Hydroxytoluene. An antioxidant.

Bioassay A test which uses animals or microorganisms for determining the biological activity of certain substances or the presence or concentration of nutrients in food.

Biodegradability Susceptibility of a chemical compound to depolymerisation by the action of biological agents.

Biological oxidation The process whereby, through the activity of living organisms in an aerobic environment, organic matter is converted into more biologically stable matter.

Biological Oxygen Demand (BOD) Microorganisms consume oxygen in their respiration. The BOD test determines uptake of oxygen by a contaminated material, e.g. sewage, water, etc. as a measure of microbial activity.

Black plate Low-carbon steel plate base for tin mill products, like tin plate.

Blanching Heating by direct contact with hot water or live steam. It softens the tissues, eliminates air from the tissues, destroys enzymes, and washes away raw flavours.

Bleaching agent Used to whiten and 'mature' flour and cheese in order to provide them with the characteristics necessary to produce an elastic, stable dough and neutralise colours which may be present in oils and fats.

Bleeders Openings used to remove air that enters with steam, from retorts and steam chambers and to promote circulation of steam in such retorts and steam chambers. Bleeders may serve as a means of removing condensate.

Bloom gelometer An instrument to measure strength or firmness of gels.

BMR Basal metabolic rate. The amount of energy utilised per unit time under conditions of basal metabolism; expressed as calories per square metre of body surface or per kg of body weight per hour.

Body hook The flange portion of the can body that is turned back for the formation of the double seam.

Body maker A machine for automatic forming of a "body" or cylindrical metal can or drum's body from a body blank (flat rectangular piece of metal). In the manufacture of tin cans, the body maker may also automatically weld the side seam.

Boiler scale Deposit left inside boilers caused by evaporation of water and precipitation of water-soluble and insoluble substances.

Bonderised black plate Also known as Chemically Treated Black Plate. This term is applied to can making quality black plate that is given a chemical treatment for the purpose of improving the adhesion of enamels and lacquers. The chemical treatment (chromate-phosphate wash) may also retard under film corrosion or, for a short time, atmospheric corrosion.

Bottom seam Also known as factory end seam. The double seam of the can end put on by the can manufacturer.

Botulism A poisoning caused by substances formed by the bacterium *Clostridium botulinum* under conditions of improper processing and storage of food. The spores of this bacterium are often found in soil and are likely to be present on soil-contaminated food.

Bound water Water chemically tied to food in the form of hydrates of inorganic salts or inorganic substances.

Bran Outer layers of the wheat kernel separated during milling.

Breakpoint chlorination Addition of chlorine to water beyond the point where chloramines are oxidised and where further increases in the dosage of chlorine will result in a proportional increase of chlorine residual.

Brine Salt, sugar, and water mixture in which most vegetables are canned. Water is not chlorinated.

Brines Salt solutions used in canning and pickling.

British Thermal Unit (BTU) The British engineering unit of heat quantity. It is approximately the quantity of heat which will raise the temperature of 1 lb of water 1°F. BTU = 0.252 Cal. = 1054 J.

Brix The measure of density of a solution, more particularly a solution containing sucrose, as determined by a hydrometer. Degrees Brix equal % sucrose in water solution at 20°C (68°F).

Brix hydrometer Hydrometer graduated in percentage sugar 20°C (68°F).

Brix/acid ratio The ratio of the degrees Brix of a juice or syrup to the grams of a specified organic acid contained in the liquid, per 100 g of the liquid.

Broasting A food service process involving pressure frying. It is more rapid than regular deep fat frying and results in less absorption of fat.

Broiling To cook by subjecting to direct radiant heat.

Broken heating curve A heating curve which shows a distinct change in the rate of heat transfer such that the curve may be represented by two or more distinct straight lines.

Bromelin Protein-digesting enzyme found in pineapple juice and stem tissues.

Broth medium A liquid medium for growth of microorganisms.

Browning reaction A reaction in foods, usually deteriorative, involving amino (e.g. from amino acids or proteins) and carbonyl (e.g. from glucose) groups; this reaction often leads to a brown discolouration and sometimes to off-flavours and changes in texture.

Buckling of cans Cans becoming permanently distorted along the double seam; caused by excessive internal pressure. See Peaking.

Buffer Any substance in a fluid which tends to resist the change in pH (hydrogen-ion concentration) when acid or alkali is added.

Bulk density Weight per unit volume of a quantity of solid particles; depends on packing density.

Bursting strength The strength of material in pounds per square inch, measured by the Cady or Mullen tester.

'C' Enamel Interior coating designed to prevent discolouration with foods containing sulphur. This enamel contains zinc compounds which react with liberated sulphur compounds to form white zinc sulphide, thus eliminating discolouration.

Caffeine An alkaloid present in coffee, tea, and cola. It is a stimulant to the heart and central nervous system.

Calciferol See Vitamin D.

Calcium The most plentiful body mineral, important for structure and growth of bones and teeth. Assists in blood clotting. Important for proper functioning of nerves, muscles, and heart. Good sources are milk products and leafy green vegetables.

Calcium propionate A mould inhibitor.

Calcium stearate An anticaking agent and emulsifier.

Calendering Subjecting a material to pressure between two or more counter-rotating rollers.

Calliper Thickness, as related to paperboard, of a sheet measured under specified procedures expressed in thousandths of an inch. Thousandths of an inch are sometimes termed 'points'. The precision instrument used in the paperboard industry to measure thickness. To measure with a calliper.

Calorie A unit of heat; the amount of heat necessary to raise the temperature of 1 g of water by 1 °C. Nutritionists use the large Calorie or kilo-Calorie (spelt with capital C), which is 1000 calories. One Calorie (kilo-Calorie) = 4184 J or 3968 BTU.

Can, cylinder A can whose height is relatively large compared to its diameter. Generally called a tall can.

Can, easy opening The end (lid) of a can which has a removable panel defined by a score line. A pull tab is mounted on this panel for opening the can.

Can, flat A can whose height is equal to or smaller than its diameter.

Can, key opening A can opened by tearing off a scored strip of metal around the body by means of a key, or any can opened by means of a key.

Can, sanitary Full open top can with double seamed bottom. Cover double seamed on by packer. Ends are gasket- or compound-lined. Used for products which are heat processed. Also known as the Canner's or Packer's Can.

Canned food Commercially sterile food in hermetically sealed containers.

Cap Form or device used to seal off the opening of the container, so as to prevent loss of its contents. See Closure.

Cap, lug A cap closure for glass containers in which impressions in the side of the cap engage appropriately formed members on the neck-finish to provide a grip when the cap is given a quarter turn, as compared to the full turn necessary with a screw cap.

Cap, screw A cylindrical closure having a thread on the internal surface of the cylinder capable of engaging a comparable external thread on the finish or neck of a container, such as glass bottle, collapsible tube, etc.

Cap, snap-on A type of closure for rigid containers. The sealing action of a snap-on cap is affected by a gasket in the top of the cap that is held to the neck of the spout of the container by means of a friction fit on a circumferential bead. Material of construction is either metal or semirigid plastic.

Cap, two-piece vacuum (two-piece vacuum cap) Standard C-T (continuous thread) or D-S (deep screw) caps, equipped with a separate disk or lid which is lined with sealing for vacuum-packing processes.

Carbohydrates Nutrients that supply energy. They help the body use fats efficiently and decrease the need for protein by furnishing energy so that protein is used for more important functions. Important sources are starches, cereal grains, rice vegetables, such as potatoes, and sugars-honey, molasses, table sugar, syrups, candies.

Carcinogen A cancer-causing agent.

Carotenoids Red and orange pigments found in carrots and other vegetables. They are precursors of Vitamin A.

Carrageenan A colloidal carbohydrate found in seaweeds. See Agar.

Case A nonspecific term for a shipping container. In domestic commerce, 'case' usually refers to a box made from corrugated or solid fibreboard. In maritime or export usage, 'case' refers to a wooden or metal box.

Catalase An enzyme which breaks down hydrogen peroxide into water and oxygen.

Catalyst Substance that alters the rate of chemical change and remains unchanged at the end of a reaction.

Catedrines Colourless flavonoids which change readily to brownish pigments.

Cation Positively charged ion such as K^+, NH_4^+.

Cationic surfactants Ionic surface-active agents in which the portion that associates with the internal phase is the cation. They include simple amine salts, quaternary ammonium salts, amino imides, and imidazolines. Cationic surfactants often have germicidal, anticorrosive, and antistatic properties.

Cellophane A colourless, transparent, flexible packaging material made of cellulose.

Celsius (°C) Temperature on a scale of 100° between the freezing point (0°) and the boiling point (100°) of water.

Centimetre (cm) A unit of measurement, abbreviated as cm. It is equivalent to 0.01 m and 0.394 in.

Centipoise Unit of viscosity equal to $1/100 \, dyne/s^2/cm^2$.

Cephalin A phospholipid whose composition is similar to that of lecithin; found in many living tissues, especially nervous tissue of the brain.

Chalaza Membranous layer holding egg yolk to thick or thin albumen.

Chelating agent A substance which forms stable bonds with metal ions. See EDTA and Sequestering agents.

Chemical Oxygen Demand (COD) An indirect measure of the biochemical load exerted on the oxygen content of a body of water when organic wastes are introduced into the water. When the wastes contain only readily available organic bacterial food and no toxic matter, the COD values can be correlated with BOD values obtained from the same wastes.

Chilling injury Colour or texture change on food surface resulting from overexposure to low temperature.

Chlorination Building up the chlorine content (as hypochlorous acid) to process or sanitise water supplies.

Cholesterol Cholesterol is a lipid or fat-like substance. A form of cholesterol is converted by sunlight on the skin to form Vitamin D. Cholesterol is found only in animal tissues and animal fats.

Chromatography A physical analytical method of separating components in a mixture.

Chuck Part of a closing machine which fits inside countersink and in chuck ring of can or lid or end during seaming operation.

Ciguatoxin Ciguatera toxin found in shellfish.

Cinnamon Barks of various species of the genus *Cinnamomum*; split-off shoots, cured and dried.

CIP Clean in place.

Clarifying agents Substances which aid in the removal of small particles of organic or inorganic matter from liquids. Vinegar often turns 'cloudy' without the use of clarifying agents.

Climacteric A critical year or period, or one of marked change.

Clinch A very loose first operation seam designed to hold the can end in place, yet allow gas to escape during double seaming.

Closing machine Also known as a double seams. Machine which double seams can end onto can bodies.

Clostridia Genus of spore-forming bacteria. *C. botulinum* is the most heat resistant of the food poisoning organisms; its growth is inhibited at pH 4.6 and below, thus it is only a problem in low-acid foods. Produces an endotoxin, botulina, highly toxic in minute doses but destroyed by heat. Destruction of this organism is generally accepted as the minimum standard of processing for low-acid and medium-acid canned food, although other *Clostridia* are more heat resistant.

Clostridium perfringens *C. perfringens welchii,* type A is a Gram-positive, anaerobe, spore-forming rod that causes a food infection, gastroenteritis, produced by the release of an enterotoxin. *C. perfringens* grows optimally at 43 °C–47 °C (110 °F–117 °F).

Closure The joint or seal which is made in attaching the cover to the glass container. Also, the type of closure, such as friction, lug, screw top, etc.

Coagulant A material, which, when added to liquid wastes or water, creates a reaction which forms insoluble floc particles that absorb and precipitate colloidal and suspended solids.

Coal tar colours Synthetic food colours.

Coccus Shape of bacteria (plural cocci). A round/spherical cell.

Cocked base plate A base plate on a double seamer which is not parallel to seaming chuck. This results in a top double seam having a body hook uneven in length.

Cocked body A can body which is not a perfect cylinder, i.e. open ends of cylinder not at right angles to body. This defect results in body hooks of uneven length at both ends. Where the body is long on one end, it will be short on the other end.

Code, can Canner's identification stamped in relief on canner's end. Also, can maker's identification stamped in relief on manufacturer's end.

Codex Codex Alimentarius (Latin for 'Book of Food') is a collection of internationally recognised standards, codes of practice, guidelines, and other recommendations relating to foods, food production, and food safety. The texts are developed and maintained by the Codex Alimentarius Commission, a body that was established by the Food and Agriculture Organization (FAO) of the United Nations and later was joined by the World Health Organization (WHO).

Co-extrusion A combination of two or more thermoplastics extruded as an entity by special dyes or made by combining extruded thermoplastics before they harden into films.

Cold break Breaking food into pieces at ambient temperatures to allow enzyme activity for a short time and then heating to halt enzyme activity.

Coliform bacteria Group of aerobic bacteria of which *Escherichia coli* is the most important member. Many coliforms are not harmful, but as they arise from faeces they are useful as a test of contamination, particularly as a test for water pollution.

Collagen Connective tissue which holds muscle fibres together. See Elastin.

Colloid Fine particles (the disperse phase) suspended in a second medium (the dispersion medium; can be solid, liquid, or gas suspended in solid, liquid, or gas).

Colloid mill Machine used to homogenise or emulsify foods.

Colloidal suspension Two-phase system having small dispersed particles suspended in a dispersant.

Colony A microscopically visible growth of microorganisms on a solid culture medium.

Come-Up Time (CUT) The time which elapses between the introduction of steam into the closed retort and the time when the retort reaches the required processing temperature.

Commercial sterility of canned food The condition achieved by application of heat which renders such food free of viable forms of microorganisms having public health significance, as well as any microorganisms of nonhealth significance capable of reproducing in the food under normal nonrefrigerated conditions of storage and distribution. Commercial sterility of equipment and containers used for aseptic processing and packaging of food means the condition achieved by application of heat, chemical sterilant(s), or other appropriate treatment which render such equipment and containers free of viable forms of microorganisms having public health significance, as well as any microorganisms of nonhealth significance capable of reproducing in the food under normal nonrefrigerated conditions of storage and distribution.

Compound (in cans) A sealing material consisting of a water or solvent emulsion or solution of rubber, either latex or synthetic rubber, placed in the curl of the can end. During seaming operation, the compound fills the spaces in the double seams, sealing them against leakage and thus effecting a hermetic seal.

Congeal To change from a liquid to a semisolid, nonfluid mass.

Consistency Resistance of a fluid to deformation. For simple (Newtonian) fluids, the consistency is identical with viscosity, for complex (non-Newtonian) fluids, identical with apparent viscosity.

Consistometer One of the several types of instruments used to measure the consistency of foods.

Contamination Entry of undesirable organisms into some material or container.

Continuous phase External phase of an emulsion.

Continuous thread An uninterrupted protruding helix on the neck of a container to hold screw-type closure.

Controlled Atmosphere (CA) storage Storage of foods in a hermetic warehouse where the concentrations of O_2, CO_2, and N_2 are controlled at specific levels.

Convection Natural or forced motion in a fluid induced by heat or the action of gravity.

Convert To change to a lower molecular weight form, as by dextrinisation, hydrolysis, etc.

Cooling (1) In a freezing plant, the process of precooling produce prior to placing it in quick-freezing chamber. (2) The process of cooling heated cans immediately after processing. Cans may be stationary or moving. In various methods, cans are immersed, partially covered, or spray cooled.

Corrosion Chemical action of dissolving the surface of a metal (e.g. tin in food medium).

Corrosion accelerator Chemical species with the ability to accept electrons, which will increase the rate of a corrosion reaction.

Countersink depth The measurement from the top edge of the double seam to the end panel adjacent to the chuck wall.

Cover Can end placed on can by packer. Also known as top, lid, packer's end, canner's end.

Critical factor Any property, characteristic, condition, aspect, or other parameter, variation of which may affect the scheduled process delivery and thus the commercial sterility of the product. This does not include factors which are controlled by the processor solely for purposes of product appearance, quality, and other reasons which are not of public health significance.

Cross section Referring to a double seam. A section through the double seam.

Crossover The portion of a double seam at the lap.

Crude fibre The remaining substance measured by weight, after food materials are rigorously extracted with the hot acid and hot alkaline solvents. These remove food components from the original sample, leaving a residue which probably reflects the cellulose and lignin content of the food sample.

Cryogenic freezing Very rapid freezing of food done by immersing or spraying with cold liquid, generally nitrogen at $-196\,°C$ ($-320\,°F$).

Cryogenic liquids Liquid nitrogen and liquid carbon dioxide.

Cryovac A flexible, transparent, heat-shrinkable food packaging material used primarily for frozen poultry.

Cryptoxanthin One of the carotenoid plant pigments. Converted into Vitamin A in the animal body.

Crystal size Grade designated for identifying the relative crystal size of nonferrous metals. For tin plate corrosion purposes, the lower the numerical grade, the better the corrosion resistance.

Cull Product rejected because of inferior quality.

Culture A population of microorganisms cultivated in a medium; pure culture – single kind of microorganism, mixed culture – two or more kinds of microorganisms growing together.

Culture medium Any substance or preparation suitable for and used for the growth and cultivation of microorganisms; selective medium – a medium composed of nutrients designed to allow only growth of a particular type of microorganism, broth medium – a liquid medium for growth of microorganisms, agar medium – solid culture medium.

Curing A food process used primarily for meat products, such as ham, consisting of the use of salts, sugar, and water to preserve food and provide certain quality attributes (desired texture, colour, and flavour).

Curing agents Salts and certain other compounds used to preserve meats such as ham, bacon, frankfurters, and bologna. Curing agents may modify the flavour and also stabilise the characteristic colour of some meats.

Curl The extreme edge of cover which is bent inwards after end is formed. In double seaming, the curl forms the cover hook of the double seam.

Cut code A break in the metal of a can end due to improper embossing marker equipment.

Cut-over Sharp bend or break in the metal at the tip of the countersink. The cut-over occurs during seaming due to excess metal being forced over the top of seaming chuck. Usually caused by heavy laps, i.e. laps containing excessive solder, but may be due to improper adjustment of the double seaming equipment.

Cyanocobalamin A most common and widely produced form of the chemical compounds that have vitamin B12 activity. Vitamin B12 is the "generic descriptor" name.

D-value Time in minutes at a specified temperature required to destroy 90% of the microorganisms in a population.

Deaeration Removal of oxygen from the product to prevent oxidation reactions.

Degradation Deterioration, chemical breakdown.

Dehydration A food processing unit operation resulting in the removal of water from the food generally to the point where spoilage is prevented.

Denaturation To alter the original state of a food substance by physical or chemical means.

Denitrification The process involving the facultative conversion by anaerobic bacteria of nitrates into nitrogen and nitrogen oxides.

Densimeter Instrument for measuring the density or the specific gravity of liquids.

Detergent Surface-active material or combination of surfactants designed for removal of unwanted contamination from the surface of an article.

Deterioration A nonbiological, physical, or chemical change in food which adversely affects quality.

Detinning The process of corrosion, where the internally plain tin coating is slowly dissolved by the food medium; rapid detinning refers to abnormally fast tin dissolution, caused by the presence of corrosion accelerators.

Dew point The temperature at which air or other gases become saturated with vapour, causing the vapour to deposit as a liquid. The temperature at which 100% relative humidity is reached.

Dextrins A polysaccharide, product of enzymatic or acid hydrolysis of starch. Used in preparing emulsions and thickening liquids and pastes.

Dextrose A widely occurring crystallisable simple sugar which contains 6 carbon atoms in contrast to 12 found in sucrose.

Dicer Equipment which cuts fruit, vegetables, and other foods into small cubes.

Dietary fibre Refers to the combined, undigested carbohydrates in food and encompasses not only the cellulose and lignin found in crude fibre but also hemicelluloses, pectic substances, gums, and other carbohydrates not normally digested by man. Crude fibre as determined is more of a refined fibre, while dietary fibre is more closely related to a true crude fibre.

Dietetic foods Those foods which comprise a diet intended to prevent or cure certain physiological conditions. Examples are low-calorie or low-sodium diets.

Differentially coated tin plate Electrolytic tin plate with a different weight of tin coating on each side, normally 'line marked' on the heavy coated side.

Diffusion Mixing of molecules or atoms by random molecular or atomic motion.

Digestion The biological decomposition of organic matter in sludge, resulting in partial gasification, liquefaction, and mineralisation.

Dill Dried ripe fruit of *Anethum graveolens*. Used in pickles and soups.

Disinfectant An agent that frees from infection by killing the vegetative cells of microorganisms.

Disinfection The reduction, by means of hygienically satisfactory chemical agents and/or physical methods, of the number of microorganisms to a level that will not lead to harmful contamination of food.

Dispersion Physical, usually temporary, mixture of two insoluble phases.

Disposal Getting rid of something, e.g. the discharge of wastewater.

Dissolving Formation of a solution by dispersion of one material (solute) at a molecular (or less) level in another material (solvent).

DNA Deoxyribonucleic acid. A long macromolecule that is the main constituent of chromosomes.

Domed A curved profile container end used for strength or appearance.

Double seam To attach an end to a can body by a method in which five (5) thicknesses of plate are interlocked or folded and pressed firmly together. A joint formed by interlocking the edges of both the end and body of a can.

Double seamed end Part of a can which is attached to the body of a double seamed can to form the top or the bottom.

DR tin plate Double Reduced Tin plate. Tin plate which is given a partial cold reduction then annealed and given another cold reduction to the final thickness. The resulting product is stiffer, harder, and stronger than a single reduced (SR) product enabling can manufacturers to use lighter thicknesses of tin plate and tin-free steel (TFS).

Drained weight Weight of the solid portion of the product after draining the covering liquid for a specified time with the appropriate sieve.

DRD Draw-Redraw. Process for making two-piece cans in which a circular blank is drawn through a die to form a shallow cup and then redrawn through two or more dies to produce a can of desired dimensions.

Droop Smooth projection of a double seam below the bottom of a normal seam. Usually occurs at the side seam lip.

Drop test A test for measuring the properties of a container by subjecting the packaged product to a free fall from predetermined heights onto a surface with prescribed characteristics.

Dry ice Carbon dioxide in solid state ($-78.5\,°C$).

DWI Drawn and Wall Ironed. Process for making two-piece cans in which a circular blank is drawn through a die to form a cup and then thinned to final dimensions by being forced through a series of two or more progressively smaller diameter ironing rings.

EDTA Ethylenediaminetetraacetic acid. A compound capable of chelating divalent metal cations. Used as a food preservative and antioxidant. Its calcium and sodium salts are used in foods to sequester traces of metallic impurities that cause food deterioration.

Effluent Wastewater or other liquid, partially or completely treated or untreated, flowing out of a process operation, processing plant, or treatment plant.

Elastin Connective tissues holding muscle fibres together. The principal component of elastic protein fibre. See Collagen.

Electrolyte A substance which dissociates into ions when dissolved in a suitable medium; hence a tin-rich electrolyte is used in tin plate manufacture.

Electrolytic Denoting a coating of tin, electrodeposited upon the base metal.

Electrophoresis Migration of the electrically charged particles towards the oppositely charged electrode.

Elemendorf test A test for measuring the tearing resistance of paper, paperboard, tape, and other sheet materials.

Emboss(-ed) (-ing) Raised design or lettering on the surface of an object.

Emulsifier (emulsion) A compound or substance which promotes and stabilises a finely divided dispersion of oil and water.

Emulsion System consisting of two incompletely miscible liquids, one being dispersed as finite globules in the other. A small amount of a third substance may render the dispersion stable. The liquid broken up into globules is the dispersed (discontinuous) phase; the surrounding liquid is the external (continuous) phase.

Enamel A vitreous or paint-like composition used as a protective coating usually baked onto the packaging material before fabrication into the finished container. On the inner surface of metal containers, its purpose is to protect either the contents or the container. On the outer surface, its purpose is to prevent corrosion or to decorate. Also called Lacquer.

Endosperm Structural component of cereal grains made up mostly of starch and some protein.

Endotoxin A toxin produced with an organism liberated only when the organism disintegrates.

Enriched A term which refers to the addition of specific nutrients to a food as established in a standard of identity and/or quality.

Enterotoxin A toxin specific for cells of the intestine. Gives rise to symptoms of food poisoning.

Enzymatic browning The darkening of plant tissues or products produced by enzymatic reactions.

Enzyme A compound of biological origin which accelerates a specific chemical reaction.

Equilibrium moisture content The moisture content of a substance at which it will neither gain nor lose moisture in an atmosphere having given relative humidity.

Equilibrium pH The pH of the macerated (thoroughly blended) contents of the product container. See Maximum pH and Normal pH.

Equilibrium relative humidity The relative humidity of the ambient atmosphere surrounding a substance when the substance neither gains nor loses moisture.

Ergosterol Pro-vitamin D. Irradiated ergosterol has served as a Vitamin D source for food enrichment.

Escherichia coli The strains of *E. coli* that produce an entero-pathogenic food poisoning syndrome.

Essential elements Those elements necessary to maintain normal metabolic functions. Some are required in trace quantities (such as iron, copper, and zinc), while others are required in larger amounts (such as calcium and magnesium).

Essential oils Flavour concentrates from spices or herbs which are generally produced by steam distillation and have no relatively high boiling constituents present.

Ester An organic compound formed by the reaction of an acid and an alcohol. Many flavouring agents are esters.

Ethyl vanillin A flavouring agent.

Ethylene oxide A gas used to accelerate ripening of certain fruits. Also produced naturally by fruits during the ripening process.

Eutectic A solution which has a melting point below that of any of the components taken separately.

Eutectic point Temperature at which a substance exists simultaneously in the solid, liquid, and gaseous states.

Eutrophication Applies to a lake or pond becoming rich in dissolved nutrients, with seasonal oxygen deficiencies.

Exhaust Heating of food in cans prior to closing the cans to produce a partial vacuum in containers.

Exhauster Equipment to heat food in cans prior to closing the cans, so as to produce a partial vacuum in the containers.

Exotoxin A toxin excreted by a microorganism into the surrounding medium.

Extended aeration A modification of the activated sludge process that employs aeration periods of 24 h or more, completely mixing, and high levels of mixed liquor solids.

Extrusion The process of forcing a material in plastic condition through an orifice.

F-value The number of minutes required to destroy a stated number of microorganisms at a defined temperature, usually 121.1 °C (250 °F), and when the Z-value is 10 °C (18 °F). 'F' value is a common term employed in the canning industry to express the lethality or sterilising value of a sterilisation process. See Z-value.

Fabricated foods Blend of food ingredients resulting in a product of special character-istics such as nutritive value or other quality attributes. Some are prepared to resemble well-accepted animal or plant foods such as soya bean burgers. Also called engineered foods.

Factory end Bottom or can manufacturer's end.

Facultative bacteria Bacteria which can exist and reproduce under either aerobic or anaerobic condition.

Fahrenheit A temperature scale in which 32 °F represents the freezing point of water and 212 °F represents its boiling point.

False seam A small seam breakdown where the cover hook and body are not overlapped, i.e. no hooking of body and cover hooks. See Knockdown flange.

FAO of UN Food and Agriculture Organization of the United Nations.

Fat A nutrient providing the most concentrated source of energy, weight for weight supplying more than twice as much energy as carbohydrates or proteins. Fats are the molecular combination of glycerol and certain fatty acids.

Fatty acids, essential Name of two fatty acids, linoleic and arachidonic. They are dietary essentials.

FDA The United States Food and Drug Administration.

Feather Beginning of a cut-over. At the top of the container's countersink, the metal is forced over the seaming chuck forming a sharp edge that may be detected with the fingernail. Commonly referred to as 'Sharp Edge'.

Fermentation The action of microorganisms upon foods. Anaerobic respiration. Usually fermentation is undesirable, but sometimes it is produced intentionally, such as in the manufacture of vinegar from apple cider.

Ficin A protein-digesting enzyme found in figs.

Fill weight The weight of the product particulates before processing. It does not include the weight of the container or covering liquid.

Finish The opening of a container shaped to accommodate a specific closure.

Firming agents Substances used to aid the coagulation of certain cheeses and to improve the texture of processed fruits and vegetables which might otherwise become soft.

First operation The first operation in double seaming. In this operation, the curl of the end is tucked under the flange of the can body which is bent down to form cover hook and body hook, respectively.

Fish meal Ground up and dehydrated parts of fish not normally used for human food. Also made from whole fish of low market value.

Flame peeling Peeling of vegetables by charring the surface by exposing it to direct flame or hot gasses in rotary tube flame peelers.

Flange To flare out the top of a can body to prepare it for double seaming to an end. Also the flaring projection around the end of a can body. The outermost projection of an end, cover, or cap.

Flange, dented A flange damaged through abuse in handling, not in manufacture. May result in false seams, lips, and breakdowns.

Flash pasteurisation Process in which the material is held at a much higher temperature than in normal pasteurisation but for a considerably shorter period.

Flat sours Thermophilic and thermoduric bacteria, facultative anaerobes that attack carbohydrates with the production of acids, but without gas formation. Flat-sour spoiled canned foods therefore show no swelling of the ends.

Flavedo The coloured outer layer of citrus fruit peel. It contains the oil sacs and fruit pigments.

Flavenoids Pigments and colour precursors commonly present in fruits and vegetables. They include the purple, blue, and red anthocyanins, the yellow anthoxanthins, and the colourless catechins and leucoanthocyanins.

Flavour Attributes of food quality which the consumer evaluates with his senses of taste and smell.

Flavouring agents Substances added to foods to enhance or change the taste. This largest group of food additives includes spices, seeds, natural and synthetic flavour concentrates, and many others.

Flexible container A container where the shape or contours of the filled, sealed container are affected by the enclosed product.

Flipper A can having both ends flat but with insufficient vacuum to hold the ends in place. A sharp blow will cause the end to become convex, but both ends may be pressed to their normal position.

Flocculation The process of forming larger masses from a large number of finer suspended particles.

Flotation Removal of solids, oil, or fat from wastewater by causing the material to float to the water surface with the aid of heat or entrained air.

Flotation grader Equipment for grading peas and lima beans in a brine solution or water.

Flour, All-Purpose Flour which can be used for making bread, cakes, or other baked 'bread-type' products.

Fluidity The ability of a substance to flow. The opposite of viscosity.

Fluming In-plant transportation of product or waste material through water conveyance.

Fluoridation Process of adding traces of sodium fluoride to drinking water to arrest or prevent dental decay.

Flux A chemical used to aid in soldering by removing the oxides.

Foaming agent Surface-active material that is used specifically to form a dispersion of a gas in a liquid or solid medium.

Folic acid The vitamin of the 'B' group, essential in the synthesis of certain amino acids. Liver, yeast, and cheese are good sources.

Food additive Any substance intentionally or incidentally added to food to protect, modify, or enhance some quality attribute or preserve the freshness of the product.

Food analogues Fabricated foods resembling well-accepted animal or plant foods.

Food chemicals codes A set of standards for purity of food chemicals in terms of maximum allowable trace contaminants and methods of analysis for the contaminants. Prepared by the Food Protection Committee of the National Academy of Sciences, National Research Council.

Food colours Synthetic or natural substances added to foods to enhance the natural colour of the food or to give the food a colour.

Food infection An illness caused by an infection produced by invasion, growth, and damage to the tissue of the host due to the ingestion of viable pathogenic microorganisms associated with the food.

Food intoxication An illness resulting from the ingestion of bacterial toxin with or without viable cells. The illness does not require actual growth of cells in the intestinal tract.

Food poisoning A general term applied to all stomach or intestinal disturbances due to food contaminated with certain microorganisms or their toxins.

Food science and technology The field of study concerned with the application of science and technology to the processing, preservation, packaging, distribution, and utilisation of foods and food products.

Food service system A facility where large quantities of food intended for individual service and consumption are routinely provided, completely prepared. The term includes any such place regardless of whether consumption is on or off the premises and regardless of whether or not there is a charge for the food.

Fortified Food to which specific nutrients have been added 'enriched'.

FPC Fish protein concentrate. A highly refined form of fish protein, white, colourless, flavourless powder. Contains approximately 85% protein.

Freeze drying A process of dehydration in which the moisture is removed by the sublimation of ice from the frozen product.

Fructose An alternate chemical name for laevulose.

FSA Food Standards Agency responsible for protecting public health in relation to food throughout the United Kingdom. It is led by a board appointed to act in the public interest.

Fumigants Substances used to control growth of insects or microorganisms on foods.

Fungicidal agent Destroys existing fungal cells.

Fungicide Any substance that destroys fungi or inhibits the growth of spores or hyphae. Legally, sometimes the term is interpreted as also including yeasts and bacteria.

Fungistatic agent Prevents growth of fungi (moulds) without necessarily killing the existing cells.

G-cap A No. 70 (70 mm) cap with abnormally deep screw. Used primarily for mayonnaise and salad dressing.

Gas chromatography A separation technique used in food analysis, involving passage of a gas through a column containing a fixed adsorbent phase. It is used principally as a quantitative analytical technique for volatile compounds.

Gas packing Packaging in a gas-tight container in which any air has been replaced by a gas that contains practically no free oxygen, such as commercial carbon dioxide or nitrogen.

Gasket In cans, a filler, usually of synthetic rubber, used in the seam for the purpose of making it hermetically tight.

Gauge Term used to designate the thickness of a plate.

Gel Semisolid system that consists of a solid held in a liquid; a more solid form than a sol.

Gelatinise To cook starch in aqueous suspension to the point at which swelling of the granules takes place, forming a viscous sol.

Gelation Solidifying, resulting in the formation of a gel.

Gelometer Instrument used to measure the time required for a fluid to gel. Also, instrument used to determine the firmness of a gel.

Germ A microorganism; a microbe usually thought of as a pathogenic organism.

Germicide Substance that will kill all ordinary microorganisms that cause disease but that is not necessarily capable of destroying bacterial spores.

Glucose An alternate chemical name for dextrose. A name given to corn syrups which are prepared by the action of acids and/or enzymes on corn starch.

Glutamate sodium Sodium salt of glutamic acid, an amino acid. Enhances the flavour of some foods. Frequently added to soup mixes, meat products, and certain other foods.

Glycerides Organic compounds resulting from the reaction of a fatty acid and glycerol. Mono- and diglycerides are used as emulsifying agents. Among the triglycerides are the fats and oils.

Glycogen A sugar stored in the liver of animals.

Goitre A condition produced by a shortage of iodine in the diet.

Gossypol A toxic yellow pigment found in cottonseed.

Grade A level or rank of quality.

Grading The selection of produce for certain purposes. Produce is sorted for size, colour, quality, ripeness, etc. May be done manually or mechanically on sizing belts, flotation graders, etc. Term also applied to finished products.

Grain Measure of weight equivalent to 0.0648 g.

Gram (g) Metric unit of weight equal to 0.035 ounces. One kilogram is equivalent to 1000 g, and one pound equals 453.6 g.

GRAS Generally Recognised as Safe.

Guar gum A stabiliser, thickener, and emulsifier.

Gum Class of colloidal substances that is exuded by plants.

Gum arabic (acacia) A stabiliser, thickener, and emulsifier.

HACCP Hazard Analysis of Critical Control Points. An inspectional approach that determines what points in the process are critical for the safety of the product and how well the firm controls these points.

Halophilic Can grow or survive in a medium with a relatively high salt concentration.

Hard swell Spoilage in which can ends are swelled too hard to be readily depressed by applying thumb pressure.

Headspace, gross The vertical distance between the level of the product (generally the liquid surface) and the inside surface of the lid in an upright rigid container (the top of the double seam of a can or the top edge of a glass jar).

Headspace, net The vertical distance between the level of the product (generally the liquid surface) and the inside surface of the lid in an upright, rigid container having a double seam, such as a can.

Heat exchanger Equipment for heating or cooling liquids rapidly by providing a large surface area and turbulence for the rapid and efficient transfer of heat.

Heat, latent Heat absorbed or liberated in a change of physical state such as evaporation, condensation, freezing, or sublimation. Expressed as BTU per lb, kcal per kg, or joules per kg.

Heat, sensible Heat that has gone into raising the temperature of steam without change of pressure or absolute humidity.

Heating curve A graphical representation of the rate of temperature change in the food throughout the heat process; this is usually plotted on semilog graph paper so that the temperature on an inverted log scale is plotted against time on a linear scale.

Heavy lap A lap containing excess solder, also called a thick lap.

Heel The part of a container between the bottom bearing surface and the side wall.

Hepatitis, infectious An infectious disease produced by a virus found in polluted waters and in shellfish growing in such waters. Also transmitted by unsanitary handling and preparation of other foods.

Hermetically sealed container A container which is designed and intended to be secure against the entry of microorganisms and to maintain the commercial sterility of its contents during and after processing, e.g. tin plate or aluminium can, glass jar, or pouch.

Herringbone score Weakening lines made in the body of a key opening can between and at an angle to the parallel scored lines. Designed to lead a tear back into the regular line.

Histidine One of the essential amino acids.

HMF Hydroxymethylfurfural. Is an organic compound derived from dehydration of certain sugars. HMF is practically absent in fresh food, but it is naturally generated in sugar-containing food during heat treatments like drying or heat processing. HMF can be used as an indicator for excess heat treatment.

Homogenisation The process of making incompatible or immiscible components into a stabilised uniform suspension in a liquid medium.

Homogeniser Mixing machine used for the preparation of emulsions of fine particle size. The emulsion is forced at high pressure through the annular space between an adjustable valve and its seat.

Hook, body That portion of the edge of a can body which is turned back or the formation of a double seam.

Hook cover That portion of an end which is turned back between the body and the body hook for the formation of an end seam.

Hook edged (side seam) That portion of the edge of the body which is turned back for the formation of a locked side seam.

Hook, uneven A body cover hook which is not uniform in length.

Hormones An internal secretion produced by the endocrine glands, secreted directly into the bloodstream to exercise a specific physiological action or other parts of the body. Many are made synthetically.

Hot dip Plate tinned by dipping into molten tin.

HPLC High Pressure Liquid Chromatography. An instrument for food chemical analysis.

HTST process Pasteurisation or sterilisation process characterised by High Temperature applied for a Short Time.

Humectant A substance that is used to help maintain moisture in foods. Humectants are added to such foods as shredded coconuts and marshmallows.

Husker Equipment for stripping husks off corn.

Hydro-cooling A process using cold water to cool foods immediately after harvest and prior to shipping to markets.

Hydrogen swell Swell resulting from hydrogen generated in the can as a result of a reaction of the product with the metal of the can.

Hydrogenated Substances that have been reacted with hydrogen. Hydrogenation of fats results in a reduction of double bonds and consequently in a higher degree of saturation and in a higher melting point.

Hydrogen-ion concentration Acidity or alkalinity of a solution measured by the concentration of hydrogen ions present. Also called pH.

Hydrolysis Process of splitting a molecule into smaller parts by chemical reaction with water.

Hydrometer densimeter Device used for measuring specific gravity or density.

Hydrophilic Attracted to water: water soluble.

Hygroscopic Absorbs water from water vapour in atmosphere.

Hyphae Thread-like elements of the vegetative part of a fungus (mycelium).

Impact strength The ability of a material to withstand mechanical shock.

Incubation Holding cultures of microorganisms under conditions favourable to their growth. Also, the holding of a sample at a specified period of time before examination.

Incubation time The time period during which microorganisms inoculated into a medium are allowed to grow.

Indicator Usually refers to a pH indicator. Various dyes change colour at specific degree of acidity or alkalinity and this colour change is used as an indicator of pH.

Inhibition Prevention of growth or multiplication of microorganisms or prevention of enzyme activity.

Initial Temperature (IT) The average temperature of the contents of the coldest container to be processed at the time the sterilising cycle begins, as determined after thorough stirring or shaking of the filled and sealed container.

Inkjet coding Use of an inkjet to print a product code or manufacturing date on the package.

Inoculate The artificial introduction of microorganisms into a system.

Inoculation The artificial introduction of microorganisms into a growth medium. This can refer to the introduction of test organisms to food, to the accidental introduction of organisms to food, or to the start of yeasts or other desirable cultures such as yoghurt.

Inoculum The material containing microorganisms used for inoculation.

Inositol A growth factor with properties similar to vitamins, generally listed with the vitamins of the B complex.

In-plant chlorination Chlorination beyond the breakpoint of water used in a food plant, usually to a residual of 5–7 ppm.

Insecticide Substance used to kill or control insects. Many are of a long-lasting nature. Care is required in the use of insecticides.

Inspection belt Conveyor belt where materials are visually inspected.

International Units (I.U.) A quantity of a vitamin, hormone, antibiotic, or other substance that produces a specific internationally accepted biological effect.

Intoxication The adverse physiological effects of an organism consuming a toxic material.

Invert, or invert sugar The mixture of equal parts of dextrose and laevulose produced by the action of acid or enzymes on solutions of sucrose.

Invertase An enzyme that catalyzes the breakdown of sucrose into glucose (dextrose) and fructose (laevulose).

Ion Electrically charged portion of matter of atomic or molecular dimensions.

Ion exchange A reversible chemical reaction between a solid and a liquid by means of which ions may be interchanged between the two. It is in common use in water softening and water de-ionising.

Iron A mineral needed in small amounts. Iron is a vital part of haemoglobin, the red substance of blood which carries oxygen from the lungs to all body tissues, and assists the body cells in releasing energy from food. Important natural sources are liver, kidney, muscle meats, dry beans, whole grains, enriched breads and cereals, and dark green leafy vegetables.

Iron chink Mechanical device used in salmon canning to automatically remove heads, tails, fins, and entrails.

Isoelectric point The pH value at which precipitation of a certain protein occurs.

Isoleucine One of the amino acids that are essential for humans.

Jam Product made by cooking to a suitable consistency properly prepared fruit with sugar, or sugar and dextrose, with or without water. No less than 45 kg fruit are used to each 55 kg of sugar or sugar and dextrose. Sometimes pectin and/or an acid are also added.

Jelly Fruit jelly is the semisolid, gelatinous product made by concentrating to a suitable consistency the strained juice or the strained water extract from fruit, with sugar, or sugar and dextrose, added. Sometimes pectin and/or an acid are also added. No less than 45 kg fruit are used to each 55 kg of sugar or sugar and dextrose.

Joule Unit of energy. One joule is equivalent to 0.239 g calories or 0.000 948 Btu.

Jumped seam Double seam which is not rolled tight enough adjacent to the lap, caused by jumping of the seaming rolls at the lap.

K-plate Electrolytic tin plate with a tin coating of No. 50 (imperial) or heavier, with improved corrosion performance for some galvanic detinning food products.

Kilogram (kg) A unit of weight in the metric system equivalent to 1000 g or 2.2046 lb.

Kilopascal Unit of pressure. One kilopascal equals 1000 Pa. 1 atm equals 1.01325×105 Pa. Abbreviation is kPa.

Knockdown flange Body hook and cover hook in contact but not tucked in.

Kraft A term derived from a German word meaning strength, applied to pulp, paper, or paperboard produced from virgin wood fibres by the sulphate process.

Kwashiorkor Term used to describe a syndrome which includes retarded growth and maturation, alterations in the skin and hair, and other changes caused by an extreme deficiency of protein intake. Occurs mostly in infants and young children.

L-Steel Base metal steel, low in metalloid and residual elements. Sometimes used for improved internal corrosion resistance for certain food product containers.

Label Any display of written, printed, or graphic matter on the container of any consumer commodity, affixed to any consumer commodity, or affixed to any package containing a consumer commodity.

Lacquer Inert organic coatings used to give additional protection to tin plate; usually applied in liquid form and 'cured' at high temperatures. See Enamel.

Lactobacillus acidophilus Bacteria used to produce buttermilk. One of the lactic acid-producing bacteria.

Lactose A white, crystalline sugar found in milk. It is less sweet than sucrose.

Laevulose A highly soluble, simple sugar containing six carbon atoms. It crystallises with great difficulty. It is sweeter than sucrose.

Lagoon A large pond used to hold wastewater for stabilisation by natural processes.

Lap Two thicknesses of material bonded together. Section at the end of side seam consisting of two layers of metal to allow for double seaming. As the term implies, the two portions of the side are seam lapped together rather than hooked as in the centre of the side seam.

Latent heat The quantity of heat, measured in BTUs or calories, necessary to change the physical state of a substance without changing its temperature, such as in distillation. A definite quantity of heat, the latent heat, must be removed from water at 0 °C (32 °F) to change it to ice at 0 °C.

Leach To subject to the action of percolating water on other liquid in order to separate soluble components.

Leavening Yeasts or a blend of approved food additives used to raise dough in baking. See Baking powder.

Lecithin A fatty substance (lipid) found in nerve tissue, blood, milk, egg yolk, and some vegetables. Used as an emulsifier.

Lethal Capable of causing death.

Lethality value (F-Value) This degree of sterility is referred to as practical or commercial sterility. The F-Value is introduced as a standard on which to base the sterilisation of food products. The F-Value is defined as the number of minutes which it takes to reduce the initial spore count of a certain microorganism to a desired safety level at a defined lethal reference temperature.

Leucine One of the amino acids that are essential for humans.

Lid Can end applied to open end of can in a cannery. Also known as top, cap, or packer's end.

Lignin A tough, fibrous material found in older plant cell walls.

Lime Calcium oxide, a caustic white solid, which forms slaked lime (calcium hydroxide) when combined with water.

Liner Generally, any linear material that separates a product within a container from the basic walls of the container.

Linoleic acid An unsaturated fatty acid occurring as a glyceride in vegetable oils. Essential in human nutrition.

Lip Irregularity or defect in double seam occurring at the lap. Due to insufficient tucking or cover hook resulting in a short cover hook and characterised by a blowing or sharp 'V' projection at the bottom of the double seam. Also known as a 'droop'.

Lipid Fats, phospholipids, waxes, and other organic compounds often containing elements other than carbon, hydrogen, and oxygen, particularly phosphorus and nitrogen.

Liquid sugar A concentrated solution of refined sucrose or of a mixture of sucrose and invert sugar.

Lock seam A seam formed by the two edges of a can body which have previously been edged or bent into hooks. The final seam is composed of four thicknesses of plate.

Low-acid foods Any food (other than alcoholic beverages) with a finished equilibrium pH greater than 4.6 (4.5 in Europe) and a water activity greater than 0.85, excluding tomatoes and tomato products having a finished equilibrium pH less than 4.7.

Lug A type of thread configuration; i.e. usually thread segments disposed equidistantly around a bottle neck (finish). The matching closure has matching portions that engage each of the thread segments.

Lug box Large box used to haul fruit from fields to cannery.

Lycopene A pigment contributing to the red of tomatoes, watermelons, and other foods.

Lye A strong alkaline solution. Caustic soda (sodium hydroxide) is the most common lye.

Lye peeling Peeling a fruit or vegetable by soaking briefly in hot dilute sodium hydroxide, then scrubbing off the softened peel.

Lypase An enzyme which promotes the breakdown of fats.

Lysine One of the amino acids that are essential for humans.

Macronutrients Nutrients which are required in relatively large amounts by humans to maintain normal growth and other body functions.

Maillard reaction A group of organic reactions, especially between amino acids and reducing sugars, producing brown colour and flavour changes in many food materials. Also known as nonenzymatic browning.

Malic acid A fruit acid found mostly in apples.

Malt Sprouted, dried barley used in the brewing industry to help digest starches into sugars.

Mammoth grader Large drum, perforated with graded holes, in which pears are graded progressively by size.

Margarine A table spread made basically of an emulsion of water in oil with milk, common salt, colouring and flavouring substances, and betacarotene (pro-vitamin A).

Maturation The process of developing quality in a product by aging under certain conditions.

Mature Fully grown and developed.

Maturity The process of coming to full development.

Mayonnaise A food product made basically of an oil-in-water emulsion with egg yolk, vinegar, common salt, and flavouring and colouring ingredients.

Mean The average value of a number of observed data.

Medium, selective A medium composed of nutrients designed to allow only growth of a particular type of microorganism.

Melting The change from the solid to the liquid state. Also the softening of harder compounds.

Mesophilic bacteria Grow best at temperatures between 23.9 °C and 40.6 °C; usually will not grow at temperatures below 10 °C or above 43.3 °C.

Methionine One of the amino acids that are essential for humans.

Metre (m) Metric unit of length, equivalent to 39.37 in or 3.28 ft.

MeV One million electron volts.

Microaerophiles Organisms which grow best in the presence of small amounts of atmospheric oxygen.

Microencapsulation The process of forming a thin protective coating around a particle of a substance.

Microgram One-thousandth of a milligram; symbol, μg.

Microlitre One-thousandth of a millilitre.

Micron One-thousandth of a millimetre.

Micronutrient Nutrients which are required by humans in relatively small or trace amounts to maintain normal growth and other body functions.

Microwave cooking Use of radio-frequency energy for cooking.

Milligram (mg) One-thousandth of a gram.

Milligrams per litre (mg/L) Equals parts per million (ppm). A term used to indicate concentration of materials.

Millimetre (mm) Equivalent to 0.001 m and to 0.0394 in.

Mineral In nutritional science, a term applied to chemical elements that act as body regulators through incorporation into hormones and enzymes. Some minerals (like calcium, phosphorus, and magnesium) are part of the body's structure.

Minimum Daily Requirements (MDR) The amount of a particular macronutrient (i.e. essential fats, proteins, carbohydrates and micronutrients – vitamins, minerals) needed by a normal, healthy adult person every day, in the absence of special metabolic needs.

Mixture Material composed of two or more substances, each of which retains its own characteristic properties.

Moisture Vapour Transmission Rate (MVTR) The rate at which water vapour permeates through a plastic film or wall at a specified temperature and relative humidity.

Molasses Syrup produced by washing raw sugar. It is boiled and as much sugar as possible crystallised out. The syrupy residue is molasses.

Molecular weight Sum of the atomic weights of all the atoms in a molecule.

Molecule The smallest theoretical quantity of a material that retains the properties exhibited by the material.

Moulds Microorganisms that belong to the fungi. The fungus body is usually composed of threads (hyphae; singular, hypha). These hyphae frequently branch in a more or less complex manner forming networks or webs, collectively called 'mycelium'. Hyphae may be one-celled or composed of many cells placed end to end. Fruiting bodies that grow from hyphae produce spores. Moulds are much less heat resistant than bacteria.

MR steel The type of steel most often used in can making due to its good corrosion resistance and high ductility.

Mullen tester An instrument for testing the bursting strength of paper, paperboard, and corrugated or solid fibreboard.

Mycotoxins Toxins produced by moulds or fungi.

Mylar A synthetic polyester fibre or film.

Myosin The protein of the muscle fibre.

Natural foods Term describing foods which are grown without chemical fertilisers or pesticides. Also foods in the preparation of which no synthetic preservatives are used.

Neck The part of a container where the bottle cross section decreases to form the finish.

Nesting Containers that fit within one another when stacked.

Net weight, minimum The weight of all the product in the container, including brine or sauce, but excluding the weight of the container.

Neutralise To adjust the pH of a solution to 7.0 (neutral) by the addition of an acid or a base.

Newtonian fluids Liquids which do not change in viscosity with a change in rate of shear.

Niacin A water-soluble 'B'-group vitamin. Important natural sources are liver, meat, whole grain, and enriched bread and cereals.

Nitrate A salt of nitric acid, usually sodium nitrate ($NaNO_3$), used to cure or preserve meats, especially ham. Saltpetre (potassium nitrate, KNO_3) has been used for many years as a curing ingredient. Nitrates occur naturally in leafy vegetables.

Nitrification The process of oxidising ammonia by bacteria into nitrites and nitrates.

Nitrite A salt of nitrous acid, usually sodium nitrite ($NaNO_2$), used in addition to sodium nitrate to cure ham or other meats. The use of nitrites allows much smaller quantities of nitrates to be used in the curing process with the same degree of protection from spoilage.

Nitrosamines Compounds which are formed from nitrates and other naturally present substances. They have been linked to cancer in laboratory test animals. Nitrosamines are also naturally occurring, normally in very small quantities. Concentration in food may increase during cooking.

Nonenzymatic browning See Maillard reaction.

Non-newtonian Materials whose resistance to flow changes with a change in rate of shear.

Nordihydroguaiaretic Acid (NDGA) Substance of plant origin used as an antioxidant for fats.

Notch To cut away small portions of a blank usually at the corners to provide for features such as beading, double seaming, tongue profile, etc.

Nucleic acids Long-stranded molecules which play a primary role in the transmission of genetic traits, in the regulation of cellular functions, and in the formation of proteins.

Nutrients Compounds that promote biological growth.

Nutrition information panel Appears on food labels to the right of the principal display panel. It provides information on the nutritional composition of the food.

Nutritional inhibitor A natural component of food which adversely affects the utilisation of a nutrient.

Oleoresins Flavour concentrates from spices or herbs prepared by extraction with volatile organic solvents.

Open lap A lap which failed due to various strains set up during manufacturing operations. Also caused by improper welding.

Open top can Another term for sanitary can.

Osmophilic Can grow or survive in a medium very low in humidity or of low water activity.

Osmosis Diffusion of two miscible fluids separated by a permeable wall.

Overlap The distance the cover hook laps over the body hook in a can double seam.

O/W Emulsion Oil-in-water emulsion; oil is the discontinuous or internal phase, water is the continuous or external phase. An O/W emulsion is dispersible (dilutable) in water but not in oil.

Oxidation The act of oxidising which is brought about by increasing the number of positive charges on an atom or the loss of negative charges.

Oxidative rancidity The deterioration of fats and oils due to oxidation.

Package Any container or wrapping in which a consumer commodity is enclosed for delivery or display to retail purchasers.

Packer's end The can end put on by the packer or canner. Also known as lid, cover, top, or canner's end.

Packing medium The liquid or other medium in which the low-acid or acidified product is packed. For example, for 'peas in brine', the packing medium is brine.

Packing test Storage and regular sampling of canned foods under controlled temperature conditions to determine internal corrosion characteristics and potential shelf life.

Palatability Sensory attributes of foods (e.g. aroma, flavour, texture, etc.), which affect their acceptability.

Pallet A low, portable platform of wood, metal, fibreboard, or combinations thereof, to facilitate handling, storage, and transportation of materials as a unit.

Palletised unit load A unitised load fixed to a pallet.

Panelling Distortion (side wall collapses) of a container caused by development of a reduced pressure (too high vacuum) inside the container.

Pantothenic acid A 'B'-group vitamin, essential for the metabolism of fats and carbohydrates. Liver, yeast, kidney, and fresh vegetables are good natural sources.

Papain A protein-digesting enzyme obtained from the juice of unripe papayas.

Paper, water-resistant Paper that is treated by the addition of materials to provide a degree of resistance to damage or deterioration by water in liquid form.

Paper, wet-strength Paper that has been treated with chemical additives to aid in the retention of bursting, tearing, or rupturing resistance when wet.

Parenchyma cell The structural unit of the edible portion of most fruits and vegetables.

Pasteurisation A heat treatment of food usually below 100 °C intended to destroy all organisms dangerous to health or a heat treatment which destroys part but not all microorganisms that cause food spoilage or that interfere with a desirable fermentation.

Pathogen Disease-producing microorganism.

Pathogenic Capable of producing disease.

Patulin A mycotoxin.

PCBs Polychlorinated biphenyls. A class of compounds known to cause cancer.

Pectin Plant tissues contain protopectins cementing the cell walls together. As fruit ripens, protopectin breaks down to pectin and finally to pectic acid under the influence of enzymes. Thus overripe fruit loses its firmness and becomes soft as the adhesive between the cells breaks down. Pectin is the setting agent in jams and jellies. The albedo of oranges and lemons and apple pomace are commercial sources of pectin. Used as a gelling agent and as an emulsifier and stabiliser.

Pectin methoxylase Enzyme in tomato juice that splits methyl alcohol from pectin leaving pectic acid, which does not have the colloidal and thickening properties of pectin. Inactivated by pasteurisation.

Pellagra A nutritional deficiency disease produced by insufficient intake of niacin and/or nicotinic acid in the diet. The disease is characterised by skin lesions, inflammation of the mouth, diarrhoea, and central nervous system disorders.

Pemmican Mixture of dried, powdered meat and fat.

Penetrometer An instrument used to determine the firmness of food.

Pepsin An enzyme found in gastric juice that promotes the digestion of proteins.

Percolation The movement of water through the soil profile.

Pericarp The plant material surrounding the seed of fruits.

Permeability The passage or diffusion of a gas, vapour, liquid, or solid through a barrier without physically or chemically affecting it.

Pesticide A chemical which kills plant or animal pests.

Pesticide residues Small amounts of pesticides remaining in foodstuff as a result of pest control operations.

Petri dish A double glass or plastic dish used in cultivating microorganisms.

pH The effective acidity or alkalinity of a solution; not to be confused with the total acidity or alkalinity, the pH scale is:
Acid solutions: 0, 1, 2, 3, 4, 5, 6
Neutral: 7
Alkaline solutions: 8, 9, 10, 11, 12, 13, 14
pH 7 is the neutral point (pure water). Decreasing values below 7 indicates increasing acidity, while increasing values above 7 indicates alkalinity. One pH unit corresponds to a 10-fold difference in acidity or alkalinity, hence pH 4 is 10 times as acid as pH 5 and pH 3 is 10 times as acid as pH 4, and so forth. The same relationship holds on the alkaline side of neutrality, where pH 9 is 10 times as alkaline as pH 8, and so on. Most meat and fish products have pH values between 6 and 7, vegetables have pH values between 5 and 7, and fruits have pH values between 3 and 5.

pH, maximum For acidified foods, the highest finished product equilibrium pH after processing. For acidified low-acid foods not controlled at pH 4.6 or below, this does not apply if the food receives a heat treatment which alone achieves commercial sterility.

pH, normal For low-acid canned foods, the pH of the product or primary ingredient (e.g. green beans) in its natural state before processing. For acidified foods, it is the pH of the primary ingredient (e.g. pimientos) in its natural state before acidification.

Phenylalanine One of the amino acids that are essential for humans.

Pheophytin A brown or olive-green plant pigment formed by the breakdown of chlorophyll.

Phospholipids Lipid compounds containing phosphoric acid and nitrogen. These compounds are important components of many cellular membranes.

Phytates Salts of phytic acid, especially sodium phytate.

Phytic acid Chelating agent used for the removal of traces of metal ions. It is of nutritional interest because it interferes with the absorption of minerals from the intestinal tract, especially calcium and iron.

Picking table The point where produce is manually inspected.

Pickle lag The time required for hydrochloric acid to reach uniform rate of attack on tin plate. It is expressed in seconds.

Pin-hole Synonym for perforation. Development of a small hole in the plate.

Polymer A very large, complex molecule formed by chemically binding together a large number of identical smaller units (or monomers).

Polyunsaturated An unsaturated bond is a chemical structure into which additional hydrogen can be incorporated. Polyunsaturated fats contain fatty acids having more than one unsaturated bond. In general, polyunsaturated fats tend to be of plant origin and liquid.

Pomace The crushed pulp of fruits pressed for juice.

Pomes Fruits such as apples, quince, and pears.

Potable water Drinkable water/fit for human consumption.

Potassium nitrate (saltpetre) A preservative and a colour fixative in meats and meat products.

Potassium sorbate See sorbic acid.

Pouch A small- or moderate-sized bag, sack, or receptacle.

Pouring agents See Anticaking agents.

Propionates Food additives having the property of inhibiting mould growth.

ppb Parts per billion.

ppm Parts per million. 1 ppm = 0.0001 percent on weight basis, 1 mg/kg = 1 ppm, and 0.032 oz./ton = 1 ppm.

Preservation Any physical or chemical process which prevents or delays decomposition of foods.

Preservatives Any substance capable of retarding or arresting food spoilage or deterioration.

Pressure ridge The pressure ridge is formed on the inside of the can body directly opposite the double seam and is the result of the pressure applied by the seaming rolls during seam formation.

Primary spoilage The spoilage due to bacterial or chemical action of product packed within the can. See Secondary spoilage.

Primary waste treatment In-plant by-product recovery and wastewater treatment involving physical separation and recovery devices such as catch basins, screens, and dissolved air flotation.

Principal display panel The part of a label on a food package that is most likely to be shown or examined under customary conditions of display for retail sale.

Process authority The person or organisation that scientifically establishes thermal processes for low-acid canned foods or processing requirements for acidified foods. The processes are based on scientifically obtained data relating to heat or acid resistance of public health and spoilage bacteria and/or upon data pertaining to heat penetration in canned foods. The process authority must have expert scientific knowledge of thermal and/or acidification processing requirements and have adequate experience and facilities for making such determinations.

Process effluent The volume of liquid discharged from a plant. It is composed of water with dissolved and suspended solids.

Process, scheduled The process selected by the processor as adequate under the conditions of manufacture for a given product to achieve commercial sterility. This process is in excess of that necessary to insure destruction of microorganisms of public health significance.

Process temperature The calculated temperature at a particular time (process time) for which a specific can size and food product need to be heated in order to achieve commercial sterility.

Process time The calculated time at a particular temperature (process temperature) for which a specific can size and food product need to be heated in order to achieve commercial sterility.

Propionates Food additives having the property of inhibiting mould growth.

Propyl gallate An antioxidant.

Propylene glycol A solvent, wetting agent, and humectant.

Protein Large and extremely complex molecules consisting of from 50 to over 50,000 amino acids. Protein is the main nutrient responsible for building and maintaining body tissues. Sources of high-quality protein are meat, poultry, fish and other seafood, milk and milk products, and eggs. Sources of fairly good protein are legumes (dried beans, peas, soya beans), peanuts, and other nuts.

Protein concentrates Food substances high in protein content obtained from natural protein-containing foods by partial elimination of nonprotein food components.

Protein isolates Protein concentrates containing over 90% protein.

Proximate analysis Determination of moisture (water), protein, fat, carbohydrates, ash, and crude fibre content of foods.

Pseudoplastic Materials whose viscosity decreases as the rate of shear to which the material is subjected increases. An example is tomato ketchup which decreases in consistency when agitated and can be poured more easily from a bottle. See Viscosity and Consistency.

PSIG Pounds per square inch gauge pressure. For absolute pressure add 14.7 lb to psig pressure.

Psychrometer An instrument for measuring the humidity (water vapour) content of air by means of two thermometers, one dry and one wet.

Psychrophilic bacteria Have an optimum temperature for growth between 15.5 °C and 21.1 °C. May grow at temperatures down to 0 °C and up to 30 °C.

Ptomaine Term that has been used to describe certain types of food poisoning known today to be caused by toxins produced by bacteria.

PUFA Polyunsaturated Fatty Acids.

Pulping Forcing soft food material through a screen resulting in a pureed food.

Puncture test A test to determine resistance of flexible packaging materials to puncturing.

Puree In food technology, a smooth, pulpy, thick fluid produced by very finely disintegrating a juicy food commodity such as a fruit or vegetable.

Putrefaction Decomposition of proteins by microorganisms, producing disagreeable odours.

Pyridoxine A 'B'-group vitamin (B_6). Meat, milk, fish, and yeast are the best sources.

QA Quality Assurance.

QAC Quaternary Ammonium Compounds (QUATS).

Quality control A system for assuring that commercial products meet certain standards of identity, fill of container, quality sanitation, and adequate plant procedures.

'R' Enamel A protective lacquer (interior) used for acid products, fruits, or coloured vegetables. Used to prevent loss of colour or discolouration of coloured fruits and contact of product with tin.

Rad A measure of energy absorbed. Defined as $1\,rad = 0.01\,Gy = 0.01\,J/kg = 100\,erg$ of energy absorbed per gram of material receiving ionising radiation.

Radappertisation Foods packed in hermetic containers and sterilised by irradiation (gamma rays).

Radicidation Exposure of food to ionising radiation at doses necessary to kill all nonspore forming pathogenic bacteria. Analogous to pasteurisation.

Rancidity An oxidative deterioration in food fat whereby a typical off-odour and/or flavour is produced.

Raw waste The wastewater effluent from the in-plant primary waste treatment system.

Recanning The transfer and sealing of a product into a new hermetically sealable container followed by a scheduled process.

Recommended Dietary Allowances The Recommended Dietary Allowances (RDAs) are quantities of nutrients in the diet that are required to maintain good health in people. RDAs are established by the Food and Nutrition Board of the National Academy of Sciences and may be revised every few years. A separate RDA value exists for each nutrient. The RDA values refer to the amount of nutrient expected to maintain good health in people. The actual amounts of each nutrient required to maintain good health in specific individuals differ from person to person.

Recycle The return of a quantity of effluent from a specific unit or process to the feed stream of that same unit. This would also apply to return of treated plant wastewater for several plant uses.

Refractometer Optical instrument that measures the percentage of soluble solids in solution by the extent to which a beam of light is bent (refracted). Soluble solids scale is based on sugar concentration in a pure sucrose solution.

Relative humidity The ratio of actual humidity to the maximum humidity which air can retain without precipitation at a given temperature and pressure. Expressed as percent of saturation at a specified temperature. See Absolute humidity.

Rendering Heating meat scraps to melt the fat which then rises to the surface, while water and remaining tissue settle below. The melted fat is then separated.

Rep Roentgen. A unit of measurement of ionising radiation absorbed by materials. It has largely been replaced by the Rad. See Rad.

Reprocessing The treatment of a canned food in its original container recovered in a salvage operation followed by a scheduled process.

Retort Any closed vessel or other equipment used for the thermal sterilisation of foods.

Retort (Retortable) pouch A flexible container in which food to be heated to commercial sterility is placed in a retort or other sterilisation system. It is made of plastic films laminated to aluminium foil.

Retrogradation Refers to reverting of starches from a soluble form to an insoluble form upon freezing or aging.

Revolving drum test A test for measuring the protection to contents, or the retention properties of a container, or both, by subjecting the packaged products to rough handling in a standard revolving drum.

Rheology Study of the deformation and flow of matter.

Rheopectic (Rheopexy) Materials which increase in consistency with an increase in rate of shear.

Riboflavin (B2) A water-soluble vitamin. Important sources are milk, liver, kidney, heart, meat, eggs, and dark leafy greens.

Rickets Bone defects caused by a shortage of Vitamin D in the diet.

Rigid container A container whereby the shape or contours of the filled and sealed container are neither affected by the enclosed product nor deformed by an external mechanical pressure of up to 0.7 kg/cm² (10 psig) (i.e. normal firm finger pressure).

Ripe Fully developed, having mature seeds, and ready for use as a fresh food or for processing.

Ripening The sequence of changes in colour, flavour, and texture which leads to the state at which the fruit or vegetable is acceptable to eat or to be processed.

Rodenticide Poisons designed to kill rodent pests such as mice and rats.

Rope A type of microbiological food spoilage characterised by bacterial colonies growing in long strands.

Rotary washer A common type of washer in which produce is tumbled and washed by sprays of water.

Saccharin A nonnutritive sweetener, approximately 300 times as sweet as sucrose (common sugar).

Sacrificial anode A metal which slowly dissolves in a corrosion reaction and, in so doing, protects a second metal from corrosion (e.g. tin behaving as the sacrificial anode to protect the coupled steel base).

Salmonella A genus of bacteria that can cause infections in man that are characteristically gastrointestinal. A common source of these organisms is faeces-soiled hands. Another is infected food that is allowed to stand in the proper growth conditions without sterilisation. Destroyed by adequate heating as in the canning process.

Salmonellosis Infectious disease caused by bacteria of the genus *Salmonella*.

Salt A chemical compound derived from an acid by replacing the hydrogen atom with a metal or a positive ion. Salts may act as buffers in solution with acids or bases. Common or table salt (NaCl) is an example.

Sanitary can See Can, sanitary.

Sanitise To reduce the microbial flora in or on articles such as food plant equipment or eating utensils to levels judged safe by public health authorities.

Sanitiser A chemical agent that reduces the number of microbial contaminants on food contact surfaces to safe levels from the standpoint of public health requirements. Sanitising can also be done by heating.

Saponification The process of hydrolysis of fats or oils of a fluid by an alkali to form soap.

Saturated–Unsaturated Saturated fat contains fatty acids with only saturated molecular bonds. A saturated bond is a chemical structure which cannot accept additional hydrogen. Saturated fats tend to be of animal origin. Most vegetable oils contain a high proportion of unsaturated fats. Most unsaturated fats (such as peanut oil) are liquid at room temperature and most saturated fats (such as butter) are solid at room temperature.

Scheduled process The ordinarily used filed scheduled process for a given product under normal conditions.

Screening The removal of relatively coarse floating and suspended solids from wastewater by straining through screens.

Scurvy A disease caused by a shortage of Vitamin C (ascorbic acid) in the diet.

Sealing surface The surface of the finish of the container on which the closure forms the seal.

Seam compound Rubber or other material applied inside can end curl to aid in forming a hermetic seal when end is double seamed.

Seam, thickness The maximum dimension measured across or perpendicular to the layers of the seam.

Seam, width The maximum dimensions of a seam measured parallel to folds of the seam. Also referred to as the seam length or height.

Seamer Machine for double seaming can ends to the body of the can.

Second operation The finishing operation in double seaming. The hooks formed in the first operation are rolled tight against each other during the second operation.

Secondary spoilage Consists of those cans rusted, corroded, and perforated (from the outside towards the inside) as a result of the action of bursting or leaking cans on them. May occur during warehousing or transport.

Secondary treatment The waste treatment following primary in-plant treatment, typically involving biological waste reduction systems.

Sedimentation The falling or settling of solid particles in a liquid, as a sediment.

Semirigid container A container whereby the shape or contours of the filled, sealed container are not affected by the enclosed product under normal atmospheric temperature and pressure but can be deformed by an external mechanical pressure of less than $0.7 \, kg/cm^2$ (10 psig) (i.e. normal firm finger pressure).

Sensible heat See Heat, sensible.

Sensory Pertaining to an impact of a food on the senses (e.g. vision, odour, taste, tactile senses).

Sequestering agent See Ethylenediaminetetraacetic acid.

Settling tank Synonymous with 'Sedimentation tank'.

Sewage Water after it has been fouled by various uses.

Shelf life The length of time that a container, or material in a container, will maintain market acceptability under specified conditions of storage. Also known as merchantable life.

Shortening A mixture of partially hydrogenated fats, generally of plant origin, used for frying and for bakery products.

Side seam The seam joining the two edges of a blank to form a can body.

Silica gel A desiccant. A substance used for drying.

Silker Usually a reel-type washer for de-silking ears of corn.

SITA (System International Tinplate Area) Unit of sale under the metric system. One sita is equivalent to 100 square metres (m^2) and equals 4.9426 base boxes.

Size grader Belts or rotary drums with graduated holes through which produce can be sized mechanically.

Skim milk Milk from which virtually all cream (fat) has been removed (0.1–0.3% fat).

Slipper A can having an incompletely finished double seam due to the can slipping on the base plate. In this defect, part of the seam will be incompletely rolled out. Term has same meaning as dead head when referring to seamers which revolve the can.

Sludge The accumulated settled solids deposited from sewage or other wastes, raw or treated, in tanks or basins and containing more or less water to form a semiliquid mass.

Sodium Sodium is an essential element. It is naturally present in foods. Most of the sodium added to food is in the form of common salt.

Sodium bicarbonate See Baking soda.

Sodium bisulphite A chemical preservative.

Soft sugars Highly refined, dark coloured, molasses-flavoured sugars which are frequently called brown sugars. They have a relatively high content of mineral and other nonsucrose materials.

Soft swell Both ends of can swelled but may be depressed fairly easily by thumb pressure.

Solid fat index A measure of the solidity of fats at various temperatures.

Soluble Solids (SS) Solids in solution largely made up of sucrose and other sugars, fruit acids, and mineral salts.

Solvents A substance which dissolves or holds another substance in solution such as common salt in water. Solvents are used in some foods as carriers for flavours, colours, stabilisers, emulsifiers, antioxidants, and other ingredients.

Sorbic acid A chemical preservative used to selectively inhibit growth of yeasts and moulds.

Sorbitol A humectant used to hold moisture in foods.

Soya bean meal The residue remaining after solvent extraction of cracked soya beans.

Soya bean milk A product made from soya bean protein, vegetable oil, and water.

Spice The bark, root, bud, flower, or fruit of plants used primarily to season foods. For example, pepper and cinnamon.

Spice oils Oil soluble extracts from spices containing, in concentrated form, the substances responsible for the flavour and aroma of those spices.

Spinner A container with a faulty double seam, caused by the container having been revolved by the seaming rolls, due to improper adjustments.

Spoilage A process whereby food is rendered unacceptable through microbial or chemical action. See Primary spoilage and Secondary spoilage.

Spores Certain rod forms of bacteria produce spores. These are not reproductive bodies, as in the case of moulds and yeasts, but are the resting stage of the organism. In the spore state, bacteria can survive extremes of cold, heat, drying, and other unfavourable conditions for long periods of time; when the environment is again favourable, the spores germinate and the organisms start another cycle of growth. Growing cells are called 'vegetative' cells. Spore-forming bacteria which can grow in the presence or absence of air (facultative anaerobes) are classified in the genus *Bacillus*, while those which grow only in the absence of air are classified in the genus *Clostridium*.

Spray drier Equipment in which material to be dried is sprayed as a fine mist into a hot-air chamber and falls to the bottom as dry powder. Period of heating is very brief. Dried powder consists of hollow particles of low density.

Springer Swelled can with only one end remaining out; on pressing this end it will return to normal, but the other end will bulge out.

Stabilisers Substances that stabilise emulsions.

Stack burn Condition resulting from placing cased cans in piles insufficiently cooled. Food may vary from over cooked to definitely burnt flavour and colour. Also, excessive corrosion of interior of container may result.

Staling A physical–chemical process in cereal products, especially bread, whereby a characteristic 'dry' texture develops.

Standard deviation Statistical measure of the scattering of data from the average; equal to the root mean square of the individual deviations from average.

Standard for grade The formulated rules by which a product will be judged to fit one of the grade categories established for the class to which the product belongs.

Standard of fill of container A statement which establishes the minimum weight or volume of a specific food which its container must hold, as determined by procedures specified in the standard, below which the food product is of substandard fill and must be clearly labelled.

Standard of quality A statement which establishes a minimum quality for a specific food product, below which it is of substandard quality and must be clearly labelled.

Standard Plate Count (SPC) Method used to determine the number of specific microorganisms present in food, other substances, or surfaces.

Staphylococci A genus of spherical bacteria (cocci) occurring in pairs, tetrads, or irregular, grape-like clusters.

Staphylococcus aureus Species of bacteria that are important as a cause of human infections and of food poisoning.

Starch White, odourless, and tasteless carbohydrates produced by plants as an energy store. Starches are primary foods for most animals and are broken down during digestion into sugars and thereby used for energy.

Starch (high amylose) A starch containing over 50% amylose (usually 55–70%).

Starter culture A culture of microorganisms used to start a fermentation process.

Steam exhausting Passing filled cans through a tunnel of steam, prior to sealing, to assist in oxygen removal from the product and headspace.

Steam flow closing machine Equipment to close containers while at the same time producing a vacuum in them by means of steam jets directed into and around the container.

Steam table Stainless-steel trays used in food service for keeping prepared food warm, over live steam.

Stearic acid A common saturated fatty acid with one of the longer lengths of carbon chain and highest number of hydrogen atoms.

Stearine The higher melting point glycerides that are separated from oils by winterisation. Stearines are used in the manufacture of vegetable shortenings. See Winterisation.

Sterile Free of living organisms. See Commercial sterility.

Sterilisation Any process, chemical or physical, which will destroy all living organisms.

Sterilisation process The time–temperature treatment necessary to render canned foods commercially sterile.

Sterilisation temperature The temperature maintained throughout the thermal process as specified in the scheduled process.

Sterilisation time The time that lapses between the moment a retort reaches sterilization temperature until steam is cut off.

Sterilisation value (Fo) The number of minutes at a reference temperature of 121.1 °C (250 °F) required to kill a known population of microorganisms with a Z-value of 10 °C (18 °F).

Sterol A complex and usually unsaturated solid alcohol compound commonly found in plant and animal lipids. Cholesterol is a sterol.

Stevia A nonnutritive sweetener/sugar substitute made from the leaves of the plant species *Stevia rebaudiana*. Steviol glycoside extracts have up to 300 times the sweetness of sugar.

Still retort A retort for the sterilisation of canned foods in batch amounts, with no agitation.

Storage life See Shelf life.

Streptococci A genus of bacteria that are spherical or oval-shaped and that divide in such a way that pairs or chains are formed.

Sublimation The physical process by which a substance passes directly from the solid state to the vapour or gas state, such as the evaporation of ice during freeze-drying.

Succulometer An instrument used to measure the degree of maturity of corn.

Sucrose A sweet, crystallisable, colourless sugar which constitutes the principal sugar of commerce. Refined cane and beet sugars are essentially 100% sucrose. Under certain conditions, sucrose breaks down to dextrose and laevulose.

Sugars (saccharides) Sweet carbohydrates obtained directly from the juices of plants or indirectly from the hydrolysis of starches. Sugars constitute the primary energy source of both plants and animals.

Sulphide staining Where naturally occurring sulphur compounds in foods or added sulphur compounds (e.g. from preservatives) react with the tin plate surface to form a blue-black stain on the tin plate or a loose black iron sulphide complex.

Supercooling Commonly referred to water freezing at a temperature several degrees below 0 °C (32 °F) before some stimulus such as crystal nucleation or agitation initiates the freezing process.

Superheated steam Saturated steam that has been heated at a constant pressure above its saturation temperature.

Supplementation See Enriched.

Surface-active agents Substances that affect the surface tension of a liquid. They include emulsifying agents, detergents, suspending agents, wetting agents, etc.

Surfactant A surface-active agent.

Suspended solids The quantity of solids, both volatile and stable, in suspension which can be filtered out by a standard filter under a specified test procedure.

Suspension A homogenous mixture of an insoluble granular or powdered material with a fluid.

Sweating If very cold cans are placed in a warm, humid place, moisture will condense on their surface. This sweating may very easily contribute to rusting of the cans.

Swell (1) (Noun) A container with either one or both ends bulged by moderate or severe internal pressure. (2) (Verb) To bulge out by internal pressure, as by gases caused by biological or chemical action.

Swell, hard A can of food which has spoiled to the point where both ends are bulged out and show no appreciable yield to thumb pressure. See Soft Swell, Flipper, Springer, Flat Sour.

Swell, hydrogen See Hydrogen swell.

Syneresis Exuding of small amounts of liquids from gels.

Synthesise To build up a compound by the union of simpler compounds or of its elements.

Syrup Water solution of sugar, usually sucrose.

Tallow Fat obtained from beef by the process of rendering.

Tannins Substances that posses astringency which influences flavour and contributes body to such beverages as coffee, tea, wine, and several fruit juices. Colourless tanning compounds, upon reaction with metal ions, form a range of dark coloured complexes which may be red, brown, green, grey, or black.

Tartaric acid An organic acid found in several fruits, particularly grapes.

TBHQ Tertiary Butyl Hydroquinone. An antioxidant.

Temper A measure of the ductility and hardness of steel plate.

Tenderometer Instrument to measure the stage of maturity of peas to determine if they are ready for canning. Measures the force required to effect a shearing action.

Teratogen An agent that causes physical defects in a developing embryo.

Terne plate Black plate coated on both sides by hot dipping in an alloy containing approximately 15% tin and 85% lead. Due to the lead content, terne plate is unsuitable for food products.

Tertiary waste treatment Waste treatment systems used to treat secondary treatment effluent and typically using physical–chemical technologies to effect waste reduction.

Texture The food characteristics that deal with the sense of feel.

Texturiser A food additive which stabilises, enhances, or changes the texture of foods, such as puddings, ice cream, and many others.

Thaw-exudate (drip) Liquid which separates from frozen foods upon thawing consisting of water with small quantities of water-soluble food components.

Thermal process The application of heat to food. Either before or after sealing in a hermetically sealed container, for a period of time and at a temperature scientifically determined to achieve a condition of commercial sterility (i.e. the destruction of microorganisms of public health significance, as well as those capable of reproducing in the food under normal nonrefrigerated conditions).

Thermocouple A bimetallic device to measure temperatures electrically.

Thermoduric Microorganisms that have the ability to withstand high temperatures, i.e. are highly heat resistant.

Thermolabile Fairly easily destroyed by heat.

Thermophiles Bacteria which grow optimally above 45 °C.

Thermophilic bacteria Describes bacteria which require temperatures between 37 °C and 82 °C for growth and grow optimally at 50–55 °C.

Thiamine A water-soluble vitamin (B-1). Important sources are pork, heart, liver, kidney, dry beans and pears, whole grain, and enriched bread and cereals.

Thickening agent A texturiser, such as starch and gelatin, which increases the consistency of a product. Gravies and soups are products that contain thickening agents.

Thixotropic Those food gels that break up (become more fluid) on being shaken and reset on standing (become thick again). Systems that show reversible alteration in their flow characteristics when work is performed on them, such as shaking.

Threonine One of the amino acids that are essential for humans.

Tin plate Steel sheet, usually of special formula and temper, coated on both sides with a controlled thickness of pure tin.

Tin plate, charcoal A type of hot-dipped, tin-coated steel plate ranging from an average of 2.2–7 lb of tin per base box. For can manufacturing, this type of plate has been completely replaced by electrolytic plate.

Tin plate, coke A class of hot-dipped, tin-coated steel plate which carries tin in the range of 1.25–1.75 lb per base box (little used nowadays).

Tin plate, differential Electrolytic tin plate having different weights of tin coatings on opposite sides of the sheet.

Tin plate, hot-dipped Black plate which has been coated on both sides with commercially pure tin by a process wherein, after pickling, the sheets are passed successively through flux, molten tin, and palm oil. The amount of coating can be varied to meet the requirements (little used nowadays).

Tin plate, type 'L' Tin plate in which the base plate is low in copper and metalloids (S, As, P, etc.). Such plate has maximum corrosion resistance to highly corrosive foods.

TLC Thin Layer Chromatography. An instrument for chemical analysis.

Tolerance A specified allowance for deviations in weighing, measuring, etc., from the standard dimensions or weight.

Tomato ketchup, catsup, or tomato sauce Product made of tomato puree, vinegar, sugar, salt, and spices.

Top double seam The double seam formed by end attached by canner. Also known as canner's or packer's end seam.

Torr A unit of atmospheric pressure equivalent to 1.0 mmHg.

Total Dissolved Solids (TDS) The solids content that is soluble and is measured as total solids content minus the suspended solids.

Total Suspended Solids (TSS) Solids suspended in solution which, in most cases, can be removed by filtration.

Toxicology The science of poisons and their antidotes.

Toxin An organic poison, a product of the growth of an organism. Some toxins are given off as waste products of a microorganism and are called 'exotoxins'. Others are contained within the cells, and are liberated only when the cell dies and disintegrates. These are called 'endotoxins'. Toxins produced by *C. botulinum* are thermolabile, that is, they are fairly easily destroyed by heat.

Trace A minute amount of a substance.

Translucent Descriptive of a material or substance capable of transmitting some light but not clear enough to be seen through.

Transparent Descriptive of a material or substance capable of a high degree of light transmission (e.g. glass).

Trichinosis A muscle infection caused by a nematode. Humans develop trichinosis by consuming improperly cooked, infected pork meat or by indirect contamination of other meats with the nematode.

Triglycerides See Glycerides.

Trim tables Area where produce are hand-cut and trimmed.

Trimethylamine A substance produced during the early stages of spoilage of fish. It gives fish its characteristic 'fishy' odour. This odour does not necessarily indicate that the fish is inedible.

Tryptophan One of the amino acids that are essential for humans.

TS Total Solids.

UHT Ultra High Temperature. Term used in reference to pasteurisation of commercial sterilisation of food by heating it for an extremely short period, around 1–2 s, at a temperature exceeding 135 °C (275 °F).

Ultraviolet irradiation A disinfection method that uses ultraviolet (UV) light at sufficiently short wavelength to kill microorganisms. It is used in a variety of applications, e.g. food and water purification. It destroys the nucleic acids in these organisms so that their DNA is disrupted. It has poor penetrating power and is thus only of value for surface sterilisation or sterilising the air.

Uperisation A method of sterilising fluid foods by injecting steam under pressure to raise the temperature to 150 °C. The added water is evaporated off.

USDA United States Department of Agriculture.

Vacuum pack In canning, the term 'vacuum pack' refers to products packed with little or no brine or water, which are sealed under a high mechanical vacuum and which require maintenance of high vacuum to assure process adequacy.

Valine One of the amino acids that are essential for humans.

Vee (vee down) Is a 'vee'-shaped deformation in cover hook associated with drastically reduced cover hook dimension. In extreme cases, it can be detected by external examination of the seam.

Vegetative cells Stage of active growth of the microorganism, as opposed to the bacterial spore.

Venting Eliminating air from a retort prior to sterilising canned foods.

Vents Openings controlled by gate, plug, cock, or other adequate valves used for the elimination of air during the venting period.

Viable Living.

Viner Equipment for removing peas, lima beans, and green beans from the vines on which they are harvested. In the case of peas and lima beans, viners also remove vegetable from the pod.

Vinyl chloride A synthetic plastic used in the manufacture of packaging materials.

Viscometer An instrument to measure viscosity.

Viscosity The internal friction or resistance to flow of a liquid. The constant ratio of shearing stress to rate of shear. In liquids for which this ratio is a function of stress, the term 'apparent viscosity' is defined as this ratio.

Vitamin Vitamins are complex organic compounds, needed in small amounts, which are essential for certain metabolic functions in humans or other animals. Vitamins act as catalysts by helping other nutrients perform their functions. See Catalyst.

Vitamin A A fat-soluble vitamin, essential for vision in dim light. Most Vitamin A is obtained from the body's conversion of carotene found in vegetables and fruits. Important sources are liver, dark green vegetables, yellow fruit and vegetables, butter, and margarine.

Vitamin B complex Folic acid, niacin, pantothenic acid, pyridoxine, riboflavin, thiamine, and biotin. The B vitamins are essential in human diets and occur naturally in meats, wheat, etc.

Vitamin B-6 See Pyridoxine.

Vitamin B-12 See Cyanocobalamin.

Vitamin C A water-soluble vitamin. Important sources are citrus fruits and juices, broccoli, brussel sprouts, raw cabbage, collards, sweet and green peppers, potatoes, and tomatoes. Also called ascorbic acid.

Vitamin D Calciferol. A fat-soluble vitamin important in the prevention of rickets. Important sources are fish liver oil, fortified milk, and egg yolks.

Vitamin E α-Tocopherol. A fat-soluble vitamin important as a natural antioxidant. Vegetable oils, especially wheat-germ oils, are important sources.

Vitamin K Vitamin necessary for proper blood coagulation to prevent haemorrhages. Good sources are green leafy vegetables, pork liver, milk, and eggs.

Vitelline membrane The membrane enclosing the egg yolk.

Vortex washer Circular tank in which produce is washed by sprays which impart a swirling motion.

Washers Equipment made in a variety of designs for washing produce prior to sizing, grading, trimming, and blanching.

Water activity A measure of water availability in food for microbial growth. The ratio of water vapour pressure of a food to the vapour pressure of pure water under identical conditions of temperature and pressure.

Water binding See Bound water.

Waxy maize A variety of corn, the starch content of which consists solely of branched molecules.

Weak lap The lap is soldered and both parts are together. However, strain on this lap, as twisting with the fingers, will cause the solder bond to break.

Whey The liquid and its dissolved lactose, minerals, and other minor constituents remaining after milk has been coagulated to separate the curd. Curd is made up of casein, most of the fat, and some lactose, water, and minerals from milk.

WHO of UN World Health Organization of the United Nations.

Winterisation The process in vegetable oil refining by which the higher melting point glycerides (stearines) are removed from oils by chilling.

W/O emulsion Water-in-oils emulsion in which the water is the internal phase and the oil is the external or continuous phase. When diluted by the addition of an oil, W/O emulsions retain homogeneity.

Wrinkle, cover hook A degree of waviness occurring in the cover hook, acting as an indication of the tightness of the seam. Several numerical rating systems are used.

Xanthophylls The yellow-orange pigments found in plant foods such as corn, peaches, and squash.

Xerophilic Can grow or survive in a medium very low in humidity.

Yeasts Microorganisms classified in the kingdom Fungi. They are spherical or more or less elongated cells. Most yeasts break down sugars to carbon dioxide and alcohol. That process is called fermentation.

Z-Value Z-value is a term used in thermal death time calculations. It is the temperature required for one log reduction in the D-value for a specific bacterium.

Index

Note: Page numbers followed by "f" and "t" indicate figures and tables respectively.

Printed by Printforce, the Netherlands